INTRODUCTION TO FOREST RESOURCE ECONOMICS

McGraw-Hill Series in Forest Resources

Avery and Burkhart: Forest Measurements
Dana and Fairfax: Forest and Range Policy
Daniel, Helms, and Baker: Principles of Silviculture
Davis and Johnson: Forest Management
Duerr: Introduction to Forest Resource Economics
Ellefson: Forest Resources Policy: Process, Participants, and Programs
Harlow, Harrar, Hardin, and White: Textbook of Dendrology
Knight and Heikkenen: Principles of Forest Entomology
Laarman and Sedjo: Global Forests: Issues for Six Billion People
Panshin and DeZeeuw: Textbook of Wood Technology
Sharpe, Hendee, and Sharpe: Introduction to Forestry
Sinclair: Forest Products Marketing
Stoddart, Smith, and Box: Range Management

Walter Mulford was Consulting Editor of this series from its inception in 1931 until January 1, 1952.
Henry J. Vaux was Consulting Editor of this series from January 1, 1952 until July 1, 1976.
Paul V. Ellefson, University of Minnesota is currently our Consulting Editor.

INTRODUCTION TO FOREST RESOURCE ECONOMICS

William A. Duerr

Formerly Distinguished Professor of Forestry
in the State University of New York, Syracuse
More recently Thomas M. Brooks Professor of Forestry
in Virginia Tech, Blacksburg

McGRAW-HILL, INC.

New York St. Louis San Francisco Auckland Bogotá
Caracas Lisbon London Madrid Mexico Milan Montreal
New Delhi Paris San Juan Singapore Sydney Tokyo Toronto

This book was set in Times Roman by The Clarinda Company.
The editors were Anne C. Duffy and Eleanor Castellano;
the production supervisor was Louise Karam.
The cover was designed by David Romanoff.
R. R. Donnelley & Sons Company was printer and binder.

INTRODUCTION TO FOREST RESOURCE ECONOMICS

1 2 3 4 5 6 7 8 9 0 DOC DOC 9 0 9 8 7 6 5 4 3 2

ISBN 0-07-017982-4

Library of Congress Cataloging-in-Publication Data

Duerr, William A. (William Allen), (date).
Introduction to forest resource economics / William A. Duerr.
p. cm. — (McGraw-Hill series in forest resources)
Includes index.
ISBN 0-07-017982-4
1. Forests and forestry—Economic aspects. 2. Forests and forestry—Economic aspects—United States. 3. Forest products industry. 4. Forest products industry—United States. 5. n-us. I. Title. II. Series.
SD393.D835 1993
634.9—dc20 92-37798

ABOUT THE AUTHOR

WILLIAM A. DUERR is a product of 60 joyous years as a member of the forestry profession.

Schooled at Williams and Iowa State colleges and at the Universities of Minnesota and Harvard, he worked during two decades of his carreer in the Branch of Research of the Forest Service, USDA. There he served in the north central, southeastern, southern, and Pacific Northwest regions.

Bill's formal experience as a teacher opened with two decades in the State University of New York and continued for another 20 years at the Virginia Polytechnic Institute and State University. He worked mainly with graduate students in forest resource management and economics, seeing more than 70 of them through master's degrees or doctorates. He says they were his prinicipal instructors.

During his teaching years, he performed also as a consultant to central and local governments and to private organizations, in the United States and abroad.

Bill is an author or editor of some 15 or 20 major works and more than 100 technical papers in forestry management and economics.

DEDICATION AND ACKNOWLEDGMENTS

I dedicate this book to my late wife, Jean, who joined me in the research for it and in the writing of it, and who gave me her guidance.

I thank the teachers and students who have used drafts of the book and given me candid criticism. I am especially grateful to Charles Lee, Douglas Richards, and James Moak for inviting me to Mississippi State University in 1981 to work on a first draft and share it with a class of senior students.

I thank the many, many other colleagues at forestry schools in the U.S. and abroad, and at the centers of learning in government and industry. These colleagues opened to me a treasury of experience and knowledge that paid my way through difficult topics.

I owe a heavy debt to my graduate students, who have served as my mentors.

CONTENTS

PREFACE

SUBJECT OF THIS BOOK

How are forest resources created? How are they used? How do they contribute to your happiness and mine?

What are some of the issues in forest resource stewardship, public and private? How should we address these issues? What principles should we follow?

Such is the subject of this book.

I call the subject **forestry economics**[1] because it treats of people and their questions about forest resource husbandry. I use this term rather than *forest* economics, a term often met. This is because I am speaking of the economics of forestry, not of the forest. The forest is a complex system, and surely it has its economics. The economics of the forest centers on the interplay of forces within the forest.

The subject of forestry, on the other hand, includes the forest and much more. Notably, it includes *people* who produce and use forest resources. It includes biological and engineering works for forest production and use, and their social setting. The boundaries of the forestry system are the limits of our world, whereas the forest system is limited by property lines and even stand boundaries.

ARRANGEMENT OF THIS BOOK

The book is divided into six parts. Part One looks at national and world forestry and so helps put the whole subject in context. The ability to understand the context of things is a mark of the educated person.

[1]Special terms are put in boldface where they are defined or their meaning is implied. They are listed in the index, which thus doubles as a glossary.

Part Two concerns consumers and their wants for forest resources. We are all consumers, with a predisposed interest in consumers.

Problems of productoin are the subject of Parts Three and Four. Part Three treats of **genetic production** (through the agency of living things such as trees and wildlife), while Part Four concerns manufacturing (with man-made machinery).

Now that demand (by consumers) and supply (by producers) have been taken up, Part Five turns to the meeting of the two in the market and to related problems of valuation. This subject matter broadening is continued and concluded in Part Six, which studies social institutions from labor organization to governmental planning.

The chapters are variously designed. In Part One, they are mainly descriptive. In Part Two, some are analytical or even philosophical. The first chapter in each part is an introduction; the last, a summary. Here and there are chapters that do inescapable chores, such as explaining the prediction process or reviewing investment criteria. The typical chapter tries its hand at problem solving, in a sense to be explained shortly.

The book is meant for use as a whole or in part. As a whole, it represents perhaps three hours of course work for an academic year. For shorter programs, the instructor may assign reading from selected chapters. Some illustrative selections are as follows:

- *General survey of forestry economics.* Chapters 1, 7, 8, 9, 13, 15, 16, 29, 30, 33, 41, 42, 51, 52, 61.
- *Macroeconomics of forestry.* Chapters 3, 4, 5, 35, 43, 45, 46, 48, 52, 53, 54, 55, 59, 60, 61.
- *World forestry.* Chapters 1, 2, 3, 4, 5, 6, 7, 9, 10, 12, 14, 25, 28, 52, 58.
- *Nontimber forestry economics.* Chapters 2, 5, 6, 7, 8, 10, 11, 12, 14, 15, 25, 27, 29, 47, 60.
- *Producer economics of forestry.* Chapters 3, 16, 17, 21, 23, 24, 25, 26, 27, 28, 30, 33, 39, 40, 41.
- *Economics of forest products.* Chapters 3, 4, 33, 34, 35, 36, 37, 38, 39, 45, 46, 48, 50, 53, 59.
- *Economics of marketing.* Chapters 13, 29, 37, 40, 42, 43, 44, 45, 46, 47, 48, 49, 50, 51, 56.
- *Consumer economics of forestry.* Chapters 1, 5, 6, 7, 8, 9, 10, 11, 12, 13, 14, 15, 53, 54, 57.
- *Forest management.* Chapters 11, 17, 18, 19, 20, 21, 22, 23, 24, 25, 26, 27, 28, 31, 32.
- *Economics of silviculture.* Chapters 2, 3, 4, 5, 11, 16, 17, 18, 21, 22, 24, 26, 27, 28, 29.
- *Microeconomics of forestry.* Chapters 18, 19, 21, 22, 23, 24, 25, 27, 28, 34, 36, 37, 38, 50, 58.
- *Forest finance.* Chapters 19, 20, 23, 24, 25, 31, 32, 44, 47, 49, 50, 55, 56, 57, 58.

HISTORY AND PROSPECT

Social science is a living thing, growing and changing all the while. In this, it resembles forests and humanity.

There was a time when social science, essentially economics, in forestry comprised mainly the financial aspects of timber growing. Its compound-interest formulas came, along with the rest of forestry, from abroad. Such topics as forest taxation, insurance, and credit found themselves at home in these financial circles.

Then came the great depression of the 1930s. The shock of prevailing unemployment and poverty, and the vision that forest wealth might contribute to recovery inspired a few adventurous American foresters to study economics and build an applied field of their own. They were taught that economics includes more than money matters. It concerns people: their behavior in the ordinary business of life.

> Do you know whether—and, if so, in what ways—American forests served to moderate the 1930s depression?

From early in their new joint profession, some forester-economists wanted to enlarge their subject matter and their viewpoint. They envisaged three directions for enlargement. One was into unfamiliar forestry questions, such as nontimber and multiple-use management. A second was into unfamiliar economic questions, such as how values, including those of the environment, come into being outside markets. A third was into areas of social science adjoining economics: political science, ethnology, law, and other institutions of the people.[2]

This book pushes into all three foregoing areas. The billing it gives to world forestry and to consumers and environmentalism nudges tradition aside. So marked are a few of its departures that, to some readers, it is scarcely forestry. Yet, by what process is a professional field defined? Surely it is defined in the work of the venturesome rather than of conformists—provided the ventures prove to be fruitful of results appropriate to new times. The work of a definer is analogous to that of any entrepreneur: to have faith, to predict, to innovate, and to take risks and the consequences.

INTENDED READERSHIP

The book is intended primarily for students in professional schools, those who have a fair idea of what forestry is and what social science is and want to put the two together.

How can students' exploration of social sciences in forestry be made rewarding? The answer on which this book pins its hope is to paint the subject in a practical light—to present it as a set of useful professional tools. You who envisage a resource-

[2]Burch, William R., Jr.: "The Uses of Social Science in the Training of Professional Social Foresters," *Journal of World Forest Resource Management,* **3** (2), 1988, pp. 73–109.

management career can, in each glimpse, envision personally using these tools and can foresee their value.[3]

Because of its stress on problems and its view of them as case studies, the book offers rewards to many who are concerned with forest resources: managers and others, and not to college students alone—indeed, to anyone who professional work is inspired by viewing it as problem solving.

PROBLEM-SOLVING PROCEDURE

What is a **problem?** How does one solve it?

A problem is a question to which there are alternative replies. Its solution consists in finding a "good" reply: one that meets agreed-upon standards. Solving a problem, then, is responding to a question—not necessarily giving "the answer," but giving a sensible response.

> You will notice throughout this section that superlative adjectives are avoided: The word "good" is used rather than "best," "sensible" rather than "perfect," "profitable" rather than "most profitable," and so on. As you read the section, think whether you approve or disapprove of this avoidance. How might you defend it or fault it?

For example, what measures shall you take to regenerate a forest on the Union Pacific tract? You propose a sensible response in four steps:

First, clarify the question. What are the alternatives at issue? They are (1) natural regeneration, (2) seeding, or (3) planting of (a) Douglas-fir or (b) western white pine by a certain corporate landowner in southwest Washington.

Second, find a procedure for working out a sensible response, a procedure for weighing the alternatives by an agreed-upon standard. In the case at hand, the standard is profit; the procedure, then, is to identify a profitable alternative.

Third, carry out the procedure, using relevant data. From the data on probable costs, production periods, and forest yields and values, it appears that to plant Douglas fir is a good (profitable) response to the question.

Fourth, follow up the question-response process to discover how it might be improved. In this case, you learn that if planting is in the picture, the source of stock is part of the question.

Notice that there are other ways to describe these same four steps, for instance: (1) clarify the question, (2) find a suitable model, (3) put data into the model and derive a reply, and (4) follow up the process.

[3]Canham, Hugh O.: "Application and Use of Forestry Economics by the Forest Service, U.S. Department of Agriculture," Rep. of a Sabbatical Study for Forest Service Policy Analysis Staff, 1991, especially pp. 3–6.

By **model** is meant the same thing as *procedure* in the first list. A model is a simplified statement of the variables to be taken into account in working on a question, and of the manner in which these variables are related to one another. A model may be treated as a hypothesis, to be tested and perhaps revised, and this is true also of the question, the data, and the response.

> Do you see how the question may be treated as a hypothesis? How the data may be so treated? —the response?

In the regeneration example, the model is as follows: Estimate the value (to the corporation) of the timber yields per acre for each tree species and reforestation method; discount the values to the present; subtract costs, discounted as necessary; and compare the resulting six net-revenue estimates to see which is satisfactorily large. And so a model is an abstraction. You might well have counted among the variables the probability of success in regeneration and the aesthetic values of the alternatives, but, in the interest of simplicity, you ruled out these considerations. This is one reason models give replies and not "the answers." (Another reason, of course, is that models are commonly used for forecasting in the face of uncertainty.)

MODELS AND THEIR ORIGIN

But where do models come from? They are derived (that is, you can *picture* them as derived) by abstracting from totality so as to create a usable statement of relationships. Totality is this: *Everything depends on everything else.* From such a mind-boggling notion, one abstracts a manageable model: *The return from forest regeneration depends, other things being equal, on whether I bring in Douglas-fir or pine naturally or by seeding or planting.* Here, from an infinity of variables, just six are abstracted. The abstraction is artistic work, and no simple rules are to be found.

Models are illustrated in a story about King Louis of France. At middle age, he commissioned the young historian, Jaubert, to write a history of the French people for him. Jaubert toiled for years and produced a gigantic work, which he brought to the palace on seven mule carts.

"Jaubert," said Louis, "I have more to do, as king, than read history. I commission you to shorten your work."

After more years, Jaubert succeeded in condensing his history of the French people to a single large volume. The king offered his compliments and then said, "Jaubert, I am growing old and won't have time to read this. Shorten it for me."

When the historian returned again with a small volume under his arm, Louis was on his deathbed. "Jaubert," said he in a thin voice, "my time is running low; you must *tell* me the history of the French people."

"Sire," Jaubert responded, "their history is this: They were born. They suffered. They died."

Into the story of Louis and Jaubert you may read two lessons. One: It shows what a model is. Each of Jaubert's four efforts was a model of the French people's history—and, incidentally, each was a model of its forerunners. The second lesson: A model's degree of abstraction must suit its intended use. As a model builder, Jaubert was successful only at the very last.

> Jaubert's models were verbal. But many models are of different sorts. There are, for example, *symbolic* models such as maps and mathematical expressions, *analogue* models such as the hands of a clock, and *scale* models such as of airplanes. What are some forestry examples of each of these three types of models?

ALTERNATIVES, THE BASIC QUESTION

Return, for a moment, to follow-up, the fourth step in problem solving. In this step, the reply can determine the question ("I can see now this isn't quite what I wanted to ask"), just as the question can determine the reply. There are alternative questions as well as replies.

In forestry economics, a study of choice, there are always alternatives. In problem solving, responding to choice, there are always alternatives. Social sciences in forestry, as problem solving, leans at every step on recognizing *what the alternatives are.*

The questioner and the replier (problem solver) in social science may be the same person, but often are not. In your study of this book, you are to think of yourself as the problem solver. You are asked a question. The questioner takes your reply and possibly others, adds a liberal dash of judgment, and comes up with a personally tailor-made answer—which, of course, may totally agree with your reply.

SUMMARY

The field of social science in forestry, once concerned largely with timber finance, has broadened with time. In this book, the field is seen as coincident with much of economics and with parts of other social sciences. The book's user who wants to focus upon particular parts of the field is given a guide to relevant chapters.

Forestry economics is presented as problem solving, replying to professional questions. The process of replying to a question is viewed as four steps: (1) clarify the question, (2) find a suitable model, (3) put data into the model and derive a reply, (4) follow up.

To clarify the question (step 1, above) is to identify the alternative replies that the questioner will consider seriously. *What are the alternatives?* To find a suitable model (step 2) is to light on a satisfactory compromise between precision and simplicity. *What is degree of abstraction?*

William A. Duerr

INTRODUCTION TO FOREST RESOURCE ECONOMICS

PART **ONE**

A SETTING FOR SOCIAL SCIENCE IN FORESTRY

CHAPTER 1

PREAMBLE TO PART ONE

A PANORAMA

The parts of this book that directly follow Part One concern bits and pieces of forestry. Not until later will you climb up for a wider view. Yet you have already been offered the thought that it's vital to appreciate the context of a subject. And so, before going on to the bits and pieces, I devote Part One to the national and world context of social science in forestry. In this way, you can hope to have the memory of a panorama that will help guide you through the small specifics of the landscape.

NATIONAL AND WORLD CONTEXT

Not only does national and world forestry offer context for forestry economics, but, what's more, it does this in the very dimension that characterizes forestry so strongly: the geographic dimension (place and space), transportation, and interarea exchange, and all sorts of other matter that can be put on maps, which are the forester's stock-in-trade.

From your study of social science, you know that the actions of any individual or group affect the actions of others. And past events influence the future. In Part One, the theme is developed that decisions made here modify those made elsewhere.

> Give some examples of how an event in a foreign country can affect forestry in some region of the U.S. For instance, what foreign event can influence the value of a U.S. timber species? the value of our forest recreation? the marketing of lumber? our consumption of forest goods or services?

EMERGENCE OF THE U.S.

The people of the United States have experienced great changes in attitude toward international affairs. Our tradition at the outset was isolation from our parent nations in the old world. Thomas Jefferson in 1801 warned us against "entangling alliances" abroad, a warning that 3,000 to 5,000 miles of intervening Atlantic Ocean enabled us to heed. But now and then, contrary tendencies appeared. We joined an alliance in World War I to "save the world for democracy," but, disillusioned by subsequent events, withdrew our international participation, built high tariff barriers, and fell back on our own varied resources and our prosperous interstate commerce to support our social development.

Then in the 1930s and especially after the peace with the Axis nations and Japan, we changed our position. In the Roosevelt and Truman administrations (1933–1953), the U.S. gave diplomatic recognition to the Soviet Union, passed a Reciprocal Trade Agreements Act that allowed drastic lowering of tariffs and led to the 90-nation General Agreement on Tariffs and Trade (GATT), established the good neighbor policy toward Latin America, collaborated with other countries in World War II and its aftermath, helped organize the United Nations, implemented the Marshall Plan for western European economic rehabilitation, and extended analogous aid to underdeveloped nations around the world (the Point-Four Program).

DEVELOPING TRADE

Along with other professionals in the years of national emergence following the mid-1940s, foresters took a growing interest in international affairs. Members of the profession were sent abroad to help with forestry projects supported by such funds as those of the Marshall Plan, Point-Four Program, and Food and Agriculture Organization (FAO) of the United Nations.

The U.S. sent abroad not only professionals, money, and ideas, but also wood products. And, too, we became a major importer of such products (Table 1-1).

TABLE 1-1
VALUE OF UNITED STATES IMPORTS AND EXPORTS OF ALL WOOD PRODUCTS AND PERCENT OF WORLD TOTAL VALUE, 1963, 1968, 1973, 1978, 1983, AND 1988

Item	1963	1968	1973	1978	1983	1988
U.S. wood-product imports:						
Million 1982 dollars	5,085	6,552	8,925	10,957	8,874	13,245
Percent of world total wood-import value	21.6	20.0	16.1	17.8	16.9	15.7
U.S. wood-product exports:						
Million 1982 dollars	1,966	3,514	5,920	5,791	5,579	9,929
Percent of world total wood-export value	9.3	12.1	12.0	10.7	12.0	12.6

Source of data in current dollars: Food and Agriculture Organization of the United Nations: Yearbook of Forest Products, Rome. (World total imports and exports reported to FAO do not agree.) Source of producer-price indices used as deflators: Survey of Current Business for December of the following year, section S (colored pages).

WORLD CHANGES

In the 1980s and 1990s, three revolutionary world changes involving U.S. forestry became evident. One was the movement of nations away from rigid central economic and social planning toward free-market democracy. Between the U.S. and the changed nations, this movement led to a lessening of military tensions and a strengthening of cooperation.

> How has the deterioration of communism in the world affected forestry in the United States?

Another revolutionary development was a profound change in the composition of the U.S. population resulting from high birth and immigration rates among Hispanics, orientals, blacks, and other minorities in comparison to the long-dominant caucasian peoples of western culture. America, once a microcosm of European nationalities, became a microcosm of the world.

The third revolutionary change in the 1980s and 1990s was the blossoming of world awareness of environmental problems that began in the 1960s and of world recognition of forests as a means for human survival. Forests became clearly perceived as carbon dioxide storage bins and as defenders of our planet's material substance against the ravages of technological development. The destruction and nonrenewal of forests, both tropical and temperate, became widely perceived as an aspect of the destruction and nonrenewal of human life.

SUMMARY

This study of social science in forestry begins with a review of the subject's national and world geographic context. Geography is a suitable setting for the study because forestry is strongly geographic in nature. World geography is a suitable setting also because the U.S. people, once isolationists, have developed an international viewpoint and concern. They exchange with other nations, not only professional resources, but also timber products. And they share with other nations the impact of changes in political systems, in ethnic distribution, and in recognition of environmental protection as a means for human survival.

CHAPTER 2

WORLD FOREST AREAS

FAO AS A SOURCE

Look, then, at the forests of the world. Almost all the information comes from the Food and Agriculture Organization (FAO) of the United Nations, with headquarters in Rome, Italy. The data are gathered in questionnaires sent by the FAO Forestry Department to the forestry official of each nation. In interpreting the results, let's recognize that:

1 Despite FAO's care in furnishing clear definitions of terms to respondents, each nation is influenced by its own conventions.
2 Different nations' units of measure are often hard to combine.
3 Many nations have only a rough idea of the quantities of their forest resources and transactions.
4 The reliability of a nation's reports and its interpretation of FAO standards may fluctuate over time.
5 Some nations do not answer FAO's inquiry, and so, if world totals are to be had, FAO must make some estimates of its own.

RECEDING OF FORESTS

When mankind was evolving, a great part of the earth's land was forested. One can imagine that during all but a fleeting fraction of subsequent time, forests maintained their dominance. It was only when most human beings ceased to be hunters and gatherers of food that the first tiny man-caused permanent changes in land use began. After that, forests were converted, along an exponential path, to farm cropland, pasture, residential space, rights-of-way, towns, reservoirs, fallow ground, and other openings.

Some corners of the earth saw widespread forest removal in ancient times. There is reason to believe that some prehistoric peoples were concerned about the shrinkage of their forests.

> What do you suppose was the basis for their concern? When you think about this question, remind yourself how primitive the transportation facilities were in prehistoric times. When a wood user's source of supply is confined to adjoining forests, what is the effect of this upon the user's concern for forest depletion?

Why did the forests recede? The quest for farm and grazing land and firewood was the prime cause. The use of timber for construction was another major cause. Some of the early Hindu scripture refers to the desirability of tree planting. Plato, about 400 B.C., wrote in these terms of land that has remained desolate to the present [2]: "There are mountains in Attica that can now keep nothing but bees, but that were clothed not long ago with fine trees. In those days, the rainfall was not lost, as it is now, through being allowed to flow over a denuded surface to the sea."

Today, forests and woodlands account for about three-tenths of the world's land area. This is roughly the same fraction of forest as in the U.S.

REGIONAL FOREST AREAS

Forest area statistics are summarized in Table 2-1. One is struck by the wide variation in percentage of forest land among the nations and regions. The lowest percentage, as we might expect, is in the Near East. The South Temperate nations are not far behind. The highest percentage of forest land, as we might not expect, is in Japan: 66 percent. Japan is an ultra-heavily populated set of islands, whose cities and farms are packed into its coastal belt, while its mountainous interiors are given to trees. The runner-up to Japan is Latin America, whose largely tropical forests have half the land.

Forest per capita, the last column in Table 2-1, is a fascinating story. Canada's 34 acres per inhabitant stand out. This forest is largely softwood, the all-purpose tree, readily usable for anything from construction lumber to paper. Canada is widely viewed as the world's wood storehouse, though Canadians themselves have misgivings. Writes one student of the subject [3]: "We have all but run out of 'back forties' in which to expand wood fiber exploitation. Very little provision has been made to restock accessible forests."

And what about the Commonwealth of Independent States (CIS, formerly the USSR), the other large northern-conifer region? The per-capita acreage statistics are less dramatic. Total forest area is greater than in Canada, but the population dependent on the forest is much greater. Acres per capita are a mere 8, less than a fourth of the Canadian figure.

TABLE 2-1
FOREST AND WOODLAND AREA: TOTAL, PER CAPITA, AND PERCENT OF ALL LAND AREA, BY NATION OR REGION, 1987–1988

			Forest and Woodland		
Nation or region	**All land, in million acres**	**Population, in millions**	**Total, in million acres**	**Proportion of all land, %**	**Acres per capita**
United States	2,316	246	669	29	2.7
Canada	2,329	26	894	38	34.4
Latin America	5,096	430	2,440	48	5.7
West, Central Europe[a]	1,195	497	397	33	0.8
CIS[b]	5,624	286	2,384	42	8.3
Near East[c]	1,717	191	119	7	0.6
Africa, except South Africa	7,177	577	1,722	24	3.0
Far East[d]	5,170	2,685	1,285	25	0.5
Japan	96	123	63	66	0.5
South Temperate[e] Zone	2,301	54	298	13	5.5
World	33,021	5,115	10,271	31	2.0

[a]Through Finland, Poland, to Black and Aegean seas.
[b]Commonwealth of Independent States, formerly the USSR.
[c]Through Afghanistan.
[d]Pakistan eastward through Pacific islands.
[e]Australia, New Zealand, South Africa.
Source: Food and Agriculture Organization of the United Nations, *Production Yearbook, 1988*, Rome, 1989; pp. 47–58, 63–79.

Besides Canada and the CIS, two regions that enjoy a significantly more extensive forest per capita than the world average of 2 acres are Latin America (5.7 acres, a figure found to be shrinking fast) and the South Temperate nations (5.5 acres), which have appreciably extended their forests through planting.

> What differences in popular attitudes and customs can one expect between regions with large per-capita forest acreage and those with little? Can you specify examples?

FOREST ASSOCIATIONS

What associations of tree species make up the world's 10 billion acres of forest and woodland? The picture, generalized to an extreme, is sketched in Figure 2-1. Here four principal types of forest are recognized:

1 *Cool conifer forest:* pine, spruce, fir, larch, birch, and related forest extending south from subarctic and high-altitude timberline to the zones where northern or tem-

FIGURE 2-1
Forest regions of the world. (*Source:* Sommer, Adrian: "Attempt at an Assessment of the World's Tropical Moist Forests," *Unasylva* 28 (112, 113), 1976, pp. 5–24.)

perate hardwoods enter the plant community or low precipitation terminates tree cover. This forest forms a belt around the earth. It crosses Canada, dipping here and there into the U.S. It extends from the Norwegian coast across Sweden, Finland, and the CIS.

2 *Temperate mixed forest:* pines and spruce, hemlock, Douglas-fir, araucaria, and other conifers; or oak, beech, maple, hickory, nothofagus, eucalyptus, or other broadleaved genera; or, most commonly, a mixture of softwoods and hardwoods. Blocks of this forest border the cool coniferous forest in northwestern and eastern Asia and in most of the southern tips of land in the South Temperate zone.

3 *Tropical moist forest:* This forest is characterized by very large numbers of tree species in mixture. Two classes may be recognized. One is the evergreen, or rain, forest, where at most a few trees in a stand will shed their leaves at any one time. The other is the deciduous forest, where the onset of a yearly dry season causes widespread leaf fall. Latin America, southern and southeast Asia, and the South Pacific islands contain most of the tropical moist forest. Central Africa also has large stretches of it.

The tropical moist forest areas of the world have been an object of concern because of the rate at which they have been deforested. Provisional estimates by FAO [1] covering 3,580 million acres that were forested in 1980 show an annual average deforestation of $41\frac{1}{2}$ million acres to 1990. The annual rate of change during the decade averaged 0.9 percent in Latin America, 1.4 percent in Asia, and 1.7 percent in Africa. All in all, the shrinkage was twice as great at the end of the decade as at the beginning.

> The tropical moist evergreen forest is said to be the most complex and to have the highest biomass per acre of any ecosystem on earth. A horizontal cross section of the forest reveals a bewildering variety of plant and animal associations. A vertical cross section displays a wide range of habitats. At the roof are sunshine and rain, encouraging flowers, insects, and birds. At the floor, dank and dim, is a different world.
>
> Can you draw a comparison between the tropical moist evergreen forest and a modern city? Is it true that the city is a less complex ecosystem? that it has a lower biomass? that its horizontal and vertical cross sections show less variety? What can we learn from the city about the forest, and vice versa?

Dry forest: In tropical regions of lower rainfall, the moist forest grades off into more open cover, with fewer stories of trees, and thence into savannah and other low, transition woodlands. Most of this is in Africa. Along with it may be classified the low, open woodlands of temperate dry areas, notably those of southwestern North America, the Mediterranean borders (including Plato's mountains of Attica), and northern and eastern Australia. The woodland types vary from stunted and twisted oaks, maquis, and other leathery broadleaves to the piñon-juniper of western American range country.

SUMMARY

Occupying just under a third of the world's land surface, forests are widely varied and unevenly distributed. The great conifer belts are in the CIS, Canada, and U.S. The most spectacular hardwoods are tropical. Sparse woodlands, too, are in the tropics and on the Mediterranean borders, in western U.S., and in the South Temperate nations.

REFERENCES

1 Dembner, S.: "Provisional Data from the Forest Resource Assessment 1990 Project," *Unasylva,* 42(164), 1991, pp. 40–44.
2 *Unasylva,* 31(125), 1979, p. 26.
3 Wetton, Frances: "Canadian Forests—Past, Present, and Future," *The Forestry Chronicle,* February 1982, pp. 11–14.

CHAPTER 3

FORESTRY AND ECONOMIC DEVELOPMENT

ECONOMIC CHARACTERIZATIONS

In this book, nations and regions are sometimes characterized as economically "developed" or as economically "developing." Here is a distinction basic to our understanding of the world, world forestry, and social science in forestry.

The economically **developed** or **more developed** nations are found mostly in earth's temperate zones; the economically developing are mostly tropical. **Developing** nations are often termed **less developed** or, occasionally, **underdeveloped** nations. They are the comparatively poor nations. A common term for them is **third-world** nations, formerly referring to the fact that they were not highly organized economies, either free-market (noncommunist) or centrally planned (communist).

> Why do you suppose developed nations are so often found in temperate zones; developing nations, in the tropics?

CHARACTERISTICS OF DEVELOPMENT

What is economic development? How does it come about? How does forestry relate to it?

Beneath the idea of **development** is **economic growth,** an increase in per-capita **real income:** income in dollars of constant purchasing power. Income is measured by gross national product (GNP) or, alternatively, by **gross domestic product** (GDP), which is GNP minus the product generated by foreign-owned assets. Development involves economic growth plus additional changes.

The additional changes, beyond economic growth, that characterize development include a reduction in absolute poverty. They include a lowering of the costs of transactions: improvements in land management and in the technology of processing, distributing, and consuming genetic and manufactured outputs and in the accompanying financial system. Changes include also the betterment of institutions for education, health, and the quality of life; provision for people's freedom of choice; and shifts in the balance among the economy's **primary sector** (raw-material production), its **secondary sector** (end-products manufacture), and its **tertiary sector** (services).

Development entails enhancement of the tertiary sector. But large tertiary activity is not in itself evidence of development. Witness the great numbers of houseworkers, shoeshine boys, and errand-runners in poor nations where growth is slow.

DEVELOPMENT AS A GOAL

After World War II, economic development became an explicit goal for many nations. People were traveling and communicating. Citizens of the poorer countries became aware of the gulf that set them apart from the materially prosperous peoples. As many of them seized upon national independence and self-determination as appropriate to the new times, so also they reached for development. They wanted to catch up with the leaders and thus enjoy what they saw not only as a better livelihood, but also as a better life. And the developed nations recognized the inequity, even danger, of wide discrepancies between their lot and that of others.

Many nations set economic-growth targets. The U.S. came to regard a 4- to 5-percent annual rate as a norm. Brazil hoped for 8 percent. In the early 1970s, the Soviet Union planned for a 7.5-percent annual growth rate. None of these nations fulfilled its hopes for long. But their goals were not unrealistic. In the years 1965–1987, Japan, a wealthy nation, enjoyed a 4.2-percent annual growth. In South Korea, a middle-income country, the rate was 6.4-percent. Botswana, in the same bracket, turned in an 8.9-percent world record. Needless to say, rates fluctuate. The 1970s saw generally high rates; the 1980s, rates not so high. Early in the eighties and again with the opening of the nineties, growth rates plunged because of widespread economic recession.

Some illustrative data on GNP and growth, along with population and its growth, are in Table 3-1. The table exhibits wide differences in material well-being. Of the six most populous nations, three generated GNP of less than $500 per capita in 1987. Sub-Saharan Africa other than Botswana and South Africa suffered a severe loss of per-capita GNP during the 1980s, while fast spreading famine in the nineties was putting 30 million Africans at risk of malnutrition or death. Meanwhile the West was said to be weary of giving aid [9].

During the 1960s, some countries began to doubt the validity of their development goal. At any rate, people came widely to question the goal of growth in GNP. Growth, they observed, is often accompanied by environmental deterioration—pollution, con-

TABLE 3-1
BASIC DEVELOPMENT INDICATORS FOR SELECTED NATIONS, 1987

Nation	Real GNP per capita		Population	
	Total, 1987, in dollars	Annual growth, %, 1965–1987	Total, 1987, in millions	Annual growth, %, 1980–1987
Switzerland	21,330	1.4	6.5	0.3
United States	18,530	1.5	243.8	1.0
Norway	17,190	3.5	4.2	0.3
Japan	15,760	4.2	122.1	0.6
Sweden	15,550	1.8	8.4	0.1
Finland	14,470	3.2	4.9	0.5
France	12,790	2.7	55.6	0.5
Australia	11,100	1.8	16.2	1.4
United Kingdom	10,420	1.7	56.9	0.1
New Zealand	7,750	0.9	3.3	1.0
Greece	4,020	3.1	10.0	0.5
South Korea	2,690	6.4	42.1	1.4
Argentina	2,390	0.1	31.1	1.4
Brazil	2,020	4.1	141.4	2.2
Mexico	1,830	2.5	81.9	2.2
Chile	1,310	0.2	12.5	1.7
Botswana	1,050	8.9	1.1	3.4
Honduras	810	0.7	4.7	3.6
Philippines	590	1.7	58.4	2.5
Indonesia	450	4.5	171.4	2.1
India	300	1.8	797.5	2.1
China	290	5.2	1,068.5	1.2
Madagascar	210	−1.8	10.9	3.3
Zaire	150	−2.4	32.6	3.1
Ethiopia	130	0.1	44.8	2.4

Source: The World Bank: *World Development Report, 1989,* Oxford University Press, New York, 1989, pp. 164–165, 214–215.

gestion, noise, ugliness, and material-resource depletion—and by institutional changes such as concentration of financial and political power and the loss of small firms.

It is among the developed economies with a noncommunist background that continuous economic growth as a national goal is most widely questioned. The developing nations, by and large, cling to the growth principle. Literally, they can't afford to do otherwise. The bulk of them, in Africa and Latin America, have a good deal of open space. "We still have plenty to pollute," as one government official puts it.

> Is it possible to enjoy economic development without suffering environmental losses?

ACHIEVING DEVELOPMENT

How is economic growth, the key ingredient of development, achieved? Primarily by making the systems of production and consumption more efficient, so that per-capita output (that is, income) can rise, and concurrently strengthening the cultural and natural resources to which the improved systems can be applied. In other words, development is achieved by enlarging the national capital per worker (**deepening** the capital) and investing it in ways that make people more productive. They may then—*given the essentials of law, order, stable government, and an efficient financial system*—use the fruits of their productivity for a better livelihood.

There are two ways of enlarging the national capital. One is to borrow abroad. Brazil is a prime example, with a total foreign debt of $124 billion in 1987. Mexico and Argentina were next in order; their cumulative total borrowings abroad were $108 billion and $57 billion, respectively, in 1987 [15]. The lenders were primarily public development organizations such as the World Bank and the Inter-American Development Bank and, secondarily, private banks such as those of U.S. and Japan. Reducing the staggering external debts is a major preoccupation in the debtor nations and in international monetary circles. Britain has led a recent move to forgive the debt of very poor nations. Japan has hung back for philosophical reasons [10]; the U.S., because of its 1990 deficit-reduction law [11].

The other way to enlarge the national capital is to save at home. Saving, in turn, can be accomplished in two ways. One is through private initiative: individuals and corporations drawing upon their income. The bulk of such saving is by the wealthiest class: the "robber barons" in the U.S., the sheiks in the Arab countries.

Since saving requires the postponement of consumption, saving through private initiative is a difficult alternative for developing economies, where people consume their meager income in order to survive. Here the nation may resort to the second way to save: through government initiative. A government can force saving by levying taxes or selling bonds, using the proceeds for investment—or reduction of external debt. Or the government can require corporations to retain more profits for expansion. Or private saving can be *forced* through monetary inflation: The people are obliged to give up some consumption as the government steps up the printing of money and uses some of the extra funds as it would use the returns from taxation or bond sales. The prime three debtor countries illustrate high rates of monetary inflation. From 1980 through 1987, Brazil's average rate was 166 percent per year (1,795 percent in 1990 alone [7]); Mexico's was 69 percent; Argentina's, 299 percent. Compare the U.S.: 4.3 percent [14].

In Latin America, Bolivia offers a dramatic example of inflation and the control of it. In the early 1980s, Bolivia's price index was rising 26,000 percent per year. Drastic government measures reduced the rate to 66 percent by 1986, 18 percent by 1990. But economic growth was a different, less successful, story. Gross domestic product rose 2.6 percent in 1990, but was outstripped by population growth [8].

How do extreme rates of monetary inflation affect the behavior of consumers? producers? the form of saving? the practice of timber and nontimber forestry?

Bolivia's population increase gives cause for thought. Here is a nation where growth in GNP, positive as a total, was rendered negative per capita, so that development was negative. This reversal of the development process by a steeply rising population is found here and there in all the world's poor regions. However, it can be argued that if a poor people can achieve development, this achievement will go far to stem the population tide and thus enhance developmental gains. Parents, more independent with rising income, need no longer rely so heavily on children for their social security. Education becomes more dominant in early life, and marriages are postponed. Women obtain freedoms—and obligations other than child-bearing. Thus a rise in prosperity changes old life patterns into new.

In the process of economic development, the national capital enlarged through borrowing or saving is put to two kinds of use and goes, also, to support a third prime requirement of development.

1 Investments in social facilities such as schools, transportation and communication systems, health institutions, and water works. Also of major importance are research projects that will promote technical change and growth in productivity. Such facilities are termed the **infrastructure** for development. They make up the bulk of the total national investment. Much of the infrastructure is versatile, even that which is created for special purposes. For example, roads built for logging come to be used for general transportation.

2 Investments in facilities not necessarily social: factories, machinery, power supplies, research and development organizations, and worker-training programs. Here we see development from the viewpoint of the firm.

3 Investments in cultural evolution. Such evolution, of course, is not primarily a function of capital investment in the usual sense. It is mainly a function of people's investment of their will and energy in behalf of development. How willing are the people to accept new ideas, live by the clock, strive for material benefits? A South Korean economist commented on the change in popular attitude: "When you have a project to propose, people used to say, 'Don't bother us.' Now they say, 'When do we start?' "

Investment in cultural evolution, and much of that in infrastructure, is investment in **human capital.** Of all the uses to which a nation's saving is directed, such investment is conceivably the most fundamental to development.

RELATION TO INCOME DISTRIBUTION

The relation of economic growth to income distribution is a persistent question. It is a big question, because from nation to nation such very different fates can befall the low-income classes while the average income per capita is growing. An accepted norm is that growth should benefit the poor: help them catch up with other households, just as the nation itself is trying to catch up with other nations. In the extreme, the income distribution in a country can be narrowed while it is being moved up the scale, so that the incomes of the poorest families are much improved and those of the wealthiest may even be reduced. Thus the Cubans redistributed incomes, though at the cost of slow growth. Planners in communist USSR held consumer spending at existing low levels while forcing economic growth.

Growth is often neutral toward income distribution. The Brazilians have believed that the only way to better the lot of the poor is to make everyone more productive and thus benefit the rich at the same time. This means moving the whole national-income distribution without much changing its shape.

But economic growth can increase the inequality of income-distribution as it did here and there after the early 1960s. A nation can grow economically without raising the income of any but the wealthiest families. And it is even possible for such growth to be accompanied by a further impoverishment of those already in poverty—the rich getting richer while the poor get poorer. In Indonesia, outstanding for its poverty and rapid economic growth, the government is concerned over the mushrooming of lavish mansions in the midst of Jakarta's prevailing squalor. In the United States, with its very modest growth rate in the early nineties, the contrast, rich vs. poor, was all too sharp: The top fifth of income earners took in nearly half of the nation's total income; the bottom fifth, less than 4 percent [10].

The relation between economic growth and income distribution remains in debate. No clear principle has been discovered. For example, the question of who has benefited from Brazil's economic development has been a subject of keen controversy among social scientists [3].

> Do you know how income distribution in the U.S. changed while our income was rising? Where, specifically, would you look for the data?

THE ROLE OF FORESTRY

Since development is resource creation, forestry is part of the story. Every one of the developed forested nations has based its growth to a significant degree upon its forests, primarily its timber inventory, at least at the start. Other forest resources have become prominent later. As for timber forestry—logging; regeneration; silviculture; and wood-products manufacture, shipment, and marketing all enter into development.

Forestry has many advantages in fostering development. Most obvious is that the mature forest is a reservoir of savings. If the trees are useful and can be made accessible, they are a highly liquid and versatile asset. They can be used at home as raw material. They can be shipped abroad in any form from logs to highly finished products and be exchanged for valuable capital goods and consumer goods.

Sweden's economic development is a classic example. Endowed abundantly with such resources as forests, waterways, iron ore, and human talent, Sweden during the mid-nineteenth century converted itself from a farming to an industrial nation. The forest was a principal source of the required capital, supplying charcoal for steel manufacture and a variety of other goods for export and home use. In the twentieth century, the Swedes repaid their loan. They rebuilt the forest from a devastated remnant to a conservatively managed asset in which capital is still accumulating as growth exceeds removals. Forestry and wood-using industry contribute largely to the Swedish economy, which provides its people with a per-capita GNP among the highest in the world (Table 3-1), growing at rates well above 3 percent in some years.

The U.S. offers interesting examples of forest-generated development. Consider the southern pine region, extending from Virginia southward and westward into Texas and Oklahoma. Between 1890 and the 1930s, the softwood timber industry mowed the forests of the region to make construction materials for the central and eastern states. In the mid-1930s, when the timber economy appeared to be at a standstill if not an end, new knowledge and initiative in pulp and paper manufacture set off an explosion based on young-growth pine forests that sent pulpmill capacity rocketing from a regional total of 5,000 tons per day to 35,000 tons by 1955, 55,000 tons in 1970, and more than 130,000 tons in 1990 [1, 2]. In another burst of activity founded on new timber growth and technology, southern pine plywood blossomed from a one-plant industry in 1964 to a multifirm, multiplant complex with an annual capacity of several billion square feet by the end of the decade. In the 1980s, reconstituted panels were capturing the market. These changes were accompanied by corresponding increases in employment, income, and the ability to save for still further ventures in development.

The point is that forests—old growth and young growth, too—are reservoirs of capital that may be drawn upon as needed to take advantage of opportunities for economic growth and development.

The opportunity for rural work is an advantage of forestry as a base for development. Logging, for instance, spurs growth in the poorer nations. It has these advantages: It can be designed to be **labor-intensive:** to use large numbers of unskilled workers per unit of product—to keep them employed, even though the return is low. In other words, logging can be **capital-extensive:** It can use sparingly the one agent of production, capital, that is in shortest supply in developing countries. Logging can be made more sophisticated and productive as labor learns to work with more elaborate capital equipment. Thus, it is a training ground, not merely a labor-user. And logging has the advantage that it keeps people on the land, and not moving into the cities. Its principal inputs, labor and land, have very low social cost—that is, if they were not used in this way, they might then be unemployed.

But when we think of the blessings of logging as an enterprise and source of employment for the peoples of developing countries, we are forced to remind ourselves of the devastating environmental impacts of the logging that is going forward in the tropical forests of the poor nations at the hands largely of emissaries from the earth's centers of wealth. The World Resources Institute reports a total of 40 to 50 million acres of tropical moist forest disappearing yearly. "It's as if we cleared and burned the state of Washington in a year." Leading targets of deforestation: Brazil, India, Indonesia, Myanmar (formerly Burma), Thailand, Vietnam, Philippines, Costa Rica [16].

Forest planting is another instrument for economic development widely used in the underdeveloped regions. Planting has much the same advantages as logging. And instead of logging's environmental faults, it has only virtues. Brazil plans to reforest 2.5 million acres of the Amazon basin over 10 years at a cost of $3 billion. Funding will come from private domestic industry and foreign contributions [6].

It can be argued that logging and planting are not effective promoters of economic growth. The argument runs that they yield low returns on investment and must occupy very large acreages if they are to have much impact on development.

> Where logging and planting make inefficient use of land and human capital, cannot development be promoted by introducing modern mechanized equipment so as to increase per-capita product? What deters the use of this remedy?

FOREST INDUSTRY AND DEVELOPMENT

While timber management has its advantages in promoting economic growth, the higher stages of production can be effective, too. Manufacturing is a case in point. The pit-sawing, or whip-sawing, of lumber (see page 300), unknown in the U.S. since the earliest colonial times, is in some nations today a significant source of employment—though admittedly a fast-dwindling source as the nations progress. Such manufacturing develops habits of work and familiarity with simple tools, just as is true of logging and planting. Once familiar with simple tools and methods, workers can graduate to more complex situations. People can be trained; schools can be improved; people can attend school longer.

Some manufacturing investments have failed because they were too far advanced for the nation where they were made. Consider the outcome of building a high-speed, large-capacity, computerized pulp-paper installation in Zaire. The infrastructure is not there yet. Conditions favor labor-intensive processing for a starter.

As a nation develops, it is fortunate if it can retain an extensive back-country that can be drawn upon for material resources and can accommodate the overflow of families from industrial centers. The U.S. is a classic example. Brazil promises to be another if it can stave off the heavy drain upon its back-country rain-forest reserves.

Consider an illustration of higher-stage timber management in Brazil: the community of Porto Santana, which has plenty of backcountry from which to draw raw material and also labor to support a logging, manufacturing, marketing, and shipping complex.

Porto Santana, at the mouth of the Amazon, gives access to ocean-going vessels and to log supplies from the interior. Built as an outlet for backcountry mines, it has docking and loading facilities. It was judged to be a good spot for a veneer and plywood plant, and a plant was built by a Brazilian–Dutch partnership.

The times smiled on the venture, for the Brazilian economy was then growing at a 10-percent annual rate. Unskilled Brazilians were employed, backed up by a few skilled foreigners. The native workers were cooperative and soon learned to arrive at the job on time. The company built a school, a road, and a medical dispensary and helped buy medical facilities. It set up a system of consumer credit that let employees buy bicycles to get to and from work.

The plant was profitable and was soon enlarged, and a sawmill was added. Supplies of *Virola,* the preferred tree genus, proved sufficient and were conserved under new government forestry requirements.

Thus the operations of this firm provided plywood and other products, employment, social capital, and some increase in the supply of foreign exchange. Further, the government's legal requirements in forestry and the corporation's good citizenship helped safeguard the timber supply.

THE FINANCIAL SYSTEM

What *is* the effect of a nation's forestry practice on its economic development? After World War II, internationalists acquired a strong faith in the efficacy of traditional forest measures: silviculture, harvesting, transportation, processing, marketing. Then around 1970, breaches began to show up in the smooth fabric of confidence. Regions that by all the rules should have been emerging from poverty were not doing so. Critical questions were argued [4] in the forestry journals. By the late 1970s, the World Bank was announcing new policy for its forestry lending: Shift emphasis from practices to people. Give high priority to forestry for environmental improvement, for rural activity, and for building human capital [12].

To grasp the bearing of forestry on economic growth and hence development, one is obliged to stand back from both the forest and the trees and to examine the whole context in which growth can readily progress.

A trend of current thinking is that an essential ingredient in economic growth is an efficient financial system [13]. Compare Switzerland, a land of meager resources, with Argentina, whose resources are varied and abundant. Switzerland, with its highly efficient financial system, leads the world in per-capita income; while Argentina lives with a mediocre income increasing at a negligible rate (Table 3-1).

A nation's **financial system** is its aggregate of institutions concerned with the use of the exchange mediums: determining their value and promoting their flow in response to supplies and demands. An efficient system reduces the cost of all transactions. It encourages trade and hence specialization and productivity. Specialization in conformity with the principle of comparative advantage (page 28) favors economic growth [5]. An efficient system promotes saving by bettering yields and liquidity. It enables financial agents to pool, evaluate, and exchange risk. And so it promotes fruitful investment, resource use, economic growth, and development.

Today's financial system is the banking system, its affiliated producers and consumers, borrowers and lenders, and its exchange mediums: currency, demand and time deposits, and other, more sophisticated instruments. These mediums of exchange have evolved from precious metals and their coinage, other valued goods used as payment, consumer goods traded in a barter system, and, ultimately, goods consumed by their producers in a self-sufficient economy.

> W. S. Jevons, writing in the nineteenth century, tells the story of Mademoiselle Zelie, a singer, who gave a concert in the Society Islands in exchange for one-third of the receipts. Her share turned out to be 3 pigs, 67 turkeys and chickens, thousands of cocoa nuts, and quantities of fruit.
>
> How could Mademoiselle Zelie have proceeded so as to maximize her reward for singing? What does your reply suggest to you about the merits of an efficient financial system?

The star performers in economic growth and development are people, not trees.

SUMMARY

Economic development is an evolutionary process characterized by two sorts of change. One is economic growth, a rising per-capita real GNP. The other is improvement in production and consumption systems, infrastructure, transactions costs, and the balance of activity among economic sectors. Development is a goal adopted by many nations and stressed by the poorer nations. It is achieved by creating resources: by enlarging the national capital and investing it to make people more productive. Forests and forestry can be instrumental in development because they provide a reservoir of capital, opportunities for rural labor to learn skills gradually, and a wide range of productive activities from logging and tree planting to end-product manufacture and marketing. However, in recent years we have come to recognize that the effectiveness of forests and forestry in the economic growth and development of a nation depends significantly on the efficiency of the nation's financial system.

REFERENCES

1 American Paper Institute: *1990 Statistics of Paper, Paperboard and Wood Pulp,* American Paper Institute, New York, 1990, pp. 4–5.

2 Duerr, William A.: *Fundamentals of Forestry Economics,* McGraw-Hill, New York, 1960, pp. 23–26.

3 Fields, G. S.: "Who Benefits from Economic Development?—A Reexamination of Brazilian Growth in the 1960's," *American Economic Review,* September 1977, pp. 570–582; Discussion, *American Economic Review,* March 1980, pp. 242–262.

4 Kromm, David E.: "Limitations on the Role of Forestry in Regional Economic Development," *Journal of Forestry,* October 1972, pp. 630–633. Gregerson, Hans M.: "The Role of Forestry in Regional Economic Development, an Alternative View, *Journal of Forestry,* February 1973, pp. 98–99. *See also* Kromm, David E.: "A response," *Journal of Forestry,* February 1973, p. 99.

5 Vollrath, Thomas L., and Paul J. Johnston: "The Influence of Commodity Composition of Trade on Economic Growth," *The Journal of Agricultural Economics Research,* Winter 1991, pp. 7–14.

6 *The Wall Street Journal,* Eastern ed., December 20, 1990, p. A12.

7 *The Wall Street Journal,* Eastern ed., December 31, 1990, p. 4.

8 *The Wall Street Journal,* Eastern ed., June 18, 1991, p. A15.

9 *The Wall Street Journal,* Eastern ed., June 19, 1991, p. A1.

10 *The Wall Street Journal,* Eastern ed., October 18, 1991, p. A11.

11 *The Wall Street Journal,* Eastern ed., November 4, 1991, p. A14.

12 The World Bank: *Forestry Sector Policy Paper,* The World Bank, Washington, 1978, pp. 5–37.

13 The World Bank: *World Development Report 1989,* Oxford University Press, New York, 1989, especially pp. 1–69, 122–132.

14 The World Bank: *Ibid.,* p. 189.

15 The World Bank: *Ibid.,* p. 205.

16 The World Resources Institute, the United Nations Environment Programme, and the United Nations Development Programme: *World Resources, 1990–91,* Oxford University Press, New York, 1990, pp. 101–102.

CHAPTER **4**

TIMBER ECONOMY OF THE WORLD

WORLD WOOD PRODUCTION

To continue our examination of world timber affairs, let's turn to Table 4-1. From columns (2) and (3), it's evident that most nations increased their total timber output between 1968 and 1988. Certainly the U.S. did, and here some figures are at hand to support interpretation.

The four decades from mid-century to 1990 witnessed a rather consistent upward trend in U.S. wood harvest. The use of commercial products was increasing, and the use of all wood was on the upswing, too, despite a two-decade downward drift in firewood use. Firewood has been bearing the brunt of the **pulverization principle,** which exerts its influence in developed economies, where labor is growing expensive while capital is becoming abundant. The principle holds that in such economies, industry turns from labor-using materials that come in chunks (firewood, other roundwood, and timbers, which are produced and lugged in **batches**) toward labor-saving materials in divided or pulverized form (plywood, particleboard, and pulp—or wood chips used as fuel—which can benefit from **flow** processing and transportation). (See pages 97 through 100 for some details.) Under the pulverization principle, firewood was being replaced by fuel oil and other easy-to-use energy sources.

But then about 1973 came the **OPEC Surge.** The Organization of Petroleum Exporting Countries decided to cut production. The price of fuel oil exploded—from perhaps 20 cents per gallon to well above a dollar. Householders by the millions, predominantly in rural areas, acquired heaters that would burn wood and turned to this cheaper source of heat, tolerating its inconvenience. Chain saws reverberated around the forest, often in timber of noncommercial quality. Nonindustrial private holdings and national forests yielded energy that was low-cost, especially if the users did the logging themselves [3].

TABLE 4-1
PRODUCTIONOF ALL WOOD AND OF FIREWOOD, ROUNDWOOD EQUIVALENT, BY NATION OR REGION, 1968 AND 1988

	All wood				Firewood			
	Billion cubic feet		Percent of world total		Billion cubic feet		Percent of all wood	
Nation or region (1)	1968 (2)	1988 (3)	1968 (4)	1988 (5)	1968 (6)	1988 (7)	1968 (8)	1988 (9)
United States	11.68	18.82	15	16	.75	4.11	6	22
Canada	3.95	6.35	5	5	.17	.24	4	4
Latin America	7.47	13.60	9	11	5.93	9.82	79	72
West, Central Europe[a]	10.64	12.87	13	11	2.26	1.98	21	15
CIS[b]	13.43	13.84	17	11	3.19	3.06	24	22
Near East[c]	1.03	1.09	1	1	.65	.65	63	60
Africa except South	10.00	16.55	13	14	9.03	15.01	90	91
Far East[d]	18.78	35.26	23	29	15.15	27.17	81	77
Japan	1.93	1.00	2	1	.22	.02	11	2
South Temperate[e]	1.28	1.77	2	1	.35	.35	27	20
World	80.20	121.15	100	100	37.70	62.41	47	52

[a]Through Finland, Poland, to Black and Aegean seas.
[b]Regions traditionally included in the former Soviet Union.
[c]Through Afghanistan.
[d]Pakistan eastward through Pacific islands.
[e]Australia, New Zealand, South Africa.

Source: Food and Agriculture Organization of the United Nations: *Yearbook of Forest Products, 1978;* FAO, Rome, 1980, pp. 73–74, 87–88. Food and Agriculture Organization of the United Nations: *Yearbook of Forest Products, 1988,* FAO, Rome, 1990, pp. 2–3, 16–17.

The OPEC Surge took the form of a $5\frac{1}{2}$-fold increase in U.S. firewood output between 1968 and 1988 [columns (6) and (7) of Table 4-1]. Such an eruption of firewood activity occurred in no other nation. Why not? Because countries other than the U.S. had not previously stretched their rejection of firewood so near the limit. They had been performing all along at a norm more nearly consistent with the accessibility of wood and the willingness of people to put up with wood as fuel.

The U.S. firewood peak was reached around the mid-1980s, and then a downward drift recommenced, reminiscent of the pre-1970s. The Persian Gulf crisis intervened, but efficient wood-burning stoves were now at hand and in use [4], and the U.S. appears to have escaped an **"Iraqi Surge."** The story is graphed in Figure 4-1. The absolute levels of the graph, especially before 1970, are only approximate.

Nowadays, firewood is still the world's dominant timber product in terms of quantity [column (9) in Table 4-1].

> How do you explain the fact that in recent years, and certainly in earlier times, the lowly firewood dominated world timber production? How do you characterize the regions where firewood was especially dominant? What place do you think the future holds for firewood?

TIMBER EMPLOYMENT AND VALUE ADDED

From Dwight Hair [1], Robert Phelps [2], and rough extensions of their work by the present writer, we have estimates of U.S. timber employment and value added cover

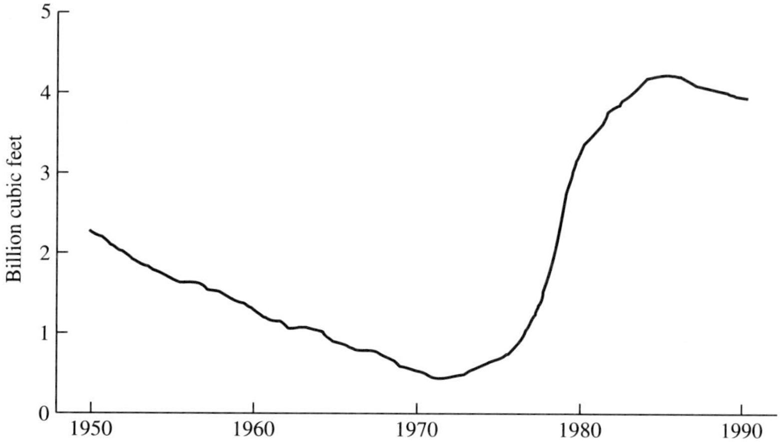

FIGURE 4-1
Annual firewood production (consumption) in the U.S., 1950–1990. (*Sources:* 1950–1975; U.S. Department of Agriculture, Forest Service: An Analysis of the Timber Situation in the U.S., 1952–2030; Forest Resource Rep. 23, 1982, p. 305. 1976–1990: Food and Agriculture Organization of the United Nations: *Yearbook of Forest Products, 1990,* Rome, 1992.)

ing much of the four recent decades. **Value added** in any sector of an economy is the value of the sector's end products minus the value of other sectors' end products used as inputs. It is the sum available for wages, salaries, interest, rent, depreciation, taxes, and profit or loss. For example, value added in stumpage production is the value of stumpage made available for harvest minus (for instance) the value of fertilizer applied to the forest, which is a starting point for figuring value added in the fertilizer industry.

> The trend of value added is an index of economic development, since the total of values added in all sectors of a national economy equals GNP. Explain why this is so and why the total entails no double-counting.

Timber's contribution to GNP in constant dollars was increasing persistently in the U.S. during the four decades, in the face of lagging contribution per cubic foot of output.

Employment in the U.S. timber economy was pointing upward. Meanwhile, employment per cubic foot of wood output was dropping off with rising labor productivity, especially in logging and in manufacturing industries other than lumber (Chap. 39).

APPARENT WOOD CONSUMPTION

Commencing with data on a year's wood production for a nation or group, if we add imports and subtract exports, we are presumably left with the total quantity of wood that remained at home for consumption that year. In some cases, we might be able to sharpen the picture by allowing for changes in stocks on hand, but here we omit this nicety. The resulting **apparent consumption** is a revealing figure because it can meaningfully be put on a per-capita basis (Table 4-2).

Just one caution respecting the figures in Table 4-2: They cover firewood, pulpwood logs, lumber, veneer, plywood, composition board, pulp, paper, and paperboard, but not more highly fabricated goods such as furniture, case goods, pallets, doors, and paper products. Consequently, the table tends to overstate apparent consumption in world regions that are net exporters of the more highly fabricated products—for example, Japan and western Europe. And it tends to understate consumption in regions, notably the developing regions, that are net importers of such products.

> Explain the foregoing reasoning.

TABLE 4-2
APPARENT CONSUMPTION OF ALL WOOD, TOTAL AND PER CAPTIA, BY NATION OR REGION, 1988

Nation or region	Production, billion cubic feet	Imports, billion cubic feet	Exports, billion cubic feet	Apparent consumption: Total, billion cubic feet	Apparent consumption: Per capita, cubic feet
United States	18.82	3.37	2.58	19.61	79.7
Canada	6.35	.41	4.19	2.57	98.8
Latin America	13.60	.43	.61	13.42	31.2
West, Central Europe[a]	12.87	8.25	6.81	14.31	28.8
CIS[b]	13.84	.11	1.34	12.61	44.1
Near East[c]	1.09	.47	.01	1.55	8.1
Africa except South	16.55	.40	.23	16.72	29.0
Far East[d]	35.26	2.02	1.72	35.56	13.2
Japan	1.00	2.83	.08	3.75	30.5
South Temperate[e]	1.77	.23	.62	1.38	25.6
World	121.15	18.52	18.19	121.48	23.7

[a]Through Finland, Poland, to Black and Aegean seas.
[b]Regions traditionally included in the former Soviet Union.
[c]Through Afghanistan.
[d]Pakistan eastward through Pacific islands.
[e]Australia, New Zealand, South Africa.

Sources: Data on total production in cubic meters; trade in roundwood, sawn and split material, and panels in cubic meters; and trade in pulp, paper, and paperboard in metric tons from Food and Agriculture Organization of the United Nations: *Yearbook of Forest Products, 1988;* FAO, Rome, 1990. Data in metric tons were converted to roundwood equivalent at the rate of 3.4545 cubic meters per metric ton. Data in cubic meters were converted to cubic feet at the rate of 35.314 cubic feet per cubic meter. Data on population from Table 2-1.

Bearing in mind the deficiencies of Table 4-2, you observe that the world's biggest consumers per capita appear to be Canada and U.S. The smallest consumers appear to be the Asiatic nations other than the CIS.

COMPARATIVE ADVANTAGE

Having brought up the subject of imports and exports, let's look further into the flow of trade in the world's timber economy. This flow between nations (or regions or firms) is explained, in the first instance, by the **principle of comparative advantage,** formulated by David Ricardo in 1817: I tend to do what I can do well (enjoyably, profitably, or the like). If my lines of competence differ from yours, we have an opportunity to trade and to gain by it. Note that I do *not* necessarily do what I can do better than you can: I don't follow my *absolute* advantage. You may outshine me in all respects; I will still choose my best alternative.

A Finnish forester and a southern pine forester were discussing the outlook for international trade. "We can always export pulp products to southeastern U.S.," said the Finn. "How so?" the American countered. "How can you maintain access to a region that can grow and process wood so much more efficiently than you can?" The Finn replied by reminding the American, "You are forgetting how many more good alternatives you have than we!"

Two nations incur the following costs of producing and shipping a unit (worth $1) of each of two goods:

	Cost per unit of	
Nation	Gammate	Deltate
Alphaland	80¢	85¢
Betaland	95¢	90¢

Where does Alphaland's absolute advantage lie? its comparative advantage? How about Betaland?

Each nation consumes 1,000 units of each good per year. If both choose to be self-sufficient, what will the net revenue of each be? On the other hand, if each nation pursues its comparative advantage and trades its surplus with the other, what will its net revenue be?

In practice, the freely competitive working of comparative advantage can be bent by deliberate institutional changes so as to manipulate the flow of trade. Quotas, tariffs, and other trade barriers can alter comparative advantage. So can changes in technology that create resources or make workers more productive. A nation can attract investors by cutting taxes, reducing regulation, or improving education. It can manipulate exchange rates so as to gain a price advantage for its goods. Looming for the 1990s are bilateral or regional trade agreements that, one hopes, will supplement but not stand in the way of such universal agreements as GATT [5].

In 1986, at U.S. urging, Canada agreed to institute a 15-percent tax on softwood lumber exports to this country, but in following years the tax withered and in 1991 was scrapped. The U.S. industry called for retaliation.

The European Community has its own plan for 1992. Some American nations are pointing to 1994 for a start in the same direction. In recent years, the U.S. wood industry favored federal control over such trade items as Douglas-fir logs shipped out to Japan and lumber shipped in from Canada. South Korea, to protect its industry, restricted the importation of wooden chopsticks from China.

DIRECTION OF TRADE IN LOGS

A word, now, about timber-trade flows between specific nations. Start with the rudimentary products, sawlogs and veneer logs.

(There are no reliable data on international trade in firewood. The world's most lavishly used—and in many a poor nation most urgently wanted—timber product is, for the most part, too heavy and bulky per unit of value to bear the cost of more than local shipment.)

And so, turn to Table 4-3. Who are the principal exporters of sawlogs and veneer logs? If you were asked to guess before looking at the table, you might well reason that the roundwood exporters must be the countries deficient in processing facilities and mechanical skills, the poor countries that have only recently emerged from the colonial system of raw-material shipment to a mother country. In Table 4-3, you do indeed find three such nations among the leading eight exporters. Malaysia is near the head of the list. But there are surprises. The U.S. is in first place and seldom, in recent years, has been below second. The CIS is third.

As for log importers, Japan is first among the 12 leaders by a wide span: more than a billion cubic feet compared to 400-odd million for China, the runner-up. Among the largest nation-to-nation flows is from U.S. to Japan, mostly Douglas-fir, true firs, and other West Coast conifers. Why this big volume of trade?

The comparative advantage of the U.S. and CIS with respect to logs arises because these countries still have old-growth conifer reserves and permit some roundwood exports, whereas British Columbia, for example, though it has reserves, does not allow such exports. The U.S. leadership is heightened by the fact that the Japanese, who are essentially the only importers of our reserves, have a processing advantage and can pay high prices for logs. The advantage arises from Japan's low (though recently rising) wage rates, high labor skills, close raw-material utilization, advanced manufacturing technology, low cost of ocean freight, active governmental support, and high end-product values. Japan is advantaged in lumber and panel, but not log, production because of the foregoing considerations and the limited output of Japanese forests.

One observes in Table 4-3 that some nations are among the heavy traders in timber which, just a few years earlier, would scarcely have been represented at all: notably members of the European Economic Community, which came prominently into the

TABLE 4-3
DIRECTION OF TRADE: MILLION CUBIC FEET OF SAWLOGS AND VENEER LOGS SHIPPED BY MAJOR EXPORTING NATIONS TO MAJOR IMPORTING NATIONS, 1988

	Major exporting nations								
Major importing nations	United States	Malaysia	Confederation of Independent States	Canada	France	Chile	Germany	Papua New Guinea	World
Japan	386.29	396.07	169.35	84.43	.07	3.11	—	24.96	1064.28
China	193.32	150.07	80.90	5.65	—	13.84	—	.78	444.56
S. Korea	108.47	112.39	—	2.40	—	14.83	—	12.92	258.40
Canada	55.19	—	—	—	—	—	—	—	114.44
Italy	2.90	—	—	1.52	19.03	—	6.25	—	96.40
Germany	6.18	—	—	—	8.51	—	—	—	60.31
Austria	—	—	8.93	—	—	—	32.31	—	58.72
Spain	.35	—	—	—	16.10	—	—	—	40.43
India	—	34.96	—	—	—	—	—	1.38	36.37
Sweden	—	—	17.23	—	—	.60	.85	—	32.38
Finland	—	—	14.79	—	—	—	—	—	30.61
Turkey	5.12	—	—	.60	—	15.85	.11	—	27.54
World	767.15	726.40	308.61	116.52	76.66	49.43	46.11	45.30	—

Source: Food and Agriculture Organization of the United Nations: *Yearbook of Forest Products, 1988,* FAO, Rome, 1990, pp. 298–299.

timber-trade scene in the 1980s through their plan for free trade, scheduled to bear full fruit in the 1990s.

DIRECTION OF OTHER TRADE

What about the trade in lumber, panels, and veneer (not tabulated here)? Lumber is far-and-away the principal item in the group. Canadian softwood is the chief export; the main target is eastern and midwestern U.S., but the Canadian output finds its way throughout the world. Other big exporters of softwood lumber are the CIS, U.S., and Scandinavian nations. Eastern and central European countries ship their product westward. Of hardwood lumber and also panels, Indonesia, Malaysia, and Singapore are large exporters. Among importers of all these products, the U.S., the U.K., and Japan lead the world.

And lastly, what about pulp and paper (also not tabulated)? Now we observe that, for 1988, all the developing countries but Brazil (with its 60 percent of total Latin American pulping capacity) have dropped out of the list of major exporters. And China and Korea are the only poor nations in the list of chief importers. The dominant trade in 1988 was again from Canada to U.S., 15 million short tons, well over half of which was newsprint paper. The next largest items in the matrix, mostly between 1 and 2 million tons, were shipments to Japan from U.S. and Canada and shipments to U.K. and Germany from Sweden and Finland. A large number of all wood-products trading nations is involved in the international markets for pulp, paper, and paperboard. This fact is partly an offshoot of the statistical compilation, in which a wide diversity of wood products is grouped together. And then, too, both raw materials and end products are included in the group: An importer of pulp may be a manufacturer and exporter of paper or board and thus be engaged in **milling in transit.**

> Japan, unlike the U.S., has avoided incurring general trade deficits and thus heavy debts abroad. How has she accomplished this, considering her strongly negative trade balance in wood?

TRADE RECAP

If one attempted to characterize the directions of wood-products trade in the world, one might say this: The flow of pulp-paper, panel, and sawn products from Canada to the U.S. constitutes the great bulk, in quantity and value, of the total international trade in forest products. However, there are large volumes of trade that serve the vital purpose of leveling out the distribution of world wood resources among their users. Nations such as the U.K., Italy, and Japan would be gravely distressed without the hardwoods of central Europe and the softwoods of Scandinavia and North America. But perhaps it is, and more surely *will be,* the tropical developing nations that stand to gain the most from trade: as exporters. If the means can be found for managing the tropical forest lands and for converting their output into useful products, especially pulp and paper, it will prove to be the means also for stimulating economic development and for lightening supply deficits through international trade. But trade is unable to repair what is, in a sense, the world's foremost timber-product deficit: in the supply of firewood.

SUMMARY

In the past few decades, U.S. wood-product output, along with that of most other nations, has been rising. Firewood has shared in the rise for tropical regions—and for the U.S., where the "OPEC Surge" of the seventies and eighties was instrumental. In the U.S., value added, employment, and apparent consumption in the timber economy reached a high level by 1990. In international trade, explainable by the principle of comparative advantage, the world's great timber and wood-product exporters are Canada, U.S., the CIS, and Scandinavia for softwood, and Malaysia, Indonesia, and Singapore for hardwood. The big importers are Japan, China, South Korea, U.S., and U.K.

REFERENCES

1 Hair, Dwight: *The Economic Importance of Timber.* U.S. Department of Agriculture Miscellaneous Publication 941, USDA, Washington, 1963.

2 Phelps, Robert B.: *Timber in the United States Economy, 1963, 1967, and 1972.* U.S. Department of Agriculture, Forest Service, General Technical Report WO-21, USDA, Washington, 1980.

3 Skog, Kenneth E., and Irene Watterson: "Residential Fuelwood Use in the United States," *Journal of Forestry,* December 1984, pp. 742–747.

4 *The Wall Street Journal,* Eastern ed., January 16, 1991, p. B2.

5 *The Wall Street Journal,* Eastern ed., October 31, 1991, p. A24; November 14, 1991, Sec. A; December 5, 1991, p. A9.

CHAPTER **5**

NONTIMBER FORESTRY

The world's forests are a reservoir, not only or even primarily of wood, but of aesthetic values, recreation opportunities, soil and water, wildlife, range, cropland, sundry material products, and other environmental benefits.

AESTHETICS: RECREATION

I ally aesthetics and recreation, though, strictly speaking, they are separate. Beauty is not a forest characteristic essential to recreation. However, recreation is enjoyable in beautiful surroundings, and such surroundings attract recreationists.

The meaning of beauty and the attractions, for the consumer, of a beautiful landscape are considered in Chap. 11. The present section of Chap. 5 looks at forests as aesthetic and recreation resources.

The person whose name is most closely associated with the economics of outdoor recreation in the U.S. is Marion Clawson. He has written a brief history of the field to 1970. His story is brought up to recent times in a supplement by George Siehl [24].

> In western cultures, forests and woodlands are widely viewed as attractive playgrounds. Why? Is the view held mainly by rural residents? urban residents? persons in particular occupations? Is the view held in cultures other than western—say, oriental?

World War II set a boundary, as it did for so many aspects of our forestry economy, in the realm of forest aesthetics and recreation. Consider the national-forest system of the United States. In the Organic Administration Act of 1897, passed just a few years after the system's creation, Congress declared as a purpose of the forests, "to furnish a continuous supply of timber for the use and necessities of citizens of the United States." During the half-century that followed, national-forest administrators gave prime attention to timber. They championed watershed management and other branches of "multiple use," but largely accepted the principle that "good management" of timber automatically brought with it the care of other forest resources.

Then in the late 1940s came a cultural revolution in the form of the environmental movement (Chap. 10). In the media and courts, forest managers were criticized as timber cutters and scolded for their neglect of forest environmental values.

> Do you agree or disagree with the "woodsman-spare-that-tree" philosophy? Why? Why is timber cutting regarded by some, but not others, as objectionable?

In 1964, a **Wilderness** Act was passed, calling for the reservation and protection of wild areas devoted to the public purpose of recreational, scenic, scientific, educational, conservation, or historical use. The Act provided for a National Wilderness Preservation System, which by 1990 comprised some 90 million acres of forest and nonforest lands in 44 states.

In the environmental era, the research arm of forestry went to work on problems of aesthetics and recreation. The forestry schools took a fresh interest in these subjects. The annual reports of the forest service chief gave increasing space to the recreation use of the national forests. These forests began to enjoy as much recreation use as did the national parks (Fig. 14-2).

In 1985, the U.S. president established a Commission on Americans Outdoors and asked the members for a program to meet U.S. needs for outdoor recreation. The commission responded in 1986 with a report and three appendices [13]. Meantime, the state forests and even some of the industrial holdings received more emphasis on recreation and aesthetic values. The subject of timber forestry was being soft-pedaled in the U.S., though our total wood output, including exports, was rising.

Trends in other developed world nations have reflected the same cultural forces as were working in the U.S. But in those other countries, timber production and recreation have commonly gone hand-in-hand on the same ground. Most of those countries did not have the tradition of forest abundance that in the U.S. allowed the belief to spread that the timber and recreation uses were incompatible.

In the poor nations, the pursuit of forest beauty and the enjoyment of forest recreation are, to a degree, unaffordable luxuries. But to cultivate the nontimber forest resource—notably for export, including on-site use by tourists—is a means for economic growth.

SOIL AND WATER

Where forest land is either steep or dry, there is a vital relation between trees on the one hand and water and soil conservation on the other. The Swiss were among the first in modern times to manage forests with this relation in view; French and Austrians were close behind. Today an awareness of forest-water-soil relations is growing in the United States, Japan, and other nations.

Forest-water-soil relations have, since ancient times, received the most attention in active, growing economies where the want of resources was keenly felt and could be vigorously challenged. Aggressive agricultural, urban, or industrial development whetted the appetite of such economies for water and productive soils. Examples from eastern U.S. are the programs of the Delaware River Basin Commission [18], the Tennessee Valley Authority [3], and the City of New York [11] to provide household or industrial water, navigation, electric power, or flood control. Western U.S. examples are the projects that center upon dams, ranging from huge to modest, built to supply electricity and water, mainly for irrigation. All in all, the nation enjoys abundant supplies of fresh water. Our rate of renewable water output amounts to about 14 times our rate of consumption [5].

Of unique interest are the systems of canals that permitted inland barge and ship transportation. They flourished in Europe and America during the eighteenth and nineteenth centuries, and some are in use today. Canals consume water by evaporation, by percolation through their banks, and in running their locks. New York's Erie Canal, completed in 1825 and replaced by the Barge Canal in 1918, was thrust into a notable role in the state's regulation of forest practices in the Adirondack and Catskill mountains. The need for safeguarding canal and urban water supplies generated in the forest figured largely in the public debates that led to establishing state forest preserves in 15 Adirondack and Catskill counties in 1885. The preserve law was put into the state constitution in 1894, with the provision that the forest should be kept forever as wild land. In 1902, a commission was established for the storage and conservation of the water supply, and in 1971 an Adirondack Park Agency was formed, to produce a master plan for state preserve lands and intermingled private lands. Chapter 54 has details.

WILDLIFE

Forest wildlife in the more developed countries is a resource primarily for viewing and for sport—and for taking comfort in its existence [2].

A recent and dramatic case is that of the northern spotted owl. This little bird's habitat, quite inflexibly, is the conifer forest of western Washington, western Oregon, and northwestern California, notably the Douglas-fir forest offering an old-growth type of environment. For some 20-odd years, the negative effects of logging upon the owl's numbers and chances for survival have been recognized. By mid-1990, there remained perhaps 2,000 pairs of the owls, and the USDI Fish and Wildlife Service, pressed by such groups as the National Audubon Society, Sierra Club, and National Wildlife Federation, declared the owl an endangered species under protection of the act of 1973. The Agriculture and Interior departments released an interim plan for implementing this declaration, relying heavily on federal old-growth acreage in parks and wilderness areas already withdrawn from timber use.

Meantime, the wood-using industry, associated labor groups, and others opposed to economic curtailment protested the threatened loss of jobs and of private and public income. Sentiment ran high in some areas against the spotted owl. "If it comes down to my family or that bird," said one person, "the bird's going to suffer. Where would we be right now if everything that lived on this earth still survived—the saber-toothed tiger, the wooly mammoth? Things adapt or they become extinct" [19]. This observation applies to the industry, too, of course—and to the environmental organizations. Frantic efforts were made during the early 1990s to save loggers' jobs, to defend the owl, and to find a compromise. There could be "no perfect solution," to quote a White House spokesperson [12]. Meantime, lumber prices were rising nationwide, contrary to the prevailing slump. Housing became more difficult. The owl was blamed [21]. And then another bird-resident of owl territory was nominated for protected status: the marbled murrelet. The wood industry called anew for congressional action to save jobs [20].

An interagency scientific committee representing the National Park Service, Fish and Wildlife Service, Forest Service, and Bureau of Land Management was instructed "to develop a scientifically credible conservation strategy." Their 1990 report called for a system of "habitat conservation areas" where no timber harvesting was to be permitted. It became clearer than ever that timber availability would decline during the next few decades [4]. Late the following year, the Secretary of the Interior appointed what became popularly known as a "God Squad," authorized to overrule the Endangered Species Act in the spotted owl case and to open several thousand acres of federal timberland. A decision was expected in 1992, but that would not be the end of the story [22]. What has been termed the "seemingly endless tussle" over the owl continues [23].

Why preserve endangered animal and plant species? Simply "because they exist and . . . this existence is . . . the present expression of a continuing historical process of immense antiquity and majesty" [16].

But preservation can be justified on financial grounds, too. Opportunity costs of endangered species management are said to be (1) easy to calculate, (2) helpful in clarifying management and policy alternatives; and the management of endangered species need not impose large costs on society [6]. The northern spotted owl and marbled murrelet hold *c* utility for human beings (page 366). And everything depends on everything else. More on this point under Sundry Nonwood Products in this chapter.

> What considerations do you think our policymakers should take into account in resolving the bird controversy? How far should they go in relying on "constructive" measures such as limiting our export of conifer logs to Japan and other Asiatic nations (Table 4-3)?

However, taken all in all, the costs of preservation have been given much more attention, and are better understood, than the benefits [17].

In some of the world's less developed regions, the prime value of wildlife is for food. These regions account for perhaps 60 percent of the total world production of game meat. In west Africa, "bushmeat" brings higher prices than beef, mutton, pork, or goat. Zaire, Botswana, Cameroon, Kenya, and Mauritius have systematic game-farming programs. Many nations export game meat, and also skins, hair, and other materials from game. The principal importers are in western Europe [10]. The nations of Africa have examined wildlife management as a more productive land use than cattle grazing [14].

However, one cannot go far with the generalization that underdeveloped countries confine themselves to bread-and-butter projects. Latin America, largely an underdeveloped region, has 30 percent of the world's nationally protected natural area, more than any other major region (Table 5-1). In Africa, with a thought-provoking 18 percent, the national parks are designed for observation of wildlife and other forms of recreation. The largest of them is the 12-million-acre Ouadi-Rimé-Ouadi Achim Faunal Reserve in Chad. Africa also contains the largest nature reserve, set aside for scientific purposes: 12.8 million acres in Botswana. Compare Yellowstone National Park, U.S.'s largest reserve: 2.2 million acres. The world's parks and reserves are, of course, only partly forested.

RANGE AND CROPLAND

Forest grazing is commonly found in the transition zones such as woodlands of the Mediterranean type and those bordering the denser tropical forest and some of the conifer forest. Where small-scale, unregulated woodland ownership abounds, as in eastern U.S., grazing, or at least the opening of the woods to livestock, is practiced. These are hardwood forests, where the activity of stock is at odds with tree reproduction and subsequent development. Here is another instance of conflict between a nontimber forest use and the timber use.

A comprehensive description of the grazing-land situation in the United States, with notes on other nations, has been put together by the federal Forest Service [9].

As for cropland, the forest has always been a prime source. In all the more humid regions of the earth, except where grass was perpetuated by fire, cropland has been made from the forest. Today, at least four kinds of cropland-forest interplay are in evidence.

One kind of interplay is the thrust of cropland against its forest boundaries, so that land use is changed somewhat permanently in favor of agriculture. Forest shrinkage occurs worldwide, despite some countervailing change, as land becomes prime for any other use.

A second kind of forest-cropland interplay is **shifting agriculture,** a method of crop production characteristic of the frontier and of underdeveloped rural regions where people are poor and fertile land is scarce. The farm family clears the trees from a forest plot and plants crops amid the stumps. The new, rich soil has been fertilized at no cost to its immediate users. The family grows successive crops until the yield falls

TABLE 5-1

NATIONALLY PROTECTED NATURAL AREAS: NUMBER, TOTAL AND AVERAGE AREA, PERCENT OF ALL LAND, AND PERCENT OF WORLD TOTAL AREA, BY NATION OR REGION, 1988

Nation or region	Number of areas	Million acres total	Thousand acres average	Percent of all land	Percent of world total
United States	247	167	676	7.4	14.9
Canada	79	57	722	2.5	5.1
Latin America	428	344	804	3.9	30.7
West, central Europe[a]	890	64	72	5.5	5.7
CIS[b]	157	45	287	0.8	4.0
Near East[c]	21	2	95	0.2	0.2
Africa except South Africa	334	204	611	2.9	18.2
Far East[d]	822	123	150	2.4	11.0
Japan	48	5	104	6.1	0.4
South Temperate[e]	924	111	120	4.6	12.0
World	3,950	1,122	284	2.9	100.0

[a]Through Finland, Poland, to Black and Aegean seas.
[b]Regions traditionally included in the former Soviet Union.
[c]Through Afghanistan.
[d]Pakistan eastward through Pacific islands.
[e]Australia, New Zealand, South Africa.

Source: The World Resources Institute and The International Institute for Environment and Development in collaboration with the United Nations Environment Program: *World Resources, 1988–89;* New York, 1990, pp. 294–295.

to a level no longer satisfactory. Then they move on to another forest plot. On abandoned plots, the forest regenerates itself as it may. Here it comes back successfully; there, as in many a Latin American tropical community, the ground long lies fallow [1].

Is there a history of shifting agriculture in any of the temperate regions of the world?

The third kind of interplay is **taungya,** a Burmese term that means, literally, hill farming. Both forest and farm crops are raised simultaneously on the same piece of land. The trees are planted at a wide spacing, and the other crop is grown in between. Or the two crops are set in alternating rows. In any case, taungya is a vertically storied mixed-cropping system of trees, shrubs, and herbs. The system is, in a sense, the forester's answer to shifting agriculture. It is used on a limited scale in a number of tropical countries. A form of taungya known as **tajesa** is a system of oaks and agriculture found in parts of Spain, Portugal, and north Africa.

The fourth kind of forest-cropland interplay is horizontal: Windbreaks of tree borders or interspersed rows guard the food crops in their lee.

The vertical wood-food systems and, by some, the horizontal systems too are referred to as **agroforestry.** An International Council for Research in Agroforestry has been established. With U.S. funds, the council is at work on a worldwide inventory of agroforestry systems and practices [7]. The survey may help answer the question of whether wood-food cooperation offers a net benefit to farmers or to the community at large.

SUNDRY NONWOOD PRODUCTS

Sundry products are harvested from forests the world over. Such harvesting is prominent in the tropics, where in some nations its earnings are comparable to those of timber [15]. The products include gums, resins, balsams, latex, drugs, edible nuts, oil-bearing nuts, essential oils, spices, and tannins. Most of these crops can be gathered near at hand by people who lack machines and elaborate equipment. Markets have expanded.

Tropical moist forests harbor millions of plant species found nowhere else. They account for perhaps half of all the world's species on about 7 percent of the total land area. A patch of forest in Borneo or the Amazon basin contains roughly 10 times as many species as a patch of similar size in New England or Montana. One-sixth of all medicines for human beings are said to be derived from tropical plants [8].

For family use, the forest yields vitamin-rich leaves and caterpillars, fungi for protein, and palm fibers for baskets, mats, and furniture. An interesting product is silk, made from the cocoons of larvae that feed on broadleaved trees.

OTHER ENVIRONMENTAL BENEFITS

There are some very general benefits of the forest that are of transcendent value to human beings. Air quality is one. Forests lighten the air's load of particulates. They gobble atmospheric carbon dioxide and thus deter the progress of the "greenhouse effect," entrapment and accumulation of the sun's heat. One observes widespread anxiety among laymen over the shrinkage, especially, of the carbon-hungry tropical moist forests.

A vital environmental benefit is the uplifting influence of forests upon the human spirit and hence upon indefinable facets of human well-being. Concern for West Coast old-growth conifers, popularly tagged as a spotted-owl issue, is very likely of this deeper nature.

In 1990, the spacecraft *Voyager I*, when it had reached the outskirts of the solar system and was 3.7 billion miles from takeoff, turned its camera in our direction. The earth showed up in the picture as a tiny pale blue dot. This mere atom is our home, the only place we know in the universe that could accommodate us. Its forests make it habitable. Our survival depends in large measure upon our care of this little speck.

SUMMARY

In this chapter, the nontimber forest resources described are aesthetic values, recreation opportunity, soil and water conservation, wildlife, range, cropland, sundry material products, and other environmental benefits, notably the preservation of planet Earth.

REFERENCES

1 Darling, Frank Fraser: "Forestry, the Environment, and Man's Needs," *Unasylva* 27(107), 1974–1975, pp. 2–8.

2 Flather, Curtis H., and Thomas W. Hoekstra: *An Analysis of the Wildlife and Fish Situation in the United States, 1989–2040.* U.S. Department of Agriculture, Forest Service Gen. Tech. Rep. RM-178, 1989, pp. iv–vi.

3 Grantham, Dewey W.: "Tennessee Valley Authority," *In The Encyclopedia Americana,* vol. 10, Grolier, Danbury, Conn., 1991, p. 802.

4 Greber, Brian J., et al.: *Conservation Plans for the Northern Spotted Owl and Other Management Proposals. . .,* Oregon State University Forest Research Laboratory, Papers in Forest Policy 1, 1990, pp. 1–4.

5 Guldin, Richard W.: *An Analysis of the Water Situation in the United States, 1989–2040.* U.S. Department of Agriculture, Forest Service Gen. Tech. Rep. RM-177, 1989, p. 2.

6 Hyde, William F.: "Marginal Costs of Managing Endangered Species, The Case of the Red-Cockaded Woodpecker," *The Journal of Agricultural Economics Research,* Spring 1989, pp. 12–19.

7 *ISTF* (International Society of Tropical Foresters) *News,* June 1983, p. 1.

8 Jablonski, David. "Extinct and Endangered Species," *In The Encyclopedia Americana,* vol. 10, Grolier, Danbury, Conn., 1991, p. 802.

9 Joyce, Linda A.: *An Analysis of the Range Forage Situation in the United States, 1989–2040,* U.S. Department of Agriculture, Forest Service Gen. Tech. Rep. RM-180, 1989, especially pp. 1–20.

10 Krostitz, Wolfgang: "The New International Market for Game Meat," *Unasylva* 31(123), 1979, pp. 32–36.
11 Merdinger, Charles J.: "Aqueduct Systems in New York City." *In The New Encyclopedia Britannica,* vol. 26, Encyclopedia Britannica, Chicago, 1990, p. 432.
12 *The New York Times,* September 22, 1990, Section 1, p. 7.
13 President's Commission on Americans Outdoors: "Report and Recommendations to the President of the United States," Washington, 1986, pp. viii plus 210 pages; "Appendix: Case Studies," Washington, 1986, pp. xii plus 173 pages; "Appendix: Literature Review," Washington, 1986, 12 sections; "Appendix: Working Papers," Washington, 1986, numerous sections.
14 Riney, Thane: "Wildlife Versus Nomadic Stocks," *Unasylva* 31(124), 1979, p. 17.
15 Robbins, S. R. J., and W. S. A. Matthews: "Minor Forest Products," *Unasylva* 26(106), 1974, pp. 7–14.
16 Rohlf, Daniel J.: *The Endangered Species Act,* Stanford Environmental Law Society, Stanford, Calif., 1989, p. 13.
17 Rubin, Jonathan, Gloria Helfand, and John Loomis: "A Benefit-Cost Analysis of the Northern Spotted Owl," *Journal of Forestry,* December 1991, pp. 25–30.
18 Thursby, J. W.: "Water Resources Management in a River Basin," *In* Duerr, William A., et al. (eds.) *Forest Resource Management, Decision-Making Principles and Cases.* O.S.U. Book Stores, Corvallis, Oreg. 1982, pp. 514–523.
19 *Time,* June 25, 1990, pp. 60–61.
20 *The Wall Street Journal,* Eastern ed., June 19, 1991, p. A14.
21 *The Wall Street Journal,* Eastern ed., June 21, 1991, p. A10.
22 *The Wall Street Journal,* Eastern ed., January 9, 1992, Section A.
23 *The Wall Street Journal,* Eastern ed., February 21, 1992, p. A3.
24 Watson, Alan H. (compiler): *Outdoor Recreation Benchmark 1988,* U.S. Department of Agriculture, Forest Service Gen. Tech. Rep. SE-52, 1989, pp. v, 3–21.

CHAPTER 6

WORLD FORESTRY CHARACTERISTICS AND PROBLEMS

The aim of this chapter is to characterize world forestry in three groups of countries: (1) the northern developed nations, (2) the developing nations, (3) the developed South Temperate nations.

NORTHERN DEVELOPED NATIONS

The northern developed nations, with their virtual corner on natural conifers, are strategic in world forestry, given today's technology of land and timber use. These countries characteristically devote explicit attention to the multiple uses of the forest. At the same time, they emphasize the timber resource and are the world's leading timber owners, developers, traders, and consumers.

Some of the most intensive timber management in the world, involving heavy per-acre investments of capital and labor, is practiced in the northern developed nations. Western Europe saw the origins of such practice in the twelfth and thirteenth centuries and its installation in the sixteenth to eighteenth centuries. Traditional conservative silviculture is still the rule in parts of western Europe. Western Germany is at the hub of it where, on many forests, diligent and expert attention is lavished upon every detail of timber management, such as the timing and method of tree pruning and of stand thinning, harvest cutting, regeneration, and protection. Scandinavia and Finland are notable for their emphasis on the timber economy. Much of their forest is held in small private tracts, and cooperation among owners is a key to forestry progress. The Finns, continually in danger of exceeding their allowable timber harvest and always mindful of timber's major contributions to national well-being, follow a very intensive form of management and achieve rates of wood growth that are amazingly high, considering the poverty of the climate.

The former Soviet Union and its eastern European neighbors have picked up the intensive conservative forestry tradition and have, in some respects, carried it further than in the west–perhaps because, when their economies were centrally controlled, they were less cost conscious than the western countries. They have devoted great effort to pushing up the GNP and have looked to their timber resources as a major means to this end. In timber development, the region has a dilemma: The bulk of its forest is in Siberia, but to open up this potential resource requires road building and calls, too, for labor relocation, because most of the population is in the south and west. The region has responded to its dilemma with a two-point policy: (1) Intensify management of the more accessible forests. (2) Mechanize timber operations, transportation, and processing. The Russians are noted for their large, well-integrated wood-utilization complexes. They are forever studying how to improve timber-conversion efficiency. Here is a paradox: extreme conservatism in silviculture; innovative performance elsewhere in forestry.

Next, Japan, the Great Britain of the Orient, the regional leader in industrial revolution. The high development that Britain took a century or more to win, Japan achieved in less than 30 years. She has emerged crowded and heavily polluted, with little of her own industrial raw materials. Those materials that she does have, such as forests, she manages intensively. But she also takes a lead among nations in importing roundwood. Indeed, she is a dominant net importer of raw materials generally. When Japan converts these materials into highly finished, often sophisticated, goods for export, she maintains her favorable balance of trade. And she ensures her economic growth. Japan has pulpmills in Alaska, forest plantations in Brazil. The Japanese buy land in the developing economies, build a port, and bring out raw materials for shipment home. These industrious people are going into Indonesia, the Philippines, and numerous other nations of the tropics, and they are going into the northern belt as well. They are not only the Britain of the Orient, but also the milling-in-transit center of the world.

> Great Britain was once the Mother Country of a widespread empire of colonial nations (on which "the sun never set"). She brought in the colonial raw materials, converted them to end products with the help of energy from her own coal, and shipped out the finished goods. The war fleet maintained to protect this vast commerce was used as a police force to keep world peace.
>
> Is Japan not only the "Britain of the Orient," but also the Mother Country of many nations, at least in a commercial sense? Does the U.S. serve as one of the "colonial" nations? For Japan, why doesn't a military responsibility accompany her commercial responsibility, as it did for Britain?

The United States and Canada are wealthy countries. They are intensifying their management of timber and other forest resources. Both have environmental problems and can afford to be concerned over them. Both have an excess of allowable cut over actual timber harvest. We find strong analogies between Canada and the CIS. Northern Canada and Siberia, the world's two great conifer reservoirs, are both inaccessible; and in Canada as in the CIS, the drive for intensive management centers on the more populous southern regions.

Where the U.S. and Canada are concerned, trade again thrusts itself into the story. The shipments of lumber and paper from Canada to the U.S. are a large fraction of all international trade in timber products. The U.S. trade deficit in wood adds to her huge general deficit, which has worried the economic world. Where, people ask, are our immense trade debts and the big surpluses of Japan and western Germany going to lead the world economy? But it was at the insistence of the U.S. forest products industry, not as a global measure, that Canada put a tax on her softwood lumber exports.

> How does such a tax affect our lumber producers? our lumber consumers, such as the housing industry? our general trade deficit? Canadian lumber producers? Canadian lumber consumers? the forestry profession?

The inaccessibility of the great conifer reservoirs and the need to manage intensively the closer-in blocks tend to force a continuous rise in the real price of wood products. The rise threatens to persist unless tropical hardwood forestry makes a big breakthrough.

DEVELOPING NATIONS

A breakthrough in management and utilization has long been awaited, not only in the developing nations themselves as a stimulus to economic growth, but also in the rest of the world as means for staving off what are seen as impending shortages. The tropical broadleaf forest is the most extensive of the forest types. Much of it is publicly owned. Here, presumably, is a vast, scarcely tapped cornucopia ready to pour out its abundance. But the fact is less straightforward than this. There are barriers:

1 Inaccessibility: Great distances; scarce roads and communications. Costly labor and capital, at least for outsiders. Wage rates are low, but employers usually must bear the expense of relocation, housing, schools, medical facilities, training, and other obligations. Tree species, at least in Latin America and Africa, are numerous; the qualities of wood as raw material vary widely and are poorly understood.

2 Silvicultural barriers: The tropical moist forests have proved to be difficult if not impossible to manage as natural growth. They are said to constitute the most complex ecosystem on earth, with many million plant and animal species. They may survive shifting agriculture so long as each abandoned site is given plenty of time to recover. But industrial utilization imposes drastic changes that the forest probably cannot endure. At least it cannot restore itself to its previous state. Some cutover lands lie idle indefinitely, while the soil oxidizes and erodes. The tropical moist forest cover of the world, some 3 to 4 billion acres, is said to be shrinking at the rate of more than 40 million acres per year. Even-aged plantation forestry looks like the only answer, with heavy reliance on exotics, for few of the indigenous species can be planted successfully [3, 7, 9, 12]. Tropical forests grow slowly and are resistant to silvicultural effort. Tropical silviculture, what little is known about it, is labor-intensive; its principal tool is the machete.

3 Utilization barriers: At nearly every step from logging onward, where capital, entrepreneurship, and management come from outside, the problem of expensive labor is pervasive. And, too, there is the problem of lagging technology. The notable case is pulp and paper manufacture. Only in recent decades under Japanese leadership has the pulping of tropical hardwoods gone ahead successfully [11]. A hopeful example of hardwood utilization emerged in the 1960s in Columbia, where Pulpapel S.A. built a plant under the auspices of a local corporation, the government, and a U.S. group. In the 1970s, the plant became one of the few in the world using tropical hardwoods on a regular commercial basis. With 30 percent of imported long fibers added to the pulp, the mill successfully produced sulphate liner board, corrugating medium, and sack paper [2].

But now, barriers aside, tropical forestry has bright aspects besides pulping advances. In southeast Asia, where the forest is more uniform than elsewhere in the tropics, silvicultural practice, including successful regeneration of indigenous trees, is well along. In Cuba, tropical pines have been planted successfully; in Kenya, pines, cypress, and *Gmelina arborea,* a fast-growing broadleaf; in Brazil, slash pine, *Gmelina,* and eucalyptus, also fast-growing. Southern pine has been established in Venezuela. Chile leads the world in its acreage of *Pinus radiata* plantations. Argentina launched a multimillion dollar project, financed through loans, for planting 250,000 acres to willows, eucalyptus, and conifers, primarily southern pine. For all the world's tropical moist forests, it has been estimated that fully 10 percent are or will be sustainably managed for timber production [8].

Everywhere in the underdeveloped economies, governments see the need for promoting home industry and exporting finished rather than raw products. Some countries are progressing. South Korea is one. The western part of Malaysia is another. True, Indonesia and Malaysia are major exporters of logs, but both nations show up as major exporters of lumber and allied products, and South Korea is not far behind. The Koreans have built a dozen or more plywood plants.

Progress is slow. Many better-situated nations still draw freely upon willing suppliers of wood in the round. Indonesia and Malaysia, as noted, are heavily tapped.

Before leaving southeast Asia, let's look in on neighboring China. The Chinese contrast with most of the poorer people in that they have allowed few outsiders to exploit their timber resources. The main recent exception came in the 1930s, when the Japanese engulfed northeastern China (Manchuria) and exploited its mineral and forest wealth.

And the Chinese themselves have engaged in forest removal aplenty. By the time of the 1949 Communist revolution, the once extensive and valuable forests had shrunk to 8 percent of the total land area. Since then, in the "greatest afforestation effort the world has yet seen" [13], planting has increased the percentage to about 15. Farmers from the communes have accomplished four-fifths of the afforestation and do much of the other woods work. The hope is to bring the forest cover up close to 20 percent by the year 2000 [5]. The central government refrains from supplying machinery, and so the forestry economy is labor-intensive—as is the economy at large. The Chinese are noted for their initiative and hard work.

DEVELOPED SOUTH TEMPERATE NATIONS

In interesting contrast to China are New Zealand, Australia, and South Africa. There, one finds central government taking initiative in forestry. The effort responds to wood shortages, present or closely threatening, and site deterioration following exploitation of indigenous forests. It consists of programs such as public land ownership, subsidies for state and private reforestation, legal control of forest land use and management practices, forestry research, and professional education.

All three countries early practiced undisciplined harvesting and burning of timber in behalf of export trade, home use, and agricultural development. Forestry in the present sense did not become established until well into the twentieth century. In 1988, South Africa's forests and woodlands occupied about 4 percent of the total land area. For Australia and New Zealand, the forest fractions were, respectively, 14 and 27 percent. In South Africa, virtually all the forest is man-made, the original tree cover having been removed. New Zealand's forests are largely man-made. In Australia, trees have been planted on both private and public lands. Most of Australia's forest and woodland receive only scant rainfall and support a stunted, sparse growth of trees.

In Australia, most afforestation is accomplished with Monterey *(radiata)* pine. An ordinary tree in its native California, the pine changed character when introduced in South Temperate lands: There it is tall, straight, and fast-growing, much prized for all products from lumber to paper. In 1986, Monterey pine plantations in Australia covered some 1.4 million acres, of which two-thirds were public lands. Three million acres was until recently the ambitious target for the year 2000. Perhaps land now allocated to other uses will be put under pine, a crop that is profitable and in the national interest.

As for hardwoods, an area of some $34\frac{1}{2}$ million acres of natural forest is officially reserved for timber production, but only 20 or 25 million acres are accessible for commercial exploitation. The best of this hardwood forest has been culled over selectively for more than a century, so that annual growth is down to about 15 cubic feet per acre. It would take a doubling of growth by 2000 if hardwoods were to contribute half of Australia's anticipated log requirement of 1 billion cubic feet. Otherwise the country would face a huge import bill [10].

In New Zealand, where the miracle pine was introduced earlier than in Australia, the vast plantations ($2\frac{1}{2}$ million acres, second only to the Chilean acreage) have had spectacular success, so that *radiata* has come to account for more than 90 percent of plantation output. The government, concerned over the nation's heavy reliance on a single tree species, has been encouraging the use of other trees, such as slash and other pines and eucalyptus. These trees are planted in South Africa and Australia, too [6].

> What are the risks in heavy reliance on a single tree species?

Australia in 1988 was a net exporter of wood. She sent pine logs to Japan, and some were shipped on to China and South Korea. New Zealand and South Africa were net importers. But in no case was the imbalance of trade very large, and it fluctuated from year to year. All three countries have a well-developed forest-based industry. Their lumber industry, as in the northern nations, has been divesting itself of the little roving mills and has made gains in productivity. Plywood and pulp-paper manufacture thrives. Australia was the leader in the pulping of eucalyptus, a process first commercialized on a full scale in the late 1930s. Some 20 species of eucalyptus are now accepted at the pulpmills.

The South Temperate nations have been giving much attention to forest resources other than timber. They have developed parks and in other ways encouraged the recreation use of forests. South Africa and Australia, because of their extensive regions of low rainfall, are concerned with forest influences on soil stability and water supply. In the drier sections of South Africa, the landowner who wants to plant trees must first obtain a permit, for trees are pumps that in such areas may put a heavy drain on the water supply.

WORLD FORESTRY ISSUES

The foregoing descriptions can be drawn together in terms of issues in the world forestry economy. Here are some, present or looming:

(The interrogative sentences that follow may be treated as questions for discussion.)

1 How shall the forestry economy respond to two somewhat opposing forces at large in the world: first, the high and rising value of human labor; second, the high value of energy? Labor cost forces a retreat from old ways, from labor-intensive silviculture and wood conversion toward capital-intensive projects, notably pulp processing and silvichemical industry. But energy cost pushes in nearly the opposite direction. Silvichemistry uses energy profusely; lumber manufacture uses it sparingly; silviculture uses solar energy. What is the relation of forestry to labor and energy supplies?

If silvichemical industry proves to be viable, then here in prospect are two kinds of use for wood. One is to extend and expand present utilization technology. The other is to convert cell-wall components into chemical feedstocks. Cell-wall polymers, which are the major part of wood, have the potential for meeting all our chemical wants in place of petrochemicals. Thus the answer to the industry's energy-scarcity problem may rest with the industry itself [1]. Yet silvichemical wood requirements are potentially vast beyond belief, far exceeding other wood requirements. Where is the wood to come from [4, 14]?

2 How will world society meet the growing wants for services of the forest? Is it possible to have more of everything from wood to wilderness and, if so, up to what point? What part can the tropical regions fill, with their wide expanse of potentially productive land? What is the likelihood that vegetative growth can outstrip population growth? How can the tropical nations establish a reasonably high priority for forest conservation, as opposed to exploitation?

3 What shall be done about the deterioration of the environment? How may forestry contribute to environmental improvement and maintenance? Most of the developed economies are engrossed in environmental concerns. When will the developing countries face the want for an environmental policy?

4 What will be the outcome of rising nationalism among the poor nations? These nations are establishing their own forestry schools and turning to their own graduates rather than foreigners to solve their forestry and wood-industry technical problems. Will they similarly dispense with foreign investors in forests and wood-industry facilities? What bearing would such action have upon their development?

5 What place will the forestry profession find in resource management? How can the profession adapt itself to changing circumstances? The old, absolute doctrines, the world's forestry heritage from western Europe, are breaking down and being rejected, even within western Europe. It is no longer possible for the forester to enforce traditional authority and traditional rules—the primacy of timber, the necessity of sustained yield, the superiority of professional over popular judgment, and so on. What authority and rules shall replace these that are outmoded?

SUMMARY

By way of condensation: (1) The northern developed countries control most of the world's strategic softwood. They tend to be conservative in silviculture, progressive in wood conversion. Most are heavy traders in wood products. (2) Nearly all the poor nations are in the tropical hardwood zone and so are faced with problems of forest inaccessibility and barriers to silviculture and wood utilization. Forestry development, if wisely administered, will be a source of better living. (3) The South Temperate nations, after exploiting their original forests, have become foremost plantation managers. They face in some cases the problem of achieving a diversity of tree species. When one views the whole forestry economy of the world, one sees fascinating issues to be resolved.

REFERENCES

1 Goldstein, Irving S.: "Chemicals from Wood," *Unasylva,* 31(125), 1979, pp. 2–9.

2 Gomez, Gustavo: "Case History of a South American Paper Mill," *Unasylva,* 30(122), 1978, pp. 7–13.

3 Gomez-Pompa, A., C. Vasquez-Yanes, and S. Guevara: "The Tropical Rain Forest, A Nonrenewable Resource," *Science,* 177(4051), 1972, pp. 762–765.

4 High, Colin, and Kenneth Skog: Current and Projected Wood Energy Consumption in the United States. *In* Klass, Donald L., ed.: "Energy from Biomass and Wastes 23," Proceedings of IGT's Conference, February 13-17, 1989, New Orleans, La. Institute of Gas Technology, Chicago, 1990, pp. 229–260.

5 Hsiung, Wen-yue, and Frederick D. Johnson: "Forests and Forestry in China," *Journal of Forestry,* February 1981, pp. 76–79.

6 Lavery, Peter B.: *Plantation Forestry with* Pinus radiata, Christchurch, New Zealand, University of Canterbury, School of Forestry, Papers No. 12, 1986, pp. 16–26.

7 Leslie, Alf: "Where Contradictory Theory and Practice Co-exist," *Unasylva,* 29(115), 1977, pp. 2–17.

8 Poore, Duncan, et al.: *No Timber Without Trees, Sustainability in the Tropical Forest,* Earthscan Publications, London, 1988, Chaps. 2–5.

9 Pringle, S. L.: "Future Availability of Wood Pulp, A World Picture," *Unasylva,* 29(115), 1977, p. 18.

10 Rule, Alexander: *Forests of Australia.* Angus and Robertson, Sydney, 1967, pp. 114–115.

11 Shimokawa, Eio: "The Japanese Pulp Industry Was First in the World to Utilize Low-Grade Hardwood," *Unasylva,* 29(115), 1977, pp. 26–27.

12 U.S. Intragency Task Force on Tropical Forests: *The World's Tropical Forests, A Policy, Strategy, and Program for the United States.* U.S. Department of State Publication 9117, 1980, pp. 12, 15.

13 Westoby, Jack: "Foresters, Forestry, and Society," *New Zealand Journal of Forestry,* 23(1), 1978, pp. 64–84.

14 Zerbe, John I., and Kenneth Skog: "Sources and Uses of Wood for Energy," *In Proceedings from International Symposium: Energy Options for the Year 2000, Contemporary Concepts in Technology and Policy,* Madison, Wisc., 1988, pp. 1.243–1.254.

CHAPTER 7

HIGH POINTS IN PART ONE

1 Appreciation for context, a mark of the educated person: the context of a material object, idea, action, assigned job, or whatever constitutes the core of a problem to be solved—a question to be answered. The context consists of those adjoining parts of a system that we include most happily in our search for a problem solution. The context of anything includes the physical circumstances, history, geography, people, and social institutions.

2 A problem may be solved in four steps: First, clarify the question. Second, find a suitable model. Third, put data into the model and derive a reply. Fourth, follow up the process.

3 Forest areas of the world. Here is a broad context of forestry as social science, with emphasis on the dimensions to which forestry is highly sensitive: space and time. Nations and regions differ greatly in their forest area characteristics.

4 Economic development and, behind it, economic growth. How do forests and forest activity contribute to development? What influence does development have on forestry?

5 Timber resources. In the U.S., how does timber contribute to income and employment? In the world, how does it augment the flow of production, consumption, and international trade?

6 Principle of pulverization. Economic development stresses labor saving and capital intensification. In timber forestry and wood industry, it stresses a diminishing use of chunks and slabs and a growing use of panels, particles, and molecules.

7 Principle of comparative advantage. Each individual, group, community, and nation tends to do what it can do best. The fact that each group has this tendency to act in accordance with its talents and the fact that talents vary from group to group make trade possible and also beneficial.

8 Nontimber forestry includes the creation and use of aesthetic values, recreation opportunities, soil and water services, wildlife, range, cropland, sundry material products, and other environmental benefits. Some of these forest resources have come into great prominence in the world and rank with or above timber as a human concern.

9 Position of tropical forests. Because the world generally is pinched for raw materials, including wood, we look everywhere for surplus timber resources. We find them in many poor nations of the tropics. There, the people want to use their resources to build a better livelihood. The other people of the world stand to gain if this use is made of the resources. How to open the tropical forests wisely, to the enduring benefit of the tropical nations and their neighbors in the temperate zones, is a vital and unanswered question.

10 Problems in world forestry. Vital and unanswered questions are pressing, not only in the tropics, but in all regions of the world. How shall wants for energy be met? wants for nontimber forest benefits? How shall the integrity of the environment be guarded? How shall poor nations become self-supporting? How can the forestry profession respond effectively to change?

11 Standing behind all foregoing high points—context, development, forestry problems of the nation and the world—is the idea of the forestry economy. The economy is a way of viewing resource creation and use. It has these characteristics [1]:

- **a** It is a *system:* a closely related set of interacting elements, which accepts inputs and generates outputs.
- **b** It is a *human* instrument, a social instrument. It is an aspect of culture. Thus it may differ from one culture to another. In any case, the function of the economy is to promote human happiness.
- **c** The economy may be described in terms of its *aspects.* It has geographic location. It has a date or dates attached to it. It operates on specified resources: persons, things, and ideas that have value.
- **d** The economy contains *stocks* of resources, measurable as quantities at a point in time: growing stocks of timber, trout stocked in a stream, a stock of technology. Resources move into and out of these stocks in the form of *flows,* measurable as quantities per interval of time: growth and cutting of timber, replenishment and removal of trout, creation of new technology and obsolescence of the old. Thus flows accomplish an essential function of the economy: change.
- **e** The economy is *measured* in a wide variety of units. One is the financial; but finance is only a subset of economics. Economic units include those used in biology and engineering and those used in philosophy. As a product of human behavior, the economy shares in the uncertainties and contradictions of its creators. As a dialectical system, it is measurable in such terms as happiness, good, conservation, utility—terms that admit simultaneously of $(X = A)$ and $(X \neq A)$, which any right-thinking logician will tell you is sheer nonsense. But then, who ever claimed that people (except, perhaps, logicians on duty) are logical?
- **f** Although both persons and things are involved in the economy, it is unlike much that foresters deal with: it is *intangible.* One can put one's hand on this

ground or this person or truck or tree, but not on the economy. The tangibles and the economy are different things, just as the substance and subject of this book are different things. The economy is a cultural institution, an idea.

g The economy is a *one-way process.* It furnishes convincing evidence that there is no such thing as perpetual motion. Seek back to the economy's very beginnings and you find that it commences with neutral stuff. This it works into primary resources and ultimately into consumer services. In bringing structure out of formlessness, it has prepared for its descent along the path described by physicists in their second law of thermodynamics, that the entropy of the universe increases continuously and irreversibly. The outputs of the pulp-paper mill sooner or later disappear, as it were, beyond capture. The noise of the mill, if by some miracle of invention it could be gathered to run a machine, could generate only a fraction of the energy originally lost in making the noise.

REFERENCES

1 Parts of the thinking under paragraph 11 depend heavily on Georgescu-Roegen, Nicholas: *The Entropy Law and the Economic Process,* Harvard University Press, Cambridge, Mass., 1971, introduction, especially pp. 4–6, 10–21 (concept of entropy; relation to social science, economics, and development); Chap. V, especially pp. 114–130 (social science lessons from physics and chemistry); Chap. VII, especially pp. 187–195 (life and development speed up the entropic process); Chap. X (value, development, social conflict; their relation to entropy); and Chap. XI (conclusions about society and social science).

PART TWO

CONSUMERS AND THEIR WANTS

CHAPTER **8**

PREAMBLE TO PART TWO

WHY START WITH CONSUMERS

One is apt to think of the economic process in forestry as commencing with land and its vegetation and animal life, continuing through their management, perhaps product harvesting and distribution, and ending with consumption. But the elements in the economic process, remember, are all related: Everything depends on everything else. The system, being thus relativistic, can be entered at any point.

For us, it's helpful to begin with consumers. The fact that we are consumers ourselves is underscored in the Introduction. Beyond this, consumers are viewed as the economy's sovereigns, and though this may be an exaggerated view, surely they are the persons whose favors are sought and perhaps fought for in the market. And when we stop to think of it, consumers are the source of our professional motivation. Yet the profession has traditionally neglected them [3]; here is a tradition worth breaking.

DEFINITIONS

Who is the **consumer?** What sort of person is the consumer? What function does this person perform?

The consumer is an individual or group that has choices and makes decisions and plans about them. Thus the consumer, by our definition, is a **firm,** in just the same sense that the producer is a firm.

The most common example of the consuming firm is the family. The firm's manager may now be one member of the family, now another, and under a skillful manager, the whole family by consensus may join in some of the consumption decisions. Besides the family, other examples of consuming firms are lone individuals, agencies of government, and private groups such as teams, clubs, and cooperative associations. These are **ultimate consumers.** And then there are the **intermediate consumers,** such as manufacturers that consume raw material in producing an end product—for instance, lumber-consuming furniture plants.

The consuming firm is commonly thought of as a receiver, a taker-in, who stands on the opposite side of the economic ball court from the producing firm. But there is a more accurate and useful way to view the consumer: as a last producer in a series. The ultimate consumer is the *very* last producer. Students of forest recreation recognize this fact; they see the forest manager as producing, not recreation, but the *opportunity* for people to have recreation. It is these people, the consumers, who produce the recreation.

Now the interesting thing is that what is true of recreation is equally true of every imaginable service, with no exception. The builder produces only the *opportunity* to enjoy a home; the artisan produces only the *opportunity* to admire and use his furniture. Consumers are the final producers.

Yes, you may say, but in such cases as the house and the furniture, material goods are involved and may be consumed in the sense of "used up." True, but of course, it's possible for recreationists to consume a forest in the same sense.

> Look through some recent issues of a professional periodical such as the *Journal of Forestry.* To what degree do the authors and readers appear to be oriented to forest-related consumers: their wants, interests, and problems? How do you explain this?

A BUNDLE OF SERVICES

Indeed, the using-up of a material good is not a necessary part of consumption, and when using-up does take place, it may be purely incidental. Things consumed are not material goods: They are services—attributes—often ideas in the consumer's mind. What is produced-consumed by the consumer is, in fact, a **bundle** of services, or attributes; this bundle's existence leads the consumer to value its source, just as one may value a well for the services of its water or as the producer in the ordinary sense values a machine for its services in production. Let me stand at my back door and try to explain the services that I experience in the firewood that I have stored just outside in preparation for winter:

> The wood will burn, at least when I get it seasoned. If I can keep it burning, it will give off heat. If I sit by the fire, I will be pleasantly warmed and will feel relaxed and able to engage in pursuits of a quiet sort that I enjoy. The fire is so tempting that I may neglect my work, to the detriment of my family and friends. If I cook over this fire, the food will taste good and give pleasure to me and my family. It will promote our good health and long life and draw welcome neighbors to our hearth. At the same time, the fire will draw the dogs to the

> hearth, and often they smell bad while their fur is warming, especially if it has been wet. The visits of neighbors will be enjoyable because it is our custom in this community to make such visits in an orderly and considerate way, not trampling down the garden, putting mud on the floors, or breaking up the furniture. In fact, the community's respect for property will enable me without conflict to maintain my home and hearth and to be reasonably secure in my possession of firewood and other belongings. The law will protect me from any gross violations of my property or privacy. I can hope to retain my personal freedom so long as I do not stray too far from the dictates of my culture.

As you see, some of the services, or values, of the firewood are positive; some are negative.

TECHNOLOGY OF CONSUMPTION

The foregoing little insight into a consumer's thoughts puts one in mind of the fact that consumption is a *process,* a *procedure,* perhaps a complex procedure. Because it involves persons, it involves, and is determined by, all sorts of cultural circumstances; place and time; capital inputs such as housing, machinery, and other equipment; materials; knowledge of how to do things (such as keep a fire going or cook over it); customs, laws, and traditions; and many subjective characteristics of the consumer, such as experience, prospects, and the mood of the moment. All these circumstances that surround the act of consumption and determine its outcome may be likened to the circumstances around the act of production and be given the same descriptive name: technology. The **technology of consumption** is its setting and procedure: the materials, methods, knowledge, and attitudes with which the firm approaches its assignment as a consumer. Here we have another analogy between consumer and producer.

In expositions on economic theory, it is common practice to lump the determinants of consumption that are not readily measured—for example, determinants other than income, price, and the like—under the heading "tastes and preferences." **Technology of consumption** clearly includes the notion of tastes and preferences, and its use helps avoid this rag-bag expression.

For an illustration of consumption technology, take the recreation service of the forest. Here is what the Appalachian Trail Conference [1] says about the technology of hiking the trail in central and southwestern Virginia:

> The ruggedness of the terrain and the exertion required to traverse it must not be underestimated. Do not undertake travel here without adequate preparation as to both physical condition and equipment. . . . Heavy rainstorms are a usual occurrence. When a tent is to be used, a thoroughly waterproof one is necessary. A full-length poncho (to cover wearer and pack) is essential. A short rain shirt is insufficient. . . .
>
> Rattlesnakes and copperheads may be encountered. A suction-type snakebite outfit should be carried. . . . Trouble may be encountered from ticks, chiggers, no-see-ums, mosquitoes, and other insects. Carry a suitable repellent. A whistle, compass, matches, fire-making material, a change of clothing, and emergency food should be in the pack of every user of the Appalachian Trail, even the day-hiker, but every effort should be made to carry the lightest possible load. . . . Every hiker must avoid cutting across switch-backs, particularly on

graded trails. Such shortcuts disfigure the trail, complicate the route, and cause erosion. The saving in time or distance is little; the damage, not apparent to those responsible for it, is great.

> Consider some act of forest-related consumption with which you are fairly familiar, such as enjoyment of viewing a landscape or using a timber product. Can you describe the technology of this act of consumption?

DIALECTICAL VERSUS ARITHMETICAL

Thinking further about consumption as production, surely anyone will tell you that these are two different events, and yet you discover that they are the same kind of event viewed in different ways. That is, you discover that the set *S* and the set *not-S* may overlap or even coincide. The subject of economics holds many examples of illogical truths such as this. "Conservation" may mean to cut timber or not to cut it. Such a contradictory subject is referred to as **dialectical** (page 51).

The distinction between dialectical and **arithmetical** subjects is not always sharp. In college, if you earn an A (four grade points) in one course and C (two points) in a second course of equal weight, your average is 3, just as though you had earned a B in each course. Yet you know perfectly well that the meaning of the letter grades is vague, that it is different for every instructor and subject, that it changes with time, and that A, B, and C are not equally spaced as are 4, 3, and 2.

This book at times works with dialectical subjects as though they were arithmetical, a common practice in social science [2]. To do so is helpful, provided you and I understand what we are doing, appreciate its limitations, and qualify the arithmetic as judgment recommends.

One last general thought about consumers: Why speak of their "wants"? Why not their needs or necessities? It is because need and necessity are absolutes, whereas social science is relativistic. The password is, "It depends." It has been attributed to Samuel Johnson that when he had been approached by a beggar woman and had turned her down and she had cried out, "But, Sir, a body's got to live!" he rejoined, "Madam, I doubt the necessity for it!"

SUMMARY

The foregoing preamble makes five points: (1) The consumer is a firm, a decision-making unit, among the last in a series of producers. (2) To view the consumer as a producer is a bit of illogic appropriate to a dialectical subject such as forestry economics. (3) What the consuming firm consumes is a bundle of services, or attributes. (4) Consumption is a process, with a technology. (5) Adopting a relativistic, not absolute, viewpoint, one speaks of the *wants,* not the needs, of the consumer.

REFERENCES

1 Appalachian Trail Conference, Inc.: *Guide to the Appalachian Trail in Central and Southwest Virginia.* Publication 25, 7th ed., Appalachian Trail Conference, Harpers Ferry, W. Va., 1974, pp. 1–9 to 1–11.

2 Georgescu-Roegen, Nicholas: *The Entropy Law and the Economic Process,* Harvard University Press, Cambridge, Mass., 1971, pp. 43–47.

3 McConaughy, Daniel B.: "Forestry's Concern for the Ultimate Consumer of Wood, A Historical Study of the *Forestry Quarterly* and *Journal of Forestry,*" Master's thesis, State University of New York, College of Forestry at Syracuse University, Syracuse, 1961.

CHAPTER **9**

CONSUMER ATTITUDES TOWARD WOOD

Now the problem at hand is to learn more about consumers' attitudes toward wood as a source of services.

WOOD VERSUS MASONRY

Among the first foresters to give scholarly attention to consumers was Joseph Zaremba [1]. He had observed that most single-family dwellings in northeastern U.S. had exterior walls of clapboard or shingles, except for a zone from Philadelphia to Norfolk, where most such dwellings had masonry walls. The question he asked himself: Why is this? His model: Look for differences in (1) the cost of the house, (2) insurance cost, (3) building codes, and (4) consumer attitudes.

How did Zaremba arrive at the first three variables on his list? He did it by sitting back and thinking about what would influence *him* if he were faced with a choice between wood and masonry. Why, he asked himself, might he prefer wood? His response came out of what he had been taught about the behavior of a **rational** human being, one who deliberately tries to maximize or minimize something, an "economic man," who strives for **economic efficiency.** This person tries to escape costs (the benefits equal) and to pursue benefits (the costs equal), or, in general, to choose the alternative that offers a favorable excess of benefits over costs. Zaremba reasoned that if people had reliable information about masonry house exteriors and about wood, they would choose the less costly alternative if the two promised equal service. And, as rational human beings, they would avoid breaking the law (the building code).

And so Zaremba gathered cost and code information from builders and insurance companies within the masonry zone and within the wood zone. He drew a blank: He found no significant difference between the two zones in respect to the cost of building and insuring a wood-exterior house or a masonry-exterior house, and he found no significant difference between the zones' building codes.

Zaremba then went after the fourth variable in his model: consumer attitudes. Data on this variable he obtained from personal interviews with a sample of homeowners. Within the masonry zone, 96 percent of respondents said they preferred masonry to wood. Why? Among their answers:

> Masonry looks better than wood.
> Masonry is a better insulator than wood.

Outside the zone, 83 percent preferred wood. Why?

> Wood looks better than masonry.
> Wood insulates better than masonry.

Don't draw the easy conclusion that consumers are lost sheep—or nitwits who kick the tires of used cars to make a choice. You are discovering again that forestry as social science is a dialectical subject. And you are observing that consuming firms are generalists. Human physiology, if nothing else, forces them in this direction: Food, drink, clothing, shelter, and a little learning and social interaction are about the irreducible minimum of their service inputs, and even here, a wide variety of judgments is represented. Consuming firms cannot be expert in all these judgments. In this, they are unlike producing firms in the ordinary sense, which, with the help of size and specialization, can be expert at the judgments they make. Zaremba recognized these things.

He then offered a reply to his original question, Why the difference between the masonry zone and the rest of the northeast? In something akin to desperation, he said that consumer preferences are a function of local custom. To rephrase his reply: Consumers' preferences are a function of their culture.

CULTURE DEFINED

In earlier chapters, and now in this one, references to the culture have come up. It's high time the term was defined. **Culture** is behavior that is learned rather than simply inherited. By virtue of our culture, most U.S. natives speak English, are governed under an administrative–legislative–judicial system, wear shoes some of the time, shake hands, and regard these trees as prettier than those. Each person is a member of a predominant culture and of larger and smaller cultural groups, too. On the large side, for example, one may belong to "western culture," possessed by all those peoples whose language, laws, art, religion, and manners can be traced to ancient Rome or Greece. And a person may belong, also, to smaller cultural groups. This person shares with the residents of a region, community, or neighborhood or with the members of an occupational or religious group a dialect, set of skills, religious convictions, and love for certain objects, places, and ideas. Each family has distinct cultural traits, and so does each individual. Each of these cultures may be regarded as a variant of a larger one: a subculture within a more general culture. The forestry profession is a subculture.

Each culture comprises at least five classes of institutions: (1) a language; (2) a social system, families, a form of government; (3) ways to make a living, an economy, a technology; (4) a philosophy of life, religious beliefs; (5) aesthetic activities and standards, ideas about what is beautiful or otherwise pleasant to the senses.

> Can you describe or give an example of each of the foregoing classes of institutions in the forestry subculture?

BENEFITS FROM CULTURE

Now is the time to ask what benefits a person's culture (subculture) bestows upon the person. What benefits did residents of the masonry zone derive from their dwellings' masonry exteriors, and what benefits did wood-zone residents get from their clapboards and shingles? It comes to this: The culture provides each member with a host of ready-made decisions; in their absence, life would be smothered in an overburden of trivia; the material to use for dwelling exteriors is such a ready-made decision. By respecting this cultural institution, the residents of each zone escaped a difficult choice. Furthermore, through their conformity they underscored the identity of the group and promoted its survival. And they assured themselves of the peace of mind that derives from harmless acquiescence.

Into which of the five classes of cultural institutions should one put Zaremba's "local custom"? Surely into class 3, technology, in this case the technology of consumption: the consumer's materials, methods, knowledge, and attitudes, including the consumer's "tastes and preferences." Consuming firms have the assignment of deriving bundles of services from their sources, and satisfactions *(net benefits)* from the service bundles. Northeastern homeowners were behaving the way "economic men" are expected to behave.

Zaremba's reply, then, gives you an example in which persons' ideas are more than just traceable back to the culture: They are, in this instance, a direct expression of cultural tenets. How did these tenets come into being? Zaremba did not explore the question. If he had done so, it would have been useful. Very often the problem solver in forestry economics can be effective by identifying and explaining the cultural tenets that underlie the question at hand. This is certainly so with questions about the why of human conduct.

> Some vocations demand or encourage specific "loyalties" from members. For instance, medical doctors commonly refrain from cigarette smoking; pulpmill managers prefer paper products to plastic products. What "loyalties," if any, are expected of foresters?

CONSUMER TERMINOLOGY

Now let's propose terminology for the set of forces that lies back of consumer behavior. First come the three forces introduced so far, in Chaps. 8 and 9. Starting with the most fundamental, or general, these three form the following hierarchy:

1 *Consumer culture,* the universe of consumers' learned behavior.

2 *Consumer technology,* the cultural institutions that embody consumers' economic behavior.

3 *Consumer want,* the part of total consumer technology that consists of attitudes toward a specific bundle of services. Such attitudes are described in the present chapter.

And while we're at it, let's list two more forces that extend the hierarchy further toward the specific. These will be featured in later chapters of Part Two:

1 *Consumer demand,* the arithmetical aspect of want, in which the flow of specific service bundles is expressed as a function of some influence such as market price or consumer income.

2 *Consumption,* the number of specific service bundles that in fact flow into consumers' hands during an interval, influences being what they are.

SUMMARY

Question: Why did the masonry people prefer masonry and the wood people prefer wood? Model: Look for differences in the cost of the house, insurance cost, building codes, and consumer attitudes. Data: From builders, insurers, and homeowners. Reply: Because of differences in local custom. Follow-up: Apparently homeowners were being rational, seeking their maximum benefits, costs being equal. Their benefit was derived from obeying the tenets of their subculture concerning an aspect of consumption technology.

A hierarchy of terms is proposed for the set of forces that lies back of consumer behavior: culture, technology, want, demand, consumption.

REFERENCES

1 Zaremba, Joseph: "The Consumer's Attitude Toward Lumber in Exterior Walls of Houses," *Journal of Forestry,* February 1959, pp. 90–94.

CHAPTER **10**

CONSUMERISM AND ENVIRONMENTALISM

CONSUMERS' DEFENSE PROGRAMS

Ultimate consumers, as generalists, are apt to be less expert than other producers in judging the quality of a consumable (page 61). But generalists have points of strength, too. They may be better fitted than specialists to appreciate the context of things (page 50)—the setting—the environment. They are born organizers for self-defense. This chapter's problem is to catch a glimpse of their work and its significance for forestry.

The theme of the chapter is this: Consumers have generated two defense programs, the consumer movement and the environmental movement. On close view, the two are seen to have much in common: consumers exercising their gift for self-protection. Consequences for forestry have been and will be heavy.

CONSUMER MOVEMENT

In the **consumer movement,** consumers band for mutual economic and political advancement. The movement may be traced to Robert Owen (1771–1858), a British textile magnate. Concerned over depression and violence in the mill towns, Owen grew convinced that human character is malleable through the surroundings and experiences of childhood. He founded model communities to embody his philosophy. None survived for long. However, his ideals of order, cleanliness, education, cooperation, and socialism have had persistent influence in many corners of the world.

An early outgrowth of Owenism was the Rochedale Society of Equitable Pioneers, a consumers' cooperative association formed in 1844. The society operated a store where its members could buy consumables of reliable quality and share in the dividends from the business. The venture pursued cooperative principles that included education of members to be successful cooperators and intelligent consumers.

The consumer movement endured and in many places prospered. In England, retail cooperative membership was, for a long time, more than doubling every 25 years. In Finland, one of the most cooperative-minded of nations, a fourth of the people are said to belong to associations. Finnish co-ops serve the interests of, among others, forest workers, owners, and small operators.

In the U.S., cooperative principles were seized, not by consumers, but by farmers. However, among these producers as among consumers elsewhere, cooperation was protection for the little fellow. Before World War II, cooperative associations were organized in the northeast and Lake States for processing and marketing timber from small holdings. In the general defense of consumers against misrepresentation and monopoly, the U.S. has enacted laws, notably those administered by the Federal Trade Commission. Private organizations such as Consumers Union have studied the quality of consumables and reported findings [4].

> One of the notable forestry co-ops was the Otsego County Forest Products Cooperative Association at Cooperstown, New York. Like the others, it was organized primarily through the effort of publicly employed foresters who had a vision of "good" forest management and profitable timber marketing from the traditionally criticized small private holdings. The Northeastern Forest Experiment Station put its expert talents into member recruitment, forest management planning, obtaining loans, and building and operating a sawmill. The venture folded in the early 1950s, and it wasn't long before the entire roster of forestry co-ops in the U.S. had been wiped out. What do you suppose was the cause?

Our first significant consumer movement was privately founded by Ralph Nader in the 1960s. He became embroiled with General Motors Corporation over his charges of automobile unsafety. The controversy led to Nader's receiving a large settlement, which he used to finance study groups oriented to consumer interests. Through books—including one on the national forests [1]—pamphlets, and campaigning, he and his colleagues rallied people and swayed legislators, administrators, and courts to safeguard consumers.

Thus the consumer movement has had these characteristics: It has worked for the consuming public's interest in safety, economy, and the good life. It has sided with the consumer, the exploitable "little guy," singly or en masse, against the allegedly exploitive "big-guy" producer.

ENVIRONMENTAL MOVEMENT: HISTORY

In the **environmental movement,** too, people join to find the good life, to escape the ungood life. They discover themselves defending the exploitable against exploiters.

The movement blossomed, as did Naderism, in the 1960s. It gave voice, across the U.S. and in other developed nations, to a long-suppressed fear: that the natural environment had been allowed to deteriorate very far. Economic growth accompanied by excessive industrialism and commercialism was seen by many as the cause of deterioration. And such growth was recognized by some as enabling nations to identify and possibly reverse deterioration.

The environmentalists pounced upon every conceivable aspect of our degeneration. They cried air and water pollution, misuse of land, defacement of scenery, magnification of noise. . . . Herein lay a supposed distinction: Environmentalism championed primarily nature; consumerism, primarily mankind.

> Can values be placed on the natural elements in our environment? Can these values be weighed against the human gains that may result from destroying natural elements? Such questions are a topic of Chap. 47.

Forest-land management and wood industry were caught up in the environmental movement to a degree scarcely matched in any other branch of American economic life. Here we find a bit of justice, for the U.S. profession had already had a foretaste of environmentalist discipline early in its history. The year was 1903; the place, an experimental and demonstration forest in the Adirondacks, administered by the New York State College of Forestry. The college's dean, Bernhard Fernow, made a plan for timber stand improvement. But influential estate-owning neighbors rose against the logging threat. The governor vetoed the college budget, thus closing the program and putting its dean out of a job [9].

Fernow wrote to Professor Adam Schwappach, at Eberswalde, Germany, for his opinion. Schwappach's reply: "Your management plan was obviously sound. Laymen ignorant of forestry shouldn't interfere. The episode reveals the shady side of the much-praised democratic system" [6].

Fernow and his successors turned their backs on the prophetic handwriting. For over half a century, the subculture inherited from central Europe successfully resisted disapproval in the U.S. Thus the faith of successive forestry generations rested on the classical tenets of our Teutonic forebears. The gist of these tenets was as follows: The value of the forest lies in its timber as a source of wood, not in its water or wildlife or scenic beauty [5]. In timber management, the forester voices standards acceptable to the forest, serving as its infallible interpreter. Laymen, ignorant of forest values, are a nuisance, especially in the woods.

ENVIRONMENTAL MOVEMENT: IMPACTS

You can readily picture the devastation wreaked in traditional forestry circles by the new environmentalism of the 1960s. The big events of the drama were enacted within a few years. Federal forestry—notably Forest Service forestry as the most widespread and clearly documented—was at target center. But other public forestry and industrial and other private forestry came in for their share of review.

The traditional forestry subculture proved slow to respond. As late as 1985, the *Journal of Forestry* [7] published an article lamenting the "rising tide of preservation" and terming politics the greatest obstacle of all to professional forestry. A commentator on the article [10] referred to "this paternalistic, elitist attitude of foresters."

The environmental movement, unlike consumerism, was led by no outstanding individuals. Its drive came from here and there as occasion prompted. It had no coordinating body; its only coordination was in the set of ideas that environmentalists found inspiring. Here a local organization took issue with resource actions that promised to affect the locality adversely. There a legislator, state or federal, seized an opportunity to head a drive against some violation of the good life.

> Are you, personally, concerned about the quality of the environment in the U.S.? How did you derive your knowledge of and interest in environmental problems?

Nationwide reputations were built by a few groups with special interests. Among organizations pointedly concerned with environmental aspects of forest resource management were the National Wildlife Federation, National Audubon Society, Sierra Club, the Nature Conservancy, and Greenpeace U.S.A., each of which counted half a million or more members [3]. Considering the diversity of its actors, the environmental movement was remarkable for the unity and concreteness of its actions.

Item Forestry judgments based on the old rules were no longer accepted as infallible. Public forest managers' decisions were no longer held exempt from judicial review. What descended upon them was a series of class-action lawsuits and court rulings, often adverse, that struck at public forestry procedures. The national forests saw early attacks on timber-harvesting methods and even on the act of timber harvesting itself [8].

Item Forestry administrative procedures that for decades had been regarded as the sole responsibility of the administering agency became invested with a public character of widened scope. Among federal laws in question was the National Environmental Policy Act of 1969, leading to creation of a federal Environmental Protection Agency to oversee a wide range of restrictions, many of them affecting forestry and forest industry. And there was the Forest and Rangeland Renewable Resources Planning Act of 1974, as amended [2], exposing the Forest Service, under floodlight, to elaborate planning and review procedures (see Social Planning for Forestry, Chap. 60). Administrators contemplating new management measures were required to file environmental impact statements for public review.

Item The classical tenet upholding the primacy of timber among forest values was rejected. Recreation values werc elevated in the national esteem. Thus the contrast between the long trends in consumption (Chap. 14) was recognized. Forestry schools that still had no major program in outdoor recreation created such programs. A number of schools amended their titles to include the term "environment" or "environmental." Wildlife, watershed, and aesthetic values were given fresh recognition. The diversity of nature received new appreciation.

Item Many environmentalists wanted forest lands preserved as wilderness. Congress passed a Wilderness Act in 1964, creating a National Wilderness Preservation System out of designated national forest, park, and other federal lands. Some of the states passed wilderness laws. (Refer again to Chap. 60.)

Item Laymen, whether in the woods or out, could be dismissed no more as a nuisance. Under the environmentalist thrust, laymen-consumers were pressing for a fresh orientation of the professional subculture: beyond timber and beyond woods and mill.

> What is your definition of forestry? Do your classmates accept your definition? Do your instructors? How does the Society of American Foresters define the field? How do you think the general public defines it?

In reviewing the prime events in the environmental movement in the U.S., we must not fail to recognize the significance of efforts abroad, some of them undertaken despite a hanging-back in the U.S. In 1992, for example, Japan and the 12-nation European Community, in the midst of talks on the environment, agreed "to take a leading role" in solving world environmental problems, notably industry's heavy carbon dioxide emissions and the tropical-forest destruction in the Amazon basin and Malaysia [11].

SUMMARY

Foregoing paragraphs have glanced at these questions: What is the consumer movement? What is the environmental movement? How are they related to forestry?

In briefest summary, the two movements are seen as expressions of consumers' genius for generalism and self-defense. Coming into force in the U.S. in the 1960s, they have called the forestry profession on the carpet to reconsider outdated tenets of its subculture.

REFERENCES

1 Barney, Daniel R.: *The Last Stand, Ralph Nader's Study Group Report on the National Forests,* Grossman, New York, 1974, especially pp. xiii–xxi.

2 Behan, R.W.: "RPA/NFMA, Time to Punt," *Journal of Forestry,* December 1981, pp. 802–805.
3 Burek, Deborah M., ed.: *Encyclopedia of Associations,* 25th ed., Gale Research, Detroit, 1991.
4 *Consumer Reports,* monthly, especially the most recent December ("buying guide") issue.
5 Fernow, Bernhard E.: *Economics of Forestry,* Thomas Y. Crowell, New York, 1902, p. 85.
6 *Forestry Quarterly,* November 1903, p. 44.
7 Heinrichs, Jay: "Pinchot's Heirs," *Journal of Forestry,* May 1985, p. 279.
8 Mock, H. Byron: "Forest Management and Litigation," *Journal of Forestry,* April 1971, pp. 200–205.
9 Rodgers, Andrew Denny, III: *Bernhard Eduard Fernow, A Story of North American Forestry.* Princeton University Press, Princeton, N.J., 1951, pp. 314–316.
10 Shanks, Bernard: "Institutional Research for a Democratic Wilderness Policy, *In Proceedings, National Wilderness Research Conference: Issues, State-of-Knowledge, Future Directions.* U.S. Department of Agriculture, Forest Service Gen. Tech. Rep. INT-220, 1987, p. 358.
11 *The Wall Street Journal,* Eastern ed., January 10, 1992, p. A10.

CHAPTER 11

CONSUMERS' EVALUATION OF LANDSCAPES

WHAT ARE LANDSCAPES?

Chapter 10 emphasizes the interest of consumers in forest values other than timber. Now let's have a closer look at this nontimber interest, taking as a case the aesthetic values of the landscape.

Landscape means one's surroundings: the part that one is aware of. The landscape is primarily what one looks at. But surely it doesn't consist of just a view. The signals that come through to the recipient are also sounds and odors and messages for the other senses. And they are filtered through the recipient's inheritance and culture and past associations. I rate this landscape high in aesthetic value because I like its looks and its quietness and because it reminds me of a pleasant experience and because I am happy this morning.

Let's address, then, the question of a landscape's aesthetic values: What attributes of a landscape may be important to consumers? What makes a landscape beautiful—or preferable to some other? How may you, as problem solver, predict whether a change in the forest landscape will meet with consumers' approval?

ATTRIBUTES OF LANDSCAPES

In an effort to make the complicated reasonably simple and straightforward, let's recognize three kinds of attributes of landscapes.

1 Elemental attributes. Examples: color (that is, hue, and its lightness or darkness, strength or weakness); texture (configuration of a surface); distance (from the viewer or between two external points); shape (of lines such as edges or boundaries), temperature; characteristics of sound, odor, taste.

The elemental attributes are those features of the landscape that lay relatively few demands upon the consumer for synthesis, for putting impressions together. The distinction between this kind of attribute and the second kind, now to be mentioned, is to a large extent arbitrary. But it's a helpful distinction, for it avoids throwing a wide range of sense signals into a single category.

2 Formal attributes: those that relate to the form in which ingredients of a landscape (elemental attributes) are brought together. Examples: color combinations and patterns, shape (of solid or complex objects), size, number, composition (arrangement of objects of different kinds, shapes, sizes, colors), combinations of simultaneous sounds or of consecutive sounds.

The foregoing definitions do not deal evenhandedly with all the senses. The characteristics of odor, taste, and touch are all classed as elemental attributes. Only sight and sound impressions are represented among the formal attributes, and sight impressions take up most of the story. It would be possible to enumerate forms and configurations of sounds, odors, and so on analogous to those of sights. But for our purposes, it would not be useful.

3 Imaginative attributes: impressions that are received from the landscape *indirectly,* as opposed to those just discussed, which are received directly. Examples: the purpose that I think a man-made feature of the landscape is intended to serve, a bed of flowers in the city that leads me to recall rural life in spring, a patch clear-cutting in the valley below that makes me think of the charming checker-boarded fields and pastures of the English countryside, a grove of trees in which I see mysterious spirits lurking.

The imaginative attributes, then, are associations and illusions that are called to a viewer's mind. They may be the most influential in a landscape's bundle of attributes. They necessarily vary among consumers as well as among landscapes. Thus my evaluation of them may be even more subjective than my response to the elemental and formal attributes.

> Is there anything that you, as a manager, could do to improve the quality of your forest's imaginative attributes?

Subjective judgments are a function largely of the viewer's culture. How very different are trees seen in different settings by different persons: (1) in a desert oasis by a native, (2) on our western range by the manager, (3) in a woodlot by a farmer, (4) in a national forest by the ranger, (5) in an industrial forest by the logger! Why the differences? They stem from differences in the way the persons associated with the trees have been taught to think. Yet botanically, the trees may be alike. By the same token, a single tree may be viewed differently by persons of different cultures. An artist and a timber forester see an old, decayed, crooked tree through much different eyes. So essential is this cultural difference that, as noted in Chap. 9, aesthetic activities and standards are recognized as a separate class of cultural institution.

Again you are confronted with the dialectical character of your subject.

WHAT IS BEAUTY?

What makes a landscape beautiful or pleasurable or preferable to some other landscape?

Philosophers have been trying for a long while to produce a clear definition of beauty. Some of the great minds in the history of humanity have addressed the question. They have not been especially successful. Nevertheless, it behooves us to give some thought to the question ourselves. There *is* such a thing as beauty! Admittedly, some consumers are unaware of the aesthetic qualities of their surroundings [6]. But in most persons' minds, beauty is preferable to ugliness. And foresters are very much caught up in the conflict between beauty and ugliness, as we learn in Chap. 10.

The Greeks were interested in the subject of beauty and tried to define it. Socrates said that the useful is beautiful. Plato said that beautiful objects are those that have harmony, unity, and proportion among their parts.

One finds in ancient Greece an attempt to be precise about beautiful proportion. The Greeks lit upon the "extreme and mean ratio," which today is called the "golden section." This is the division of a length such that the smaller part, *a,* is to the larger part, *A,* as the larger part is to the whole:

$$\frac{a}{A} = \frac{A}{a + A}$$

The value of a/A turns out to be about .618, or about $\frac{5}{8}$. With its nice symmetry, it holds an appeal for the mathematically minded. A golden rectangle is sketched herewith.

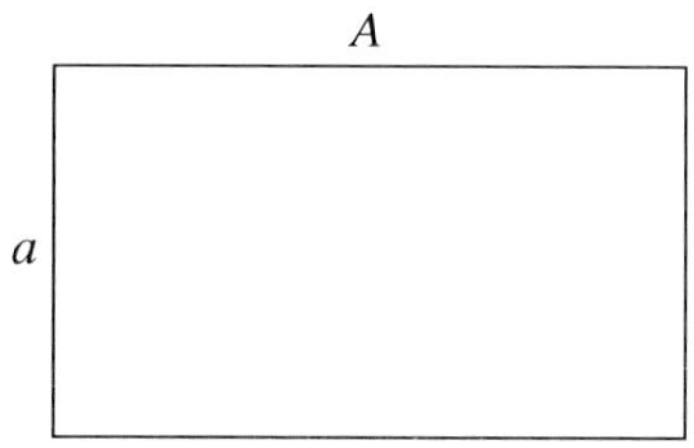

> Is the golden rectangle pleasing and harmonious to your eye? Would you consider a rectangle with a smaller *a/A* ratio more beautiful? a larger ratio? Why?

In modern times, the German philosophers have devoted thought to the subject of aesthetics. Immanuel Kant, who worked in the eighteenth century, proposed an interesting postulate: Aesthetic enjoyment must be disinterested. When I contemplate a landscape aesthetically, I cannot be concerned with its practical significance or material value. I cannot, for example, consider whether I would like to own the land. Kant's philosophy led him to the conclusion that it is impossible to develop a rule for recognizing beauty, such as "All objects possessing thus and such qualities are beautiful." The rule would involve a logical, not an aesthetic, judgment. Aesthetic judgments, Kant believed, are necessarily subjective: "I think this scene is beautiful."

A Darwinian definition of beauty has been proposed: Beauty is the quality in living things that leads to their success in natural selection: to their conservation and survival as a species or tribe.

Landscape architects have given attention to rules for recognizing or creating a pleasurable scene. Among desirable qualities in a landscape or structure are symmetry, order (as opposed to randomness), definiteness (clarity of form, outline, and so on), naturalness (the quality of being untouched, unchanged), distinction (based upon the presence of pleasing contrasts or strikingly prominent objects), vividness (again a matter of contrast), unity (made up of parts that belong together), diversity or variety (as of a rock-strewn stream bordered by low-hanging trees), congruity (of an object that is appropriate to its surroundings, blends well).

One can appreciate a list such as the foregoing and use it to advantage, but in fact the list taken alone is ambiguous. Consider congruity. Clearly, you would not want to minimize the congruity of elements in a landscape. But you would not want to maximize their congruity, either, for this would mean to minimize contrast. Get in a small plane and fly over an extensive, even-aged, well-stocked, pure pine forest on level land. The congruity of the landscape elements is a maximum, and the inherent aesthetic appeal of the landscape is essentially nil. After you have assembled a list of qualities desirable in a landscape, you are still as far as ever from defining beauty, because you have not been able to specify the optimum for any of the qualities. Kant would say it's impossible, that the optimal degree of any quality is a subjective judgment, that there cannot be a rule that fits all landscapes and viewers.

Every one of us has been *taught* (in our family, at school, among our other associates, and by ourselves) beliefs about beauty and the pleasurability of the scene. We have been taught to think that this kind of landscape is preferable to that kind. That is, as noted earlier, our aesthetic tastes are a function of our culture. Remember what people said in the "masonry zone" (Chap. 9): Masonry looks better than wood. And outside the zone: Wood looks better than masonry. If I can examine my landscape preferences and search for their origins, I may be surprised to discover how many of my notions are hand-me-downs from persons whose views I respected.

GUIDES TO MANAGEMENT

Still, the old search continues for a "science" of aesthetics: for rules, or "laws," preferably with numbers attached. In the U.S. in recent years, the search has been concentrated to some extent on predicting how viewers will respond to alternative scenes or changes in a scene: how viewers will rate them or rank them. A pioneering effort of this sort was reported by Elwood Shafer [7] in vol. 1, No. 1 of the *Journal of Leisure Research.* In the lead article, he featured the phrase, "It seems possible to quantify aesthetics" and went on to describe a method for predicting consumers' preferences for landscapes. But his equation, though rather reliable statistically, employed variables (perimeters and areas on black-and-white photos) that lack significance for the land manager. Cannot variables be found that clearly *mean* a course of action for the manager: that help *instruct* the manager what to do?

Much of the more recent quantitative effort has gone into the quest for independent variables that are meaningful and instructive to the manager. Louise Arthur [2] in 1977 described her search for variables that could be used "to manage for scenic beauty." Studying ponderosa pine landscapes in Arizona, she found that three sorts of forest attributes are usefully related to consumer response: (1) physical features such as downed wood and ground cover, (2) design features such as extent of the view and direction of lighting, (3) timber-cruise data such as tree-diameter distributions and stand density.

Robert Brush [4], in a 1979 report on Massachusetts forest owners' landscape preferences, linked these preferences to spatial qualities of the forest. He found that large enclosed spaces and spaces created by thinning heavily stocked stands were more attractive to his respondents than unbounded openings or dense stands. Linda Anderson [1], in a 1981 Arizona study, concluded that viewers' liking for a landscape can be significantly affected by the designation applied to it, such as commercial timber, leased range, national forest, national park, recreation area, or wilderness area. The 1980s saw more efforts [for example, 3, 5] to discover landscape variables that have the dual quality of being controllable by the resource manager and strongly correlated with viewers' preferences.

> Anderson's findings bear out the experience of the creators of effective advertising: Names for consumables can be devised that appeal to consumers and increase sales. Can you devise a term or slogan for some technically sound forestry practice currently in public disfavor which would enhance the public image of the profession?

PREDICTING CONSUMERS' EVALUATION

How may you, as problem solver, predict whether changes in a forest landscape will meet with viewers' approval? For example, the manager shows you a forest tract, describes a few alternative procedures that will change the appearance of the tract, and asks which of the procedures, if any, people will approve of. The question is approachable by the usual steps:

First, what sort of question is this? It comes down to forecasting consumer preferences among landscapes, ranking or weighing the landscapes in terms of aesthetic appeal. If A is the existing scene, and B, C, and D are alternative scenes, you will try to learn how the four scenes compare in people's eyes and then assume that the preferences you find today will hold for tomorrow also. Remember that you are *not* being asked to *explain why* people prefer this landscape to that. You will try simply to discover *what* their preferences are.

Second, what is a suitable model? It's this: Select a sample of consumers, arrange for them to view the alternative scenes, and ask them to evaluate the alternatives.

Third, what is involved in working the model and deriving a reply? Some jobs are involved:

1 One job is statistical. You try to identify the universe of consumers whose preferences will be studied and to select a representative sample of sufficient size to bring errors within acceptable limits.

2 Another job is to find or simulate the landscapes to be compared. Bring together photographs, drawings, or scale models. Or ask consumers to picture the alternatives in their mind's eye. For instance, would they prefer to have more people in the landscape, or fewer?

3 Still another job is to design the questions to be put to the consumer sample. You may ask them merely to rank alternatives. Or you may ask them for ratings on a scale, as from 0 to 10. It may be helpful to ask them some things about themselves that may bear upon preferences.

Fourth, what about the follow-up? This, understand, is *your* follow-up, not the decision-maker's: not the latter's consultations, hearings, explanatory meetings with consumers, or whatever. Your follow-up may take the form of repeating your analysis with different interviewees, materials, or procedures so as to test the sensitivity of your conclusions to your approach. If studies have been made of the principles governing preference for landscapes of the type you are working with, you will want to check your results against those studies' findings.

Let's recognize some common and not always escapable flaws in such an analysis of consumer preferences as is sketched here.

One is that the spectrum of consumers' aesthetic reactions is apt not to be well represented. If you ask me to look at a picture or other model of a landscape, I will give you my evaluation of the model, not necessarily of the landscape. My evaluation will be solely visual; the other senses will not be represented. Furthermore, my responses to the imaginative attributes of the landscape may well be muffled. I may fail to see mysterious spirits lurking in the modeled grove.

You may find it difficult to identify the universe of consumers, especially if you face such a question as "What would the viewers of this area think about changes or developments of this or that sort?" You don't know who the viewers would be if the changes were made. Beyond this difficulty is one more perplexing: The universe of viewers and the universe of persons in whose behalf aesthetic decisions are made may be different.

> Is it conceivable to you that most of the consumers whom the forest manager is interested in are absentees, not viewers? Is it possible? Describe a case in point.

Is aesthetic enjoyment essentially a social experience? Some say it is. As a rule, one tends to indulge one's aesthetic tastes in the company of others. Kant said that if alone on a desert island, he would adorn neither his hut nor his person. Much analysis of aesthetic preferences fails to account for their social character.

COMPARING MODERN AND ANCIENT THOUGHT

What is the connection between the classical effort to define beauty and the current effort to predict how people will react to objects? The first response may well be that these are two very different pursuits: Today's philosophers are not trying to explain what beauty is; they are trying to explain whether people will like what they see, or how strongly they will like or dislike it. Although Louise Arthur entitles her article, "Predicting scenic beauty. . . ," she soon settles down to a reworded phrase, "Explaining people's evaluations of scenic beauty."

If you can understand the difference between the classical and current pursuits, you will have come a long way toward grasping the subject of this book, the economics of forestry as social science.

To take the ancient Greek philosophers as the classical case, they delighted in the discovery of universal truths—for example, that the square on the hypotenuse equals the sum of squares on the other two sides of a right triangle. Most of us accept this proposition as universally true in the sense that it holds everywhere on earth. The natural sciences deal in such universal truths.

However, the inquiry about beauty and essentially all the other inquiries in this book are not natural science, but social science. The question of what is beauty is a question, if you like, in cultural anthropology. If someone asks you this question, you are obliged to reply, "That depends." It depends on the object, the time, the beholder, and the other circumstances of beholding.

And so, how do contemporary analyses of beauty differ from those of Plato? There is no fundamental difference at all. Contemporary analysts are simply giving more attention than Plato did to the independent variables other than the object itself.

SUMMARY

Consumers evaluating a landscape are responding to three kinds of attributes: elemental, formal, and imaginative. The qualities of the attributes taken as a bundle determine the degree of beauty of a given landscape for a given consumer or group at a given time. As a general proposition, you can think of each attribute as capable of taking on any quality within a range. For the given case, the optimal quality (that which contributes to a top evaluation) invariably lies between the extremes of the range, at some point that can be identified only subjectively at best. Universal truth respecting beauty has therefore been impossible to formulate. However, it *is* possible to predict with some reliability the preferences of given consumers for given landscapes at a given time, using variables over which the resource manager has some control.

REFERENCES

1 Anderson, Linda M.: "Land Use Designations Affect Perception of Scenic Beauty in Forest Landscapes," *Forest Science,* February 1981, pp. 392–400.

2 Arthur, Louise M.: "Predicting Scenic Beauty of Forest Environments, Some Empirical Tests," *Forest Science,* February 1977, pp. 151–160.

3 Brown, Thomas C., and Terry C. Daniel: "Modeling Forest Scenic Beauty, Concepts and Application to Ponderosa Pine," USDA Forest Service Res. Pap. RM-256, 1984.

4 Brush, Robert O.: "The Attractiveness of Woodlands, Perceptions of Forest Landowners in Massachusetts," *Forest Science,* March 1979, pp. 495–506.

5 Hull, R. Bruce, IV: "Forest Visual Quality Management and Research, *In* Watson, Alan H. (compiler): *Outdoor Recreation Benchmark 1988.* U.S. Department of Agriculture, Forest Service Gen. Tech. Rep. SE-52, 1988, pp. 485–498.

6 Knudson, Douglas M., and Elizabeth B. Curry: "Campers' Perception of Site Deterioration and Crowding, *Journal of Forestry,* February 1981, pp. 92–94.

7 Shafer, Elwood L., Jr., John F. Hamilton, Jr., and Elizabeth A. Schmidt: "Natural Landscape Preferences, A Predictive Model," *Journal of Leisure Research,* Winter 1969, pp. 1–19.

CHAPTER **12**

CONSUMERS' SUBJECTIVE VALUES: RESOURCES

This chapter asks two questions. First, what is value and how is it measured? Second, what is a resource? The replies will be useful for problem solving in later chapters.

Chapter 9 works with consumer preferences for building materials; Chap. 11, for landscapes. Both concern consumers' subjective values. Let's go now to consumables in general. The consumables of interest are the services of the forest.

Our value system comes out of our culture. Every institution of the culture is influential. Language, for instance, through the connotation of words, colors our view of the things the words represent. As cultural institutions, values are different from one culture to another and thus often from one person to another; and they are continually changing.

How is the foregoing point related to the one attributed to Linda Anderson on page 74?

You can think of your subjective values as given to you simultaneously with your goals. Values and goals are not always readily distinguishable.

VALUE DEFINED AND MEASURED

Earlier chapters work with instances of value that lend themselves to ordinal treatment: Which do you prefer, other things equal: *j* or *k?* For subjective valuation, this *ranking* approach is inherently appealing and believable. Surely I can expect you, as a consumer, to be able to express a preference between *j* and *k*, and thus between *k* and *m* and among any two or more alternatives (unless, of course, you conclude that the one and the other are equal, so which one you possess is a matter of indifference to you).

The problem now is to explain cardinal subjective value. Other things equal, how much is *k* worth to you? How much is *m* worth?

What is value, and what is the cardinal measure of it? The firewood *m* that I have stored behind my barn has value in your eyes, and the evidence of this is that you are prepared to give up possessions in order to own some of it. **Value** is the characteristic of a thing that makes it exchangeable with other things *in the sphere of possession.* The **measure of value,** then, is found in the other things that you stand ready to exchange. What are these other things, and how much of them are you prepared to give up?

The foregoing definitions give the term **possession** a special meaning. They require the term to include more than legal title to such items as firewood. They require the term to include any sort of command over the services of objects or persons or ideas. Hence the possession of a landscape is the opportunity to view it; and the measure of its subjective value is what one stands ready to sacrifice for the view, perhaps the net cost of traveling to the scene. Here one is not thinking about the value of the land. Indeed, Kant has warned (Chap. 11) against clouding one's judgment of an object's beauty with thoughts about ownership of the object.

The "other things" that are exchangeable as a measure of your subjective valuation can be various. If you are a subsistence farmer and are prepared to acquire a piece of furniture by making it yourself, the "other things" are those that you think you could produce in *h* hours, the time you expect will be required to make the furniture. To get the furniture, you have to exchange these other things—not with someone else, but with yourself.

Or if you are involved in barter, then the subjective value you place on my firewood is measured by the goods you stand ready to offer me in exchange—say, your shotgun for a cord of wood.

Or, of course, you can form your subjective valuation in financial terms: $100 for a cord.

CARDINAL VALUE: PRICE-DEMAND

What the consumer stands ready to give up—the time, the shotgun, the money—for the wanted consumable is its **price** (if per unit) or **value** (if fractional or multiple units) in the consumer's eyes. This price or value P of a stated quantity Q of consumable m (a cardinal measure) can be thought of as found by ranking (ordinally) P_m against Q_m, varying P_m until a value is reached where the choice between P_m and Q_m is a matter of indifference.

> Explain in your own words how a relative (ordinal) approach to valuation can be converted into an absolute (cardinal) measure.

Subjective value, then, is the price or value that a consuming firm stands ready to pay for a given quantity of something, such as firewood, under given circumstances. When it comes to acquiring the wood in actual fact, you and I may bargain and come up with an *objective value,* or a **market value (price),** of less than a shotgun or \$100 for a cord of wood. Or you may find in the community a well-established market for firewood, in which you can buy a cord for \$80, and this is what you will offer me. In the transaction, you will enjoy a **consumer surplus** of \$20 (Fig. 12-1). Part Five of this book looks at marketing and market value. Here, however, let's stick with subjective value, being reminded as we go along that although the *amount* of it will in some instances equal market value, its *nature* is different. It is what economic theorists have termed **value in use,** in contrast to **value in exchange.** It is what consumers *stand ready* to pay, as opposed to what they *actually* pay.

And so, to generalize, we have a given population of consuming firms contained in a given location *L.* This population stands ready to acquire a quantity *Q* of consumable *m* per time period *t,* this quantity being a function of *d* of the firms' subjective valuation of *m* (that is, P_m) under stated conditions. The stated conditions, aside from *L,* include the technology of consumption *T,* the firms' **disposable income *Y*** (available for consumption and saving, after income taxes) during *t,* and the market prices *P* of services other than *m.* Letting Q_m vary with P_m, we have the **price-demand function** of economic theory:

$$Q_m/t = d(P_m) \mid L, T, Y, P_1, P_2, \ldots, P_{m-1}, P_{m+1}, \ldots, P_{n-1}, P_n \qquad (1)$$

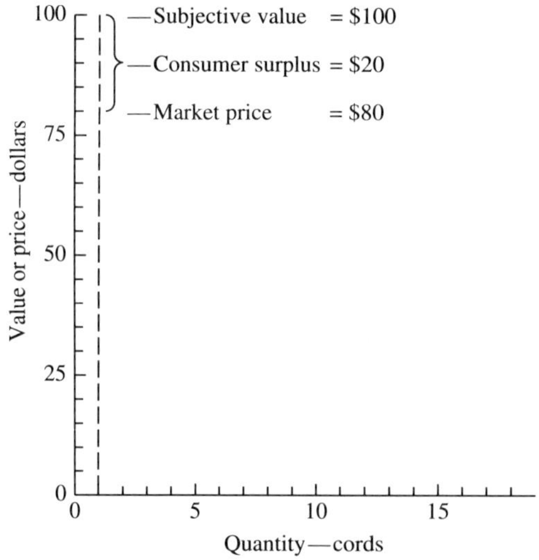

FIGURE 12-1
Your subjective value, the market price, and your consumer surplus for a cord of the firewood stored behind my barn. Hypothetical example.

CONDITIONS OF DEMAND

To be more specific, one can call expression (1) the **short-run demand** for *m,* using the qualifier in the same sense as the theorist uses it in referring to the short-run *supply.* Short-run demand is pictured as an intended, or planned, flow through the consumption process, given a goodly measure of fixed inputs into consumption: The firms' technology is fixed, and thus much of their durable consumer capital is fixed, such as dwellings, means of transport, and lesser equipment for living. Short-run demand contrasts with **long-run demand,** where in the extreme case none of the inputs to consumption is fixed and the firms' demand is a set of plans for consumption constrained only by prospects that have not materialized. For example, a plan for forest resource use might look to regional or national policies for an enunciation of forest stewardship in the social interest. Consumers might be directed toward wants consistent, say, with environmental enhancement. What, then, would be consumers' response to changes in price or other variables given the planned constraints upon demand? When prospects materialize, constraints become fixed and the demand is short-run. Hold these responses in mind when we come to Fig. 12-3 and accompanying discussion.

Of the foregoing **conditions of demand,** this book works almost exclusively with the first, the short-run demand. The other condition is introduced here because, when it comes to Parts Three and Four, both short-run and long-run *supply* will figure in the analysis. The preceding paragraph seizes the opportunity to remind you again of the close family resemblance between consumption and production.

A PRICE-DEMAND CURVE

A price-demand function such as that in expression (1) can be useful. It tells about subjective evaluations by a group of consumers—say, in a community or wider region. If the function is derived from real data, which is possible after a fashion, it explains how people actually behaved during a past period and may help predict how they *will* behave.

If a price-demand function is plotted on arithmetic scales and a price-demand curve drawn, the result might resemble Fig. 12-2. You can look upon this graph as symbolizing the short-run demand. The curve need not be concave. But it is apt to have a negative slope because the lower the level of P_m, (1) the greater the disposable income available to firms for all their purposes, including *m* (the **income effect**); (2) the more attractive *m* is compared to other items, whose prices are fixed (the **substitution effect**); and (3) the greater the probability that firms will anticipate a price rebound and thus view P_m as a "limited-time offer" (the **bargain-sale effect**).

Subjective values P_m are shown along the curve in Fig. 12-2 from left to right in descending order of magnitude. Individual firm's evaluations are scattered along the curve. They reflect the firms' diminishing interest in *m* as their shopping list calls for acquiring larger quantities. Some firms are not represented at all: They have no interest in *m* in period *t;* indeed, the implication is that its subjective value to them is negative.

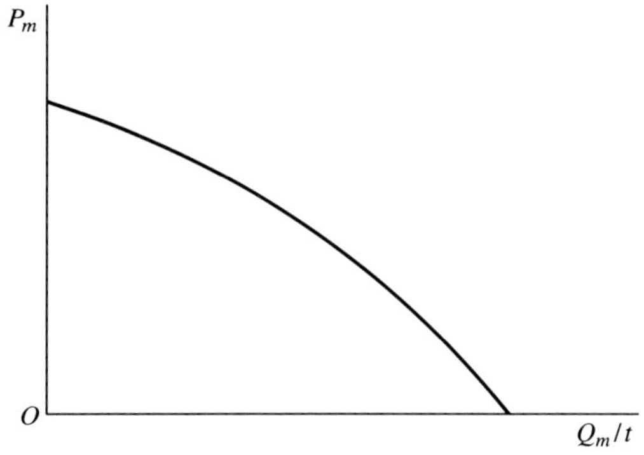

FIGURE 12-2
Hypothetical aggregate price-demand curve for consumable *m:* quantity demanded per unit of time in relation to price, or subjective value per unit of measure.

> Why does it follow that firms not represented on the curve in Fig. 12-2 place a *negative* value on consumable *m*?

COMMENTS ON THE CURVE

Here are three points about demand, such as that represented in Fig. 12-2:

First, the price-demand says that if the price of m is P_m, then consumers stand ready to acquire the quantity Q_m per unit of time. The independent variable is P. But here tradition is broken; P is put on the vertical axis. The justification for this is that P is not only price, but also subjective value, a surrogate for marginal utility, a function of Q. Thus the arrangement of axes in Fig. 12-2 is best for most purposes.

The price-demand function (marginal utility in relation to quantity per unit of time) is a **consumption function.** It may be viewed as derived from the basic function, *total* utility in relation to quantity. Our interest hereafter lies in the marginal-utility function, not the total.

Second, it will help to mention again the conditions of demand. Let's focus on the distinction between short and long run and take for illustration the opportunity to swim, for a fee, at Forest Park. The fee is lowered, and the community responds by using the facilities more heavily. A new point is found on the short-run demand curve, representing a **change in the quantity demanded.** As time goes on, the low fee attracts more swimmers more frequently to Forest Park, away from alternative recreation. And nonswimmers learn to swim. Visits to the park continue to increase, not along the short-run curve, but to a new point on a *long-run* curve that allows for change in the technology of consumption. The old short-run demand no longer exists: There has been a **change in demand** for the short run. The evidence of this is that when the former fee is reinstated, visits do not drop all the way back to their former rate. The rate is higher than before, because the technology has changed and short-run demand has increased. Prospects, initially little constrained, have now materialized. See Fig. 12-3.

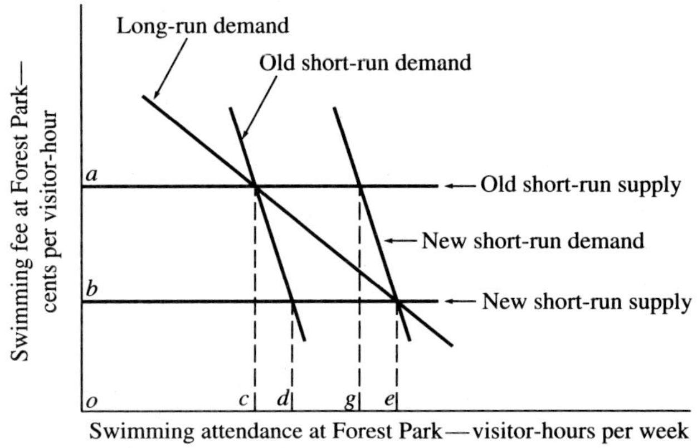

FIGURE 12-3
Short-run demand and supply, and long-run demand, for swimming opportunity at Forest Park (*oa* = old fee; *ob* = new fee; *oc* = old attendance; *od* = intermediate attendance, incorporating an increase in amount demanded in the short run, on account of new fee; *oe* = new attendance, incorporating an increase in short-run demand, or an increase in amount demanded in the long run, on account of new fee; *og* = new attendance when old fee is reinstated). Hypothetical data.

How do you interpret the long-run demand in this case? It is not a statement of plan. Yet it is conjectural, existing only by virtue of the short-run demands and supplies.

Third, it is essential to distinguish between demand in the economics sense—its sense in this book—and **"demand"** in the sense intended by the federal Forest Service in such statements as, "Demand for timber products will double by the year 2030." Hence, let's interpret demand as a functional relationship of quantity per unit of time to some independent variable such as price. Let's look at "demand" in Forest Service terms later on. For the time being, you can think of it as somewhat related to consumption, a topic taken up in Chap. 14. In terms of the present chapter, it is most nearly the quantity demanded.

RESOURCES DEFINED

Having talked about value, let's say something about resources, a closely related idea broached early and mentioned frequently in this book. Like the idea of value, that of resources will be useful to you as a problem solver. A **resource** is any bundle of attributes that has subjective value. The bundle may be embodied in an object such as a landscape or in an act such as to transport the viewer to the scene or in an idea such as how to make the scene accessible. Further, the bundle may be embodied, not directly in a consumable, but in a producer's instrument capable of giving birth to a consumable that has subjective value: timberland, a sawmill, a forest range, a forest [1].

Of all kinds of resources, knowledge (technology) is very possibly the most significant, because it is, so far as one can imagine, boundless. Knowledge is the only such resource. Forest resources may be renewable, but they are not boundless.

Just as value is a creature of the culture, so is a resource. Consider one of the principal forest resources, water. How greatly the significance of the water resource varies from one culture to another! Depending on its quality and abundance and on the technology that the culture bestows on its members, water may be used for drinking, washing, flushing away wastes, navigation, irrigation, baptism, brewing, viewing, flood control, power generation, manufacturing, cooling, heating, and so on. You can picture different cultures that have equal access to water but make vastly different uses of it.

So it is. Consider forest land. In America before the European immigration, the forest held an immense quantity and variety of material that had physical and spiritual significance, but, as a resource, it was small and mostly inaccessible to the primitive cultures in which it was set. Later, in the early years of European colonization, the forest formed a barrier to migration and agricultural development. Thus some of its values were negative: The resource was diluted with **resistances.** Today, the extent and stature of the American forest are comparatively puny, but its weight as a resource is far, far greater than in the days of the Native American and the settler, simply because of the difference in culture. The value of the resource has been increasing all the while, not because trees grow, but because knowledge grows.

What are these cultural institutions that are capable of making resources out of stuff that formerly was neutral? They are transportation and communication, social organizations, ideals and value systems, aesthetic appreciation, and the understanding to build and use tools and machines, to manage forests, to use their services, and to safeguard our environment to the extent that our circumstances permit.

Clearly, resources come into being by the agency of man. Something was there before, but man was required to make a resource of it. A resource is like a painting: The materials are at hand, but to make a painting of them, an artist with an idea and a technology must be brought in. Behind the artist are the artisans who created the materials, and behind them, still other artisans in succession, until we come at length to the quiet and unpeopled earth. Are resources, then, man-made? Well, at least their existence is the outcome of collaboration in which the human agent took a large part.

> What analogies can you find between the ideas about forest resources just expressed and the ideas about forestry and development in Chap. 3?

SUMMARY

Values arise from the culture. A consumable has subjective value if someone is prepared to sacrifice possessions for it. The measure of this value is found in the kind and quantity of these possessions.

The quantity of a resource that consumers stand ready to acquire in a time period can be stated as a function of its subjective value, other things equal. This is the price-demand function of economic theory, a consumption function.

Resources arise from the culture and are thereby, in a sense, man-made. They are items that have value, measured by human willingness to sacrifice for them. Knowledge is the foremost resource. Resource creation is essentially the same process as economic development.

REFERENCES

1 The ideas about resources in this book are taken largely from the classic work on the subject: Zimmermann, Erich W.: *World Resources and Industries,* Harper & Row, New York, Evanston, and London, rev. ed., 1951, especially pp. 3–20.

CHAPTER **13**

CONSUMER DEMAND FOR FOREST RESOURCES

This chapter addresses the question, What are some of the characteristics of consumer demand for services of the forest? It uses three models: the model of price-demand derived in Chap. 12, and models of income-demand and of demand elasticity given in the present chapter. The data come from studies made in the U.S. The follow-up consists of interpreting the findings.

ELASTICITY OF DEMAND

As you know, a useful measure of price-demand is its **elasticity,** ϵ, the sensitivity of quantity, Q, to price, P. Since price-demand is a statement of how Q is related to P, this question of sensitivity is of central interest. One *could* ask how much difference in Q is associated with a unit difference in P:

$$\Delta Q \div \Delta P \tag{1}$$

which is a question of slope.

But a moment's reflection suggests that slope as in expression (1) is not a useful measure, because dividend and divisor are expressed in different units. To derive a useful measure of sensitivity, one works with *proportional* differences in Q and P, or elasticity:

$$\epsilon = \frac{\Delta Q}{Q} \div \frac{\Delta P}{P} \tag{2}$$

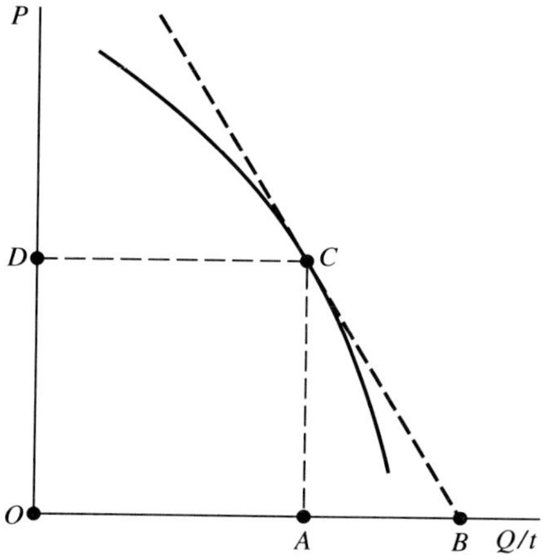

FIGURE 13-1
Measuring elasticity at point C on a price-demand curve, or the income-demand curve for an inferior good; ϵ, a negative quantity, equals $BA \div OA$.

You can think of elasticity as the percentage of difference in quantity accompanying a 1-percent difference in price.

Now consider the **point elasticity** of demand at point C in Fig. 13-1. The elasticity of the curve at C is the elasticity of the tangent, CB. The price at C is OD; the quantity, OA. Now let's examine the case where $\Delta P = OD$, so that $\Delta Q = BA$:

$$\epsilon = \frac{BA}{OA} \div \frac{OD}{OD} = BA \div OA \tag{3}$$

The sign, like the slope of the curve at C (the direction B to A) is negative.

Figure 13-1 illustrates the superiority of elasticity over slope as a measure of Q-to-P sensitivity: The magnitude of elasticity is independent of the scales on the axes; that of slope is not.

When a price-demand function was developed in Chap. 12, income was held constant. But surely one can take firms' income as the variable on which quantity per unit of time depends, hold all prices constant, and obtain an **income-demand** function. The appropriate rearrangement of expression (1) in Chap. 12 is

$$Q_m/t = d'(Y) \mid L, T, P_1, P_2, \ldots, P_m, \ldots, P_{n-1}, P_n \tag{4}$$

Observe that in expression (4) Q_m/t is the quantity of m that firms stand ready to take (or have actually taken, if this is historical) per unit of time, where Y is their disposable income. As in the price-demand function of Chap. 12, firms or groups of firms are found strung out along the range of quantities, but there is no unique supply function to fix a quantity for the market. Every occupied point along the range is a fixed quantity. This observation will become clear as we move through the next section of this chapter.

Quantity may be a negative function of income per firm, so that the income-demand curve has a negative slope and elasticity like the price-demand curve. The consumable with such an income-demand curve (Fig. 13-1) is termed an **inferior**

good—because it is shunned by the better-heeled consumers, who have the freer choice among goods. Or the quantity may be a positive function of income, as in Fig. 13-2, which represents a **superior good.** Whatever the sign, the rule for measuring elasticity still holds. Notice in Figs. 13-1 and 13-2 that points *A* and *B* are located on the *Q* axis.

ILLUSTRATIONS OF INCOME-DEMAND

While we are on the subject of income-demand, let's pursue it with some forestry illustrations. Consider forest recreation. Or rather, consider *outdoor* recreation, for neither the data nor the reasoning about outdoor recreation is divisible into forest and nonforest categories.

Although no income-demand functions as such have been derived from outdoor recreation, a fair bit is known about the incomes of participants. Without doubt, outdoor recreation opportunity is a superior consumable. The Survey Research Center [4] found that the more active U.S. recreationists are the younger, better-educated caucasians who live in suburbs or in towns smaller than 50,000. At a time when the median annual income of all families, recreationists or not, was $8,660, the median for participants in walking, nature study, and pleasure driving was more than $9,000; for hunters, picnickers, and campers, more than $10,000; for hikers, above $11,000; and for those who drove off-road vehicles, over $12,000.

By analogy, one may speak of the income-*supply* of forest-based recreation opportunity. In a study of factors influencing the recreation use of private woodland in Vermont [1], the Forest Service learned that land held by younger owners with more formal education or those reared in large cities was most likely to be used for recreation.

If I were to analyze the incomes of ultimate consumers of wood-product services, would I discover relationships similar to those for outdoor recreation? Let me speculate about a reply, even though I have little data.

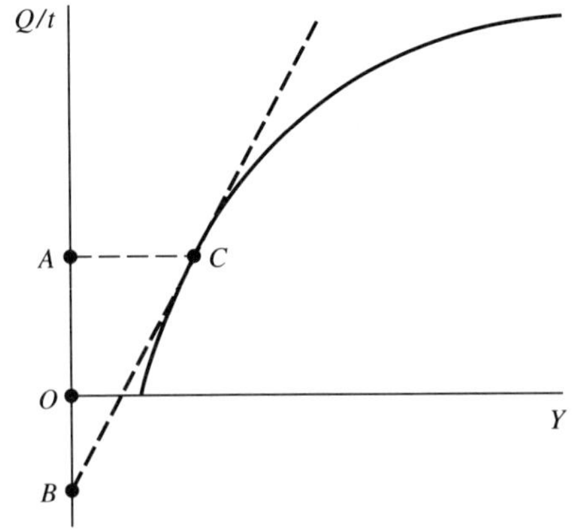

FIGURE 13-2
Measuring elasticity at point *C* on the income-demand curve for a superior good; $\epsilon = BA \div OA$, positive.

There was a time in the U.S. when wood, by and large, was an inferior good. The poor family on the frontier was wood's big user. Such a family in a forested region, and even in parts of the prairies and plains, made use of wood for housing, other buildings, fuel, fences, tools, wagons, utensils, furniture, toys, machinery, and goodness knows what else. Roundwood was abundant and cheap, and it could be worked and shaped by individuals with simple devices and skills. It was one of the prime essentials of rural life.

Higher-income, urban families used wood, but with a few minor exceptions in the furniture category and the glaring exception of their frame houses, they were proud to demonstrate their independence of the humble substance.

In those days, wood as an inferior good served as a leveler of the people's living. It infused a measure of equity into the economy's treatment of its ultimate consumers.

Then, one by one, wood-derived consumables lost their inferior status and became superior goods. Following our industrial revolution after the Civil War, the cities found their population swelled by migrants from rural regions and from abroad. Poverty became more an urban than a rural phenomenon. Tenement dwellers in want of such fuel and other material as might be useful to them had no woodland to draw upon. Their stair bannisters were a poor and scarcely renewable resource. The frame house, banned from downtown as a fire hazard, became a symbol of middle-class suburbia. Wooden furniture and fixtures, now factory made, became a securer expectation for those who were better off financially, and such persons could be counted upon to acquire such items in larger quantities as time went on.

Today, those who take advantage of access to a wide variety of forest products are, notably, the class that we previously identified as outdoor recreationists: the better-educated, higher-income caucasians who live in rural residences, suburbs, and small towns.

> Surely wood for the fireplace is a superior good. But what about wood for cooking and heating in the U.S.? The pulverization principle tells us it's an inferior good. Does the U.S. line in Table 4-1 tell us anything?

Among major commodities, the one in which wood probably has longest been a superior good is the single-family dwelling. Some time ago, the same man whose work is reviewed in Chap. 9, Joseph Zaremba [9], made a survey of all families in New Haven, Conn., who had built or bought a new house during the year. In each case, he determined (1) the family's income per year and (2) the quantity of wood in the house structure: boards, dimension, millwork, plywood, shingles, and so on. He found a clear positive relationship, little changed after a period of years (Fig. 13-3). He concluded that high income allows the family (1) to fulfill its want for a large house,

(2) to install more wood paneling, and (3) to have a ranch-style dwelling, which uses more wood than other styles, per square foot of floor space.

If you sketch or picture the tangents in Fig. 13-3, you find that they strike the vertical axis above the origin. This means that *BA* is less than *OA,* so that $\epsilon < 1$. Thus Zaremba's demand curves are **inelastic:** A 1-percent difference in income is accompanied by less than a 1-percent difference in quantity. Elasticity appears to be close to **unitary** ($\epsilon = 1$) at very low incomes and to diminish steadily with higher incomes. Wood use is most sensitive to income in those brackets where much wood is just beginning to be afforded; in higher brackets, incomes are reached that allow the consuming firm to buy about as much wood as it cares to use. Nowhere are the curves **elastic** ($\epsilon > 1$).

PAPER CONSUMPTION AND INCOME

In many respects, pulp and paper products are seen to differ from other wood products. Consider income-demand and related forms of demand.

To start with, it's necessary to recognize that pulp products comprise a great variety of consumables, which are used in such very different ways that they cannot always be lumped for analysis. In this section, the discussion is limited to *household papers,* the kinds used directly by ultimate consumers: paper for printing and letter writing, napkins, towels, tissues, and tableware.

People commonly take it for granted that paper consumption has strong positive correlation with income—and so it has over time. In the history of the U.S., income and paper consumption have risen together. But this tells us little or nothing about income-demand: whether in a given period, high-income firms use more paper or less than firms with lower incomes. The historical statistics tell how consumption changed,

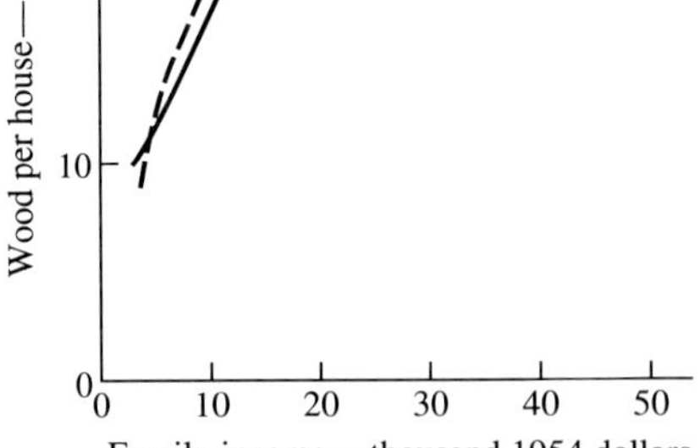

FIGURE 13-3
Income-demand for wood in single-family dwellings, New Haven, Conn., 1940 and 1954. (*Source:* Zaremba, Joseph: "Family Income and Wood Consumption," *Journal of Forestry,* June 1961, p. 445.)

but they don't say what this change is related to. If you were planning a paper-sales program in a region, you would be foolish to aim the program at higher-income families on the strength of the historical statistics.

> What light do the above observations shed on the question about Table 4-1 posed on page 89?

Another piece of conventional wisdom is that paper consumption has a positive net relationship to education [5].

The truth, apparently, is that paper-demand functions are complex, and what has been learned about them does not, in general, lend strong support to our preconceived ideas. In a very intensive sample of families in Liverpool, New York, then a town of 8,000 households, Thomas Wood [8] reported the following findings about household paper consumption in relation to (1) family income, (2) education of the family head, (3) age of the family head, and (4) size of family:

Purchase of *daily newspapers:* Relation to anything but family size and age of family head is negligible.

Stationery consumption: Age and education are the strongest predictors. Relation to family size is negative.

Purchase of *magazines* and *books* and the quantity of *mail* received: These are positively related to income, but related to education only as it also relates to income.

Consumption of *towels, napkins, tissues,* and *tableware:* These are negatively related to education, and little related to anything else.

PRICE-DEMAND FOR LUMBER

Let's study a few more demand functions, concentrating on the relation of quantity demanded to price or unit cost. The raw material for the study is a set of historical data for each commodity, data on the quantity that changed hands and the price paid in each of as many different market transactions as possible within place and time limits. You can picture plotting these P and Q data on a graph sheet and finding that the points are scattered.

Each point represents a supply-demand intersection under one set of circumstances. If I run a regression analysis and derive a function, Q_m in relation to P_m, the function's **identity** may be unclear: whether it represents demand or supply or something else. Somehow, using appropriate statistical methods, I must find some way of holding demand steady while letting supply run its course, so as to come up with at least an approximation of expression (1), Chap. 12. Thus empirical price-demand functions are a bit harder to construct than income-demand functions, and one approaches them with humility.

Work has been done on the price-demand for lumber. The results indicate pretty clearly that demand is inelastic, falling between −1 and zero, and in most cases between −0.5 and zero. Inelastic price-demand, generally speaking, is characteristic of services that (1) are used mostly by upper-income consumers; (2) represent a small fraction of consumers' total outlays; (3) represent a small fraction of consumers' outlays for a particular purpose, such as for nails in building; (4) are "necessities" rather than "luxuries;" or (5) have no close substitutes—or some combination of these characteristics that outweighs contrary influences.

Lumber is a **producer good** used, not for final consumption, but as raw material for making final consumables **(consumer goods).** Housing and other light construction are softwood lumber's principal end uses. It follows that the demand for such lumber is a **derived demand:** derived mainly from housing demand. In view especially of characteristics (2) and (3), foregoing, we infer that derived price-demand tends to be less elastic than the final consumer price-demand from which it derives.

Why does one not speak of derived income-demand?

Inelastic price-demand produces this result: the higher the price, *P,* the greater the consumers' total outlay, *PQ.*

Lee James [2], working before 1940 and reporting in 1946, was one of the early students of lumber demand. He derived a number of elasticity estimates, but gave most credence to the lowest of them (the one nearest zero), −0.5. The Forest Service [7] in 1973 published an estimate that a 30-percent higher lumber price would be accompanied by a 15-percent lower annual apparent national consumption, giving us, as did James, an elasticity of −0.5. However, **interval elasticity** such as this is more difficult to interpret than the point elasticity with which we have worked so far: If we turn the statement around, then the Forest Service is saying that a 23-percent lower price (1 ÷ 1.30 = 0.77) means an 18-percent higher quantity (1 ÷ 0.85 = 1.18), which makes elasticity equal to −0.78.

There is some argument about whether the price-demand elasticity of lumber is changing and, if so, in which direction. It is pointed out that the rise in both wood and nonwood substitutes for lumber may tend to increase the elasticity. But it is emphasized that so far as nonwood substitutes are concerned, the environmental movement is highlighting the superiority of renewable, reusable, biodegradable materials and thus tending to hold elasticity down. Henry Spelter [6], in a study of softwood lumber, estimated that its price-demand elasticity had fallen from about −0.3 in 1950 to −0.1 in 1980.

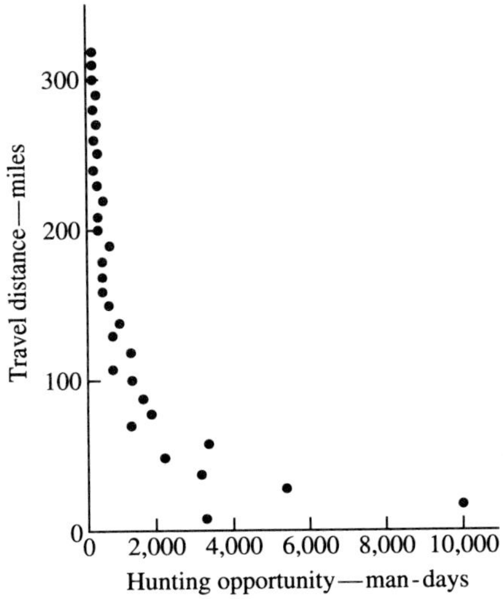

FIGURE 13-4
Price-demand for opportunity to hunt small game, 1970–1971. (*Source:* King, William E.: "An Economic Analysis of the Spatial Behavior of Hunters," Ph.D. dissertation, Washington University, St. Louis, 1975, p. 82.)

PRICE-DEMAND FOR HUNTING

Finally, let's have a look at price-demand for a quite different bundle of services: the opportunity for hunting small game. The work is that of William King [3]. The locus is the state of Missouri. The period is the 1970–1971 hunting season, which, one notes, antedated the energy crisis. King's results are graphed in Fig. 13-4. Quantity demanded is measured in hunter days, while price (unit cost) is represented by travel distance to the hunting site. We are to understand that the length of stay at the hunting site is not markedly related to travel distance. We are to disregard the fact that longer distances take a longer time to cover and that the journey itself is part of the hunting experience.

A glance at the plotted points leads one to conclude intuitively that the correlation between hunter-days and travel distance is remarkably high. One can also see that the elasticity of demand is low at high prices, rises as price falls, reaches −1.0 at a price of about 50 miles, and thereafter climbs toward high negative amounts.

> Can you offer any explanations for the variation in elasticity along King's demand function?

High price-elasticity of demand is characteristic of services that (1) are used mostly by lower-income consumers, (2) represent a large fraction of consumers' total outlays, (3) are "luxuries" rather than "necessities," (4) have close substitutes, (5) are durable or repairable, or (6) have a variety of uses. If the price-demand for a service is elastic, then the higher the price, *P*, the lower the consumers' total outlay, *PQ*.

Harken back to the *identity* problem mentioned on page 91: To get an empirical demand function, the trick is to hold onto demand while letting supply vary, so that the intersection points will define demand. Clearly, King does not turn the trick to complete perfection. But by taking a given hunter population in a given season and by choosing a region with wide variations in supply, he carries off the empirical operation with a notable degree of success.

SUMMARY

This chapter looks at two kinds of demand, quantity in relation to price and quantity in relation to consumer income, and develops a rule for measuring elasticity. Landmark studies of income-demand in forestry include work on recreation opportunity, construction materials, and paper. Some notable studies of price-demand concern lumber and hunting opportunity.

REFERENCES

1 Dennis, Donald F.: "Factors Influencing Recreational Use of Private Woodland," U.S. Department of Agriculture, Forest Service Res. Note NE-341, 1990. 3 pages.

2 James, Lee: "A Forecast of Lumber Demand, *Journal of Forestry,* July 1946, pp. 477–487.

3 King, William E.: "An Economic Analysis of the Spatial Behavior of Hunters, Ph.D. dissertation, Washington University, St. Louis, 1975.

4 Mandell, Lewis, and Robert W. Marans: *Participation in Outdoor Recreation, A National Perspective,* Survey Research Center, Institute for Social Research, University of Michigan, Ann Arbor, 1972.

5 Rich, Stuart: *Marketing of Forest Products, Text and Cases,* McGraw-Hill, New York, 1970, p. 45.

6 Spelter, Henry: "A Product Diffusion Approach to Modeling Softwood Lumber Demand," *Forest Science,* March 1985, pp. 685–700.

7 U.S. Department of Agriculture, Forest Service: *The Outlook for Timber in the United States,* Forest Resource Rep. 20, 1973, p. 182.

8 Wood, Thomas H.: "Some Effects of Education and Other Selected Variables on Consumption of Paper," M.S. thesis, State University of New York, College of Forestry, Syracuse, 1972, pp. 26–39.

9 Zaremba, Joseph: "Family Income and Wood Consumption," *Journal of Forestry,* June 1961, pp. 443–448.

CHAPTER **14**

CONSUMPTION OF FOREST RESOURCES

AGGREGATE CONSUMPTION

Let's have a look at the trends in consumption of forest resources in the United States. The ideal way to express the quantity of consumption, if we can believe Chap. 8, would be in bundles of attributes. But, alas, these are units that can't be either measured or counted, so it's necessary to fall back on proxies, such as cubic feet, user-days, and numbers of participants.

Remember the distinction drawn in the Terminology section of Chap. 9: *Demand* is the functional relation between (1) the quantity that consumers stand ready to take and (2) some determinant such as price or income. *Consumption,* on the other hand, is the quantity that consumers do in fact take, the determinants being given.

The best-documented set of forest resources is wood products. Annual data are available on *apparent consumption* (page 26) expressed in cubic feet of roundwood equivalent, a convenient common measure. In Fig. 14-1, these data are converted to 13-year moving averages for the period 1900–1986. The consumption plotted for each year is $\frac{1}{13}$ of the total for the 13-year period, which includes the six earlier years and the six later years. Moving averages are used to smooth out the short-term ups and downs in consumption and give a clearer view of the general trend.

At this point, please review the two paragraphs about the OPEC Surge on pages 23 and 25 so that you will have in mind why the functions in Fig. 14-1 turn up their tails so oddly.

Total wood consumption, shown by the uppermost line in Fig. 14-1(a), rose briefly after 1900 to a 1909 peak, fell away disastrously into the early 1930s, and then revived gradually until the mid-1970s to about the level of the early peak. Thus wood held its own in terms of total consumption during the first three-fourths of this century.

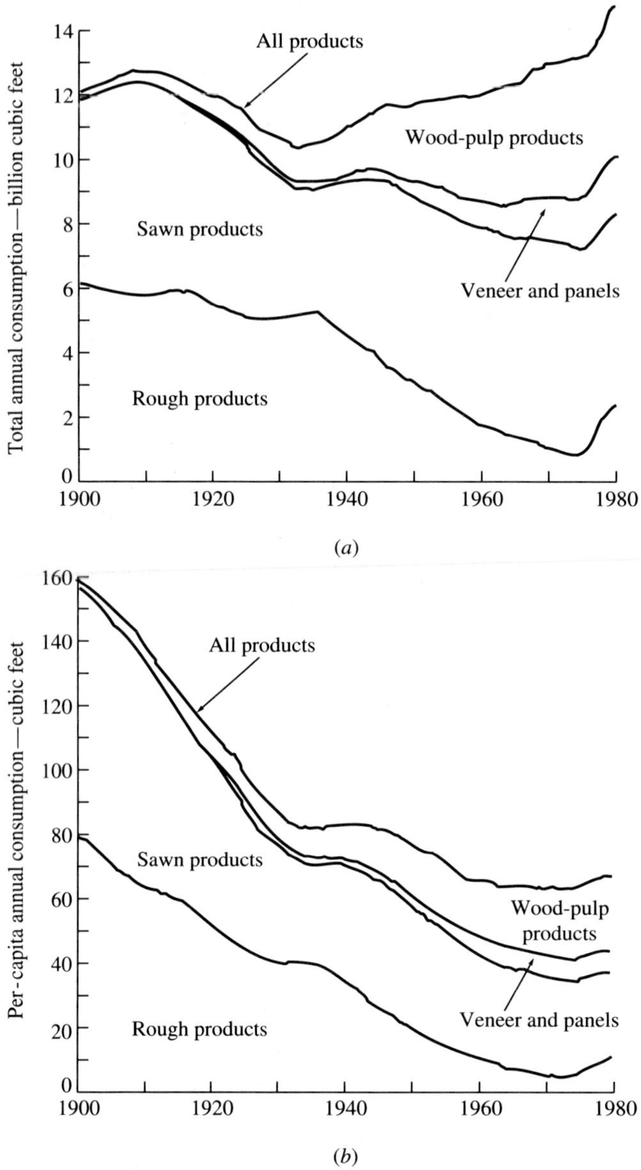

FIGURE 14-1
Apparent annual consumption of wood products in the U.S., 1900–1986: 13-year moving averages, roundwood equivalent. (a) Total consumption; (b) per-capita consumption. (*Source:* Federal government historical series.)

> Why the "disastrous decline" in wood consumption into the early 1930s?

However, in the whole scheme of things, wood lost ground. About the turn of the century, wood made up nearly half the value of all U.S. industrial raw material. By 1975, the figure had fallen well below 20 percent [3], and if wood prices hadn't been rising relative to other raw-material prices, wood's loss of position would have appeared more dramatic yet. The place of wood as a raw material has been taken over principally by minerals: metals, petroleum-based plastics, glass, brick, stone, sand, concrete, gypsum, and so on.

Taking population trends into account, one finds that consumption per capita [the uppermost line in Fig. 14-1(b)] underwent a general decline throughout $7\frac{1}{2}$ decades, the rate of decline slackening off as the trend progressed. After 1975, there appears to have been an increase in per-capita consumption.

Mainly three forces account for the 86-year trends in U.S. wood consumption: (1) changes in the technology of wood-product consumption and manufacturing production, (2) changes in wood-product prices relative to other prices, and (3) the OPEC Surge. A sketch of the history of some specific wood products will help explain these comments.

CONSUMPTION OF WOOD-PRODUCT GROUPS

Wood products divide themselves into four greatly different groups. Let's study the apparent consumption of each group, starting with the most elementary products and working our way up the ladder. You will recognize here some new thoughts about the principle of pulverization, which was introduced in Chap. 4.

First, rough products: round or split chunks and firewood from industrial leavings. The consumption trends are traced out by the lowermost lines in Fig. 14-1. Firewood is the principal member of the group, accounting for three-fourths of the group total in 1900, half in 1975, and then in the 1980s, near the height of the OPEC Surge, back up to three-fourths and well beyond. Other members are posts, poles, piling, and hewn ties and timbers. Both total and per-capita consumption fell consistently during the first 75 years. The total fell slowly before World War II, speedily thereafter. The **real price** (in dollars of constant purchasing power) of the cumbersome chunk commodities rose steeply, and, in most cases, substitutes readily supplanted them in consumers' favor.

> Will the old trends be resumed? Or will the OPEC Surge prove to have been a permanent interruption?

But apparently the most promising use of firewood is for heating and power generation in wood-using industry. The firewood in this case is not our primitive round or split chunks, but factory residues that are readily reducible to small pieces and transportable by belt or pipe. Such residues qualify as pulverized products, and we may think of their contribution to the *surge* in recent years as a manifestation of the pulverization principle. However, the chunks that make up household firewood enjoy no such stimulus. Almost any household that can afford electricity, gas, oil, or even coal will prefer it over wood for heating and cooking. Consumers would rather push a button or flip a switch than lug dirty billets of wood in and bucketsful of ashes out. Wood for the fireplace, as noted earlier, is a different bundle of attributes: The price-demand for it appears to be quite inelastic; the income-demand, positive and elastic.

I remember visiting friends in New Orleans one steamy summer evening. These former northerners had the air conditioning turned on full tilt in their expensive home and a fire burning merrily in the fireplace. What do you suppose was the elasticity of this family's price-demand for firewood?

Second, sawn products: slabs, including lumber, both boards and dimension, and sawn timbers and railroad ties. The consumption trends are represented by the interval between the lowermost line and the line above it in Figs. 14-1(a) and 14-1(b). The group long dominated the field of wood products, and its entire history resembles the 1900–1975 history of the whole field. Total consumption sagged during the middle decades and then came back. Per-capita consumption dwindled slowly but persistently. The real price of sawn products rose steadily—not so steeply, perhaps, as those rough products that were bought and sold, but still about 2 percent per year. At the same time, the technology of lumber *use* was not able to keep up. For example, several boards—requiring separate handling, trimming, placement, and fastening—are required to cover the same 32-ft^2 space that can be carpentered in a jiffy with a single sheet of plywood or a reconstituted panel, and the panels are more stable dimensionally.

True, the lumber industry has, especially since 1950, enjoyed technological changes that have undoubtedly buoyed lumber consumption by slowing price increases. All sorts of devices have been introduced that allow the industry to practice close utilization of logs, making pulpwood, fuel, or reconstituted panels out of nearly all its raw material that doesn't become lumber. Electronic devices have been invented for guiding each of the principal processes in the mill, from log sorting and channeling through saw operation, to grading the end product. Further innovations are in prospect, such as a method for making studs from hardwood trees that have been going unused [2], and even the substitution of a log-splitting process for sawing. What effect the near extinction of the little, inefficient, family-operated mill has had upon consumption is less easy to say. On the one hand, the industry has become less competitive than it used to be. On the other hand, its product has been improved.

Third, veneer and panels: plywood, particleboard, fiberboard, and the like. The long trend in total consumption is up; in per-capita consumption, up or steady. These trends are represented by the intervals marked with arrows in Fig. 14-1. Apparently the advantage of these products rests more on consumption technology than on price. They are the first products we have come to in the list that are standardized and interchangeable to a large degree. One need state only four specifications for ordinary softwood plywood in order to be reasonably assured of the outcome: species, thickness, grade, and whether for use indoors or out. Reconstituted panel classification is at least as straightforward. The grades are simpler than for lumber. The amateur carpenter can cut pieces of panel of almost any requisite width and length, which lend themselves more easily than lumber to building shelves, cabinets, and other conveniences.

Fourth, wood-pulp products: particles, fibers, and molecules, including paper, **paperboard** (paper that is 0.012 in or more in thickness), and pulp for rayon and acetate fibers and the like. Consumption is represented by the uppermost interval in Fig. 14-1(a) and 14-1(b). This is the most aggressive and successful wood-product group in consumers' eyes.

One speaks of *wood*-pulp products because, in the first decades of the study period, a large share of all paper was made from cotton, linen, and other nonwood fibers. In the U.S., no paper at all was made from wood until the 1860s. The pulp products are therefore unique among the four groups of products as age-old consumables, long rare and expensive, that were quite suddenly brought within reach of everyone for a myriad of uses by virtue of the introduction of wood as an abundant raw material.

The wood-pulp products are all reconstituted. Mostly they are chemically rather than mechanically processed. Both their total and their per-capita consumption have tended to rise over time. For the most part, by virtue of advances in manufacturing technology and labor productivity, they have held the line in terms of real price. They enjoy advantages of uniformity and wide adaptability, flexibility, and convenience in use. They have made inroads on lumber and veneer in packaging and on cotton and linen in household use.

At 25-year intervals up to the OPEC Surge, the percentage of total wood consumption in the U.S. accounted for by each product group was as follows:

Product group	1900	1925	1950	1975
Rough products	52	45	27	8
Sawn products	47	47	49	48
Veneer and panels	—	1	3	10
Pulp products	1	7	21	34

Here is more evidence that the pulverization principle was working.

It is fascinating to observe in how many different respects the four groups of wood products fall into a hierarchy that helps explain their history and perhaps foreshadow their prospects. This and earlier chapters have touched upon their characteristics: their technology of consumption and production; their consumption trends, both absolute and relative to general economic development; their price trends; and other characteristics. With respect to each characteristic, the four groups appear in the hierarchy in the same order, the order of their listing in this chapter. There is still another characteristic, which must be mentioned here because it relates to consumption. This is the use of man-derived energy, as from minerals, wood, water power, and atomic reaction. In the course of their production and consumption, rough products make no net use of energy; in fact, they *produce* energy. The other product groups *consume* energy, sawn products in modest amounts, pulp products most lavishly, a function of their extreme pulverization and their heavy use of water that must be removed before manufacture is complete. Energy consumption is the one characteristic in which the hierarchical order of advantage is reversed, with rough products at the top, pulp products at the bottom.

> In order for the last statement to be true, how must the term *advantage* be defined? Can you devise a definition that will be valid for all the characteristics of the wood-product groups?

CONSUMPTION OF RECREATION SERVICES

Let's turn now from the consumption of timber services to that of outdoor recreation in the U.S., remembering that scenery, wildlife, water, and other nontimber resources are subsumed, to a degree, under recreation.

The long trend in consumption is divisible into three eras. Before a time rather early in the twentieth century—let's say, before World War I—outdoor recreation was a minor element in the forestry economy. Sixty percent or more of the people lived in rural areas. Much of their work was done outdoors. For recreation, they commonly went indoors.

Then, as the century developed, there came a phenomenal upburst of outdoor-recreation activity that continued with little slackening until perhaps the mid-1970s. We can get a rough idea of what was happening by looking at estimates of activity on the national parks and national forests (Fig. 14-2).

People were enjoying more leisure time. The U.S. workweek averaged 60 hours in 1900, 50 hours in 1930, 40 in 1960, and was still falling in the seventies. The proportion of leisure time given to outdoor recreation rose from maybe $\frac{1}{4}$ of 1 percent in 1900 to perhaps 3 percent by 1950 and was thought to be heading for 7 percent by century's end. During much of the period, labor was becoming more productive, not only in factories and farms and business offices, but also in the home. Women were being emancipated. The traditional Calvinist work ethic was relaxed. The middle classes were growing apace. As incomes rose, people came to enjoy greater freedom to move about by automobile. Larger proportions of families became urbanized and for recreation turned to the outdoors, many of them indulging their rising curiosity about their natural environment. A cultural revolution was in progress.

But now we have strong evidence that the mid-1970s were a turning point, when the exponential growth of outdoor recreation gave way to a general leveling off, marked by downs as well as ups.

Our ethnic composition was in flux. Urban centers drew families into tightening confinement. The workweek stopped shrinking and even threatened to reverse its trend. The Calvinist ethic did not reappear in its old form, but there did appear and spread a new breathless drive to be busy, which fostered an environment friendly to television viewing as a replacement for formerly attractive firsthand activities.

ACTIVITIES OF RECREATIONISTS

Who are the outdoor recreationists, and what are their activities? Chapter 13 glances at this question. A little more information will be useful.

In its 1990 analysis of the outdoor recreation and wilderness situation in the U.S. [1], the Forest Service reported on a study of the relative popularity of various outdoor activities. The report not only ranks the activities, but also gives us a look at changes in activity and in ranking between 1977 and 1987, thereby checking our impression

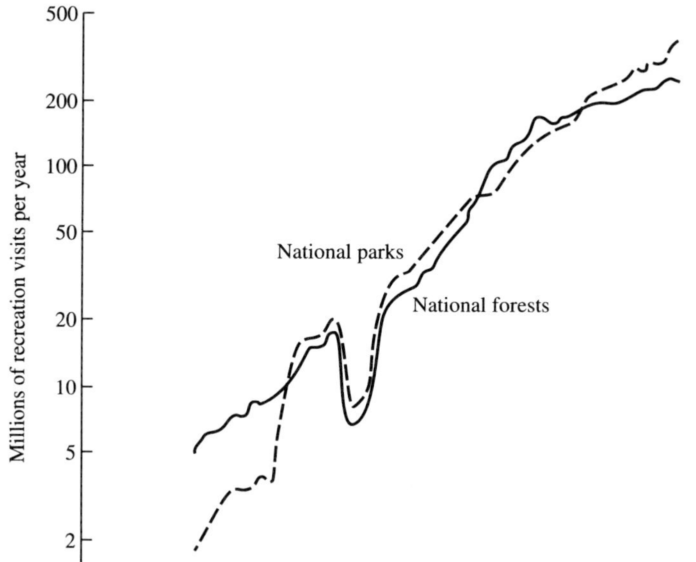

FIGURE 14-2
Recreation visits per year to U.S. national forests and national parks, 1924–1982. (*Source:* Clawson, Marion: *Outdoor Recreation: 25 Years of History; 25 Years of Projection,* Virginia Polytechnic Institute and State University, Blacksburg, April 19, 1984.)

that the mid-1970s were an outdoor-recreation turning point. Have a look at Table 14-1.

SUMMARY

The question in this chapter is *What has been happening to the use of U.S. forests during the twentieth century?* The model is the time series, expressed as total consumption and per-capita consumption. Wood products have, as a total, held their own at best. In conformity with the pulverization principle, the use of pulp products has grown, while firewood chunks and other rough products have until recently lost their importance as consumables. Lumber and veneer products have been intermediate.

TABLE 14-1
PERCENTAGE OF POPULATION PARTICPIATING IN OUTDOOR RECREATION ACTIVITIES IN TWO TIME PERIODS, 1977 AND 1987

	Percentage participating at least once annually	
Activity group and activity	1977 (households)	1987 (individuals)
Land-based		
Camping (developed)	30	20
Camping (dispersed 1977) (primitive 1987)	21	11
Driving off-road vehicle	26	12
Hiking	28	16
Horseback riding	15	10
Nature study/Photography	50	13
Picnicking	72	54
Pleasure driving	69	54
Sightseeing	62	51
Water-based		
Canoeing	16	9
Sailing	11	7
Other boating	34	7
Swimming outdoors	61	—
Outdoor pool swimming	—	48
Nonpool swimming	—	35
Water skiing	16	10
Snow/ice-based		
Cross-country skiing	2	4
Downhill skiing	7	7
Ice skating	16	7
Sledding	21	5
Snowmobiling	8	4

Source: Cordell, H. Ken, et al.: *An Analysis of the Outdoor Recreation and Wilderness Situation in the United States, 1989–2040,* U.S. Department of Agriculture, Forest Service Gen. Tech. Rep. RM-189, p. 33.

Meanwhile, a cultural revolution has created recreation resources on much forest land. Picnicking, driving, walking, swimming, and nature study are major recreation activities.

REFERENCES

1 Cordell, H. Ken, et al.: *An Analysis of the Outdoor Recreation and Wilderness Situation in the United States, 1989–2040,* U.S. Department of Agriculture, Forest Service Gen. Tech. Rep. RM-189, 1990.

2 National Forest Products Association: *Forest Industries Newsletter,* Washington, February 4, 1983, p. 4.

3 U.S. Department of Commerce, Bureau of the Census, and U.S. Department of the Interior, Bureau of Mines. "Raw Materials in the United States Economy, 1900–1969," Working Paper 35, 1972.

CHAPTER 15

HIGH POINTS IN PART TWO

1 Emphasis on the consumer. It is the consumer toward whom forestry is directed. The consumer provides the reason for forest resource management and for the existence of the forestry profession. Our study of consumers reminds us that much of economics is dialectical, not arithmetical.

2 What consumption is. Consumption is, philosophically speaking, the final phase of production, for producers in the ordinary sense merely provide consumers with the opportunity to produce satisfactions from the services they consume. Consumption is the act of drawing upon the bundles of attributes that constitute consumables: the act of deriving satisfactions from the bundles. Satisfactions are ordinarily positive, but can be negative (that is, dissatisfactions).

3 Technology of consumption. As a process analogous to production, consumption is carried out in accordance with a technology. This technology is the setting and procedure of consumption: the materials, methods, knowledge, and attitudes with which the consuming firm approaches its assignment. Consumption technology includes what are sometimes termed the "tastes and preferences" of the consumer.

4 Consumers' culture. Culture is behavior that is learned rather than simply inherited. Each culture comprises at least five classes of institutions, ranging from language to aesthetic standards, and each culture is distinguished from others by the character of these institutions. It is their culture that determines consumers' technology, just as it is the professional culture that determines foresters' behavior.

5 Consumer behavior. Indeed, we can arrange in a hierarchy the forces that lie behind consumer behavior. Starting with the most fundamental, or general, force, the hierarchy is (a) consumer culture, (b) consumer technology, (c) consumer want, (d) consumer demand, and (e) consumption.

6 Consumer movement and environmentalism. The consumer movement is traceable to Robert Owen (1771–1858) and to the Rochdale consumer cooperative, formed in 1844. Spiritually akin to consumerism is environmentalism, which gained force in the U.S. in the 1960s and has had profound influences on forestry.

7 Subjective value. Value is the characteristic of a thing that makes it exchangeable with other things in the sphere of possession. Value is measured by the quality and quantity of these other things. Demand functions are statements of subjective value.

8 Landscape beauty as a subjective value. It is not possible to generalize about the characteristics that make for beauty, because they are subjective judgments of individual consumers or groups, a manifestation of culture. One can list the attributes that relate to beauty, but cannot identify the optimal state, or quality, of these attributes. However, it is possible to predict approximately how a given kind of consumer group under given circumstances will rank alternative scenes in terms of beauty.

9 Price-demand. A price-demand function is a statement of the quantity of a consumable that a specified group of consuming firms stands ready to acquire per unit of time, depending on the consumable's price. Or the function may be viewed as a statement of the subjective marginal unit value that the group places on any given flow of the consumable per unit of time. Price-demand for a producer good is derived from the demand for the consumer goods that are end products.

10 Income-demand. An income-demand function is a statement of the quantity of a consumable that a member of a specified group of consuming firms stands ready to take per unit of time, depending on the firm's accustomed disposable income.

11 Elasticity of demand. A revealing characteristic of any demand function is its elasticity, the sensitivity of quantity to some determinant such as price or income.

12 Man-made resources. Resources are the attributes of things or ideas that have value. They are man-made in the sense that their value is created within human culture.

13 Pulverization principle. A study of timber-products consumption in the U.S. since 1900 reveals trends consistent with the pulverization principle: Pulp-products consumption, both total and per capita, has risen steadily. Firewood and other rough-products consumption has fallen steeply, except for an "OPEC Surge" after the mid-1970s. The trends for other products are intermediate.

14 Cultural revolution in outdoor recreation. Both total and per-capita outdoor recreation, measured in hours per year, increased exponentially during the first three-fourths of the twentieth century. The increase is attributed to continual shortening of the workweek and to rising trends in income and urbanization.

PART THREE

GENETIC PRODUCTION

CHAPTER 16

PREAMBLE TO PART THREE

TIMBER IN GENETIC PRODUCTION

Genetic production is that in which living genes other than human genes have a major part. The factory consists of plants and animals. For us, genetic production means using the forest to grow timber, forage, or wildlife or to produce watershed or aesthetic or recreation values—or some combination of these services.

The foregoing list of services notwithstanding, only one, timber, is heavily represented among the problems in Part Three. Indeed, timber is the dominant topic in all but a couple of the questions analyzed. The reason for this unequal treatment is that Part Three is designed to offer a set of guides to forest management—that is, guides in the sphere of economics, of financial value, of net monetary contribution to human income—and it is timber that lends itself most readily to this purpose. The principles developed for timber may be extended, to the degree possible, toward other forest resources.

In all the forested world, only a few communities can afford the luxury of harboring untouched resources. Most peoples must call upon their forests for some of their life's support. The mission of forestry science is to help supply the material as well as the spiritual wants of human beings. To discover and practice methods for perpetuating the forest and sustaining its contribution to people's livelihood—and this emphasizes the use of timber—is a principal obligation of the forestry profession. (For more on this vital point, see Sustaining the Supply, pages 223 and 224.)

GENETIC IDENTITY

A characteristic of genetic production is found in the **principle of genetic identity:** The living factory and its product are the same. The factory that produces forest forage is the forage itself. The forest scene derives from the forest, as does the forest watershed. Where the product, to be used, must be removed from the forest, the identity principle takes on added force: If one wants wood from the wood factory (the tree), one must take the factory. If one wants venison, there is no way to have it but to harvest the deer, thus liquidating the factory. The identity principle has interesting economic implications that will make themselves evident as Part Three moves along.

Factory and product are *commonly* identical in genetic production. There are exceptions. One is the goose that laid the golden egg. Here, identity worked in just one direction, as with any sort of production: Taking the product left the factory unaffected, but not vice versa.

> Can you think of some examples of genetic production in forestry where the principle of genetic identity does not hold? How does one characterize such examples?

CHOICES AND GUIDING CRITERIA

The genetic producer in forestry has a multitude of choices, covering such problems as *where* to locate, what *products* to produce and for what *outlets,* the *scale* of production, *whom* to employ in the management program, the *technology* (for example, which silvicultural system for timber), and the *intensity* of management (for instance, the amount of variable input per unit of fixed input, such as per-acre investment in growing stock and improvements). Thus the problems in genetic production are numberless, just as they are in every branch of forestry. But the models for solving these problems are comparatively few, and this is what allows you to see the forest despite the trees, so to speak.

What criteria are to be used by the genetic producer for making a choice from numberless alternatives? We can recognize two: (1) economic efficiency and (2) economic equity. Let's introduce these criteria now and return to them in more detail in Chap. 23, after we've studied some useful background.

1 Economic *efficiency* holds the same meaning in production as in consumption (page 60): Choose the high-benefit alternative, costs equal. Choose the low-cost alternative, benefits equal. In any case, choose the alternative with the most favorable relationship of benefit to cost. If circumstances change, shift the program so as to pursue benefit and escape cost. *Benefit* and *cost* include those pluses and minuses that are normally weighed in money and also those that aren't—for example, the utility of a pleasant scene, the disutility of an eyesore.

Where much of the benefit or cost cannot be measured and assigned numbers, what constitutes a "favorable relationship" between the two is left to judgment, just as the homeowner in Chap. 9 judges what weight to place on local custom. The problem solver replies to the questioner in these terms: Here is the monetary net revenue that you will most likely derive from alternative A. Here is the corresponding amount for B. In your judgment, do the other net benefits of A outweigh its poorer showing in money terms? If so, choose A; if not, choose B [1].

Even where much of the benefit and cost *can* be measured and assigned numbers, the problem solver is obliged to come back to the questioner with some diffidence because of uncertainty about the value (positive or negative) that the questioner may attach to the small unquantified leftovers. Here is still another reason the problem solver must expect to produce replies and not answers.

And if the problem solver is working with a totally quantified question where there are no leftovers, there may still be some difficulty in identifying the most favorable relationship of benefit to cost unless the questioner has stated the firm's aims (criteria for choice) very clearly. In this book, a choice is viewed as rational if, in consideration of both efficiency and equity, it aims for the greatest excess of benefit over cost per unit of time or the greatest value of prospective net benefits all brought to some point of time.

In any case, the problem solver is often found working with interest rates, or what may be called guiding rates of return. This is because, first, forest resources are capital; the cost of holding them takes the form of interest; and sometimes the revenue (benefit) from them may be looked upon as interest; and, second, forest resources commonly yield benefits and entail costs that extend over a period of years, so that the analyst is obliged to do some compounding and discounting, and this means working with interest rates. Chapters 19 and 20 give thought to the choice of interest rates and to their use in compounding, discounting, and related operations.

2 If the firm is a nation or group of nations or public agency, considerations of economic equity are a second criterion for making choices. And equity is involved, too, in the choices made by private firms. Equity considerations are those concerned with the distribution of benefits and costs among individuals and groups in society. Equitable distribution is applauded or at least sanctioned by the culture. It may tip the scales in favor of persons who are poor or well-educated or aggressive or handicapped or possess some other characteristic that concerns the community. More on this in Chap. 23; and meantime we set the equity criterion on the back burner and deal with economic efficiency.

DEFINITIONS

Whenever you use a criterion for guiding a questioner's choice, you are obliged to consider whether the questioner is interested in the firm as a whole or only in some one segment of the firm's activity—some one enterprise or project. An **enterprise** is a major part of a firm's affairs that, the manager thinks, is best viewed separately for planning or accounting. Type of output and location are two common bases for recognizing enterprises. Thus a forestry firm may have a wildlife enterprise, and it may have one timber enterprise at Digby and another at Denby. A **project** is a unit of investment, a part of an enterprise. The forestry firm may have the sale of hunting rights as one project, the use of fertilizer at Digby as another. Usually the firm's interest lies in maximizing the net benefits to the firm as a whole and not necessarily to any one enterprise or project. Following this **principle of unity of the firm** ensures that the firm avoids increasing its revenue from one activity at the sacrifice of even greater returns from some other. For example, before one can soundly recommend that the farmer plant pines in a back pasture, one must make sure that both the land and the farmer's time are best used for trees. Forthcoming chapters illustrate problem-solving methods that conform to the principle of unity of the firm. The firm in question may be a private individual, a forest-products corporation, or a government agency such as the Forest Service—or a nation or group of nations.

The term **capital** has been used and *will* be used a good bit; so here is a place to define it. The term has a dual meaning. The first is **capital goods:** durable equipment or machinery for manufacturing a product. The tree, deer, and goose are capital in this sense. The second meaning is **financial capital:** that which is in **liquid** form—for example, money. The proceeds from liquidating the tree are capital in this sense. A related term is to **capitalize:** to express the financial value of a productive instrument as the sum of net proceeds expected from it, all brought to a single point of time, commonly the present. Or you may think of capitalizing as estimating the investment required to earn the expected net proceeds in the form of interest.

Before we leave definitions, we owe it to ourselves to recognize a realm of possibilities that we'll term **microgenetic production,** which is still only at the research stage. Its application in our economy may lie generations ahead. This is the manufacture of materials, not by macroorganisms such as trees, animals, and grass, but by microorganisms such as bacteria. Wood is a complex substance, the potential raw material for a wide variety of useful chemicals. Microorganisms can be bred to make such chemicals from wood by genetic processes safer and less costly than those crude methods now followed in the chemical industry. You may live to see such developments.

PHILOSOPHICAL NOTE

The chapters that follow often work with neat problems and precise solutions—for example, the optimal harvest age of a tree, in years; the best stocking of a timber stand, in board feet; the most efficient fire-management outlay for a forest region, in dollars. The purpose of the neatness and precision is not to mislead you into supposing that professional life is always orderly and predictable. Rather, the purpose is to make good on a procedure promised at book's beginning: to work with models and thus escape the bewildering complexity of affairs. Models, as abstractions, furnish replies, not answers. The forestry firm that calculates a 60-year timber rotation will discover countless necessities for departing from plan. For this book, it's the plan that's vital. The departures are mostly relegated to the indefinable realm of judgment. Plans are seldom followed to the letter, but they are still indispensable. Hidy et al. [2] give us insights into the scope and degree of abstraction in genetic production by a wood-industry firm.

> Although judgment is referred to as "indefinable," there surely must be some helpful way of explaining what the term signifies. What does it signify to *you?*

SUMMARY

This preamble makes four points. First, genetic production is that performed by plants and animals. In genetic production, "factory" and product are commonly identical. Second, in its effort to solve genetic-production problems, this book uses as a criterion the economic efficiency of the firm as a unit. It also recognizes economic equity as a criterion. Third, problem solving in genetic production, as in other realms of forestry economics, involves the use of capital and the rates of return to capital. Fourth, as you study precise plans for problem solving, recognize that plans are seldom precisely accomplished.

REFERENCES

1 Canham, Hugh O.: Break-Even Benefit-Cost Model, A Useful Tool for Forestry Decision-Making, *Northern Journal of Applied Forestry,* March 1986, pp. 14–16.
2 Hidy, Ralph W., Frank Ernest Hill, and Allan Nevins: *Timber and Men, The* Weyerhaeuser Story, Macmillan, New York, 1963, pp. 490–511.

CHAPTER **17**

BEST STOCKING OF DIVISIBLE TIMBER CAPITAL

CHOICE OF THE BEST-STOCKING QUESTION

Let's turn first to a fundamental question in genetic production: Taking as the case a stand of timber, what is the most efficient quantity of growing stock? In other words, if you are making a plan for such a stand, what stocking will you aim for, so that by thinning or accumulating stock you can achieve the goal you have set? This chapter elaborates on principles first enunciated in 1952, with pine at Crossett, Ark., as an example [1].

> Name some of the forest-management tasks that would be guided by an estimate of best stocking.

As we'll see, the stocking question is tied to a great number of interrelated questions. We need to be alert to the folly of taking some little question in isolation, ignoring the other questions in the interrelated system. This book attempts to make the reader aware of interconnections. Occasionally, it studies how to cope with interconnections. But it also bows to the need for taking its subject a problem at a time.

The best stocking of a stand is a good question for a starter because the firm's alternatives are simple: Either it cuts more of the timber or it cuts less. This choice among intensities of cutting is open to the firm because our timber stand is **divisible**

capital: It can be *divided* into two sets of trees, one to be harvested and one to be left growing.

Divisibility is not a necessary characteristic of genetic capital. There is **indivisible capital.** Here the simplest example is the tree. The firm either cuts the entire tree at a given time or it cuts none of it. There are no halfway measures, such as peeling off the tree's outer rings and leaving the core to produce a new envelope. For the tree, therefore, the harvest question is *when.* We'll have a look at this question in Chap. 18.

> Are forest resources other than timber classifiable into divisible and indivisible capital? Can you cite examples?

A DOUGLAS-FIR ILLUSTRATION

Good illustrations of divisible timber capital are many-aged stands managed under a single-tree or group selection system—say, northern hardwoods or ponderosa pine. But any one-aged timber stand can be managed as divisible capital, and such a stand is taken here. It is a 50-year-old Douglas-fir stand on site I (index 200 ft at 100 years) in western Oregon, managed under a thinning regime. The problem: How much growing stock should be left after the thinning that is coming up this year? That is, what is the best stocking of this 50-year-old stand? The model is this: The financially best combination is found in the region of diminishing return, where marginal revenue and cost are equated and net revenue therefore maximized (or loss minimized).

Table 17-1 tells the story. Let's study first the upper part of the table, which gives **production functions** in relation to **input.** These are analogous to the consumption functions defined in Chap. 12.

First, the physical production function. Column (2) displays a range of **variable inputs** (the **growing stock,** or stocking, of merchantable wood) from 26,000 to 38,000 board feet (bd ft) per acre at 2,000 bd-ft intervals. Column (5) shows the corresponding **outputs** of **current annual growth.** The **fixed input,** land, is recorded in column (1), and all else in the problem, including silviculture technology and the unit value and age of the timber, is fixed.

The fact that age is constant deserves emphasis. Column (5) does *not* say what happens to the stand's growth as the growing stock builds up over time. All the stands represented in columns (2) and (5) are 50 years old. Age and other things are deliberately held constant so that the net effect of stocking upon growth can be examined. The figures respond to the question, "If I have a 50-year-old stand that has been brought along through life under a thinning regime, and this stand has thus-and-so-much growing stock per acre, what will its current annual growth be?"

Since only the input of stock and the output of growth are variable, and everything else is fixed, including land and technology, this is called a **short-run production function.** The implication is that time is not allowed for innovations to be adopted that

TABLE 17-1

ANNUAL PRODUCTION FUNCTIONS FOR A 50-YEAR-OLD DOUGLAS-FIR STAND ON SITE I IN WESTERN OREGON[a]

Quantities and Values in Relation to Input (stock)								
Total fixed input (land)	Total variable input (growing stock)		Total cost (5% of stock)	Total output or revenue from growth		Total net revenue	Marginal values per unit (2 M bd ft) of variable input	
(1) Acres	(2) M bd ft	(3) Dollars	(4) Dollars	(5) Bd ft	(6) Dollars	(7) Dollars	Revenue, (8) Dollars	Cost, (9) Dollars
1	26	2,600	130	1,535	153.50	23.50		
							11.50	10.00
1	28	2,800	140	1,650	165.00	25.00		
							11.20	10.00
1	30	3,000	150	1,762	176.20	26.20		
							10.50	10.00
1	32	3,200	160	1,867	186.70	26.70		
							9.60	10.00
1	34	3,400	170	1,963	196.30	26.30		
							8.50	10.00
1	36	3,600	180	2,048	204.80	24.80		
							7.20	10.00
1	38	3,800	190	2,120	212.00	22.00		

Quantities and Values in Relation to Output (growth)

Total fixed input (land)	Total output or revenue from growth		Total variable input (growing stock)		Total cost (5% of stock)	Total net revenue	Marginal values per unit (100 bd ft) of output	
(1) Acres	(2) Bd ft	(3) Dollars	(4) M bd ft	(5) Dollars	(6) Dollars	(7) Dollars	Cost, (8) Dollars	Revenue, (9) Dollars
1	1,600	160	27.15	2,715	135.75	24.25		
							8.60	10.00
1	1,700	170	28.87	2,887	144.35	25.50		
							8.90	10.00
1	1,800	180	30.65	3,065	153.25	26.75		
							9.75	10.00
1	1,900	190	32.60	3,260	163.00	27.00		
							11.00	10.00
1	2,000	200	34.80	3,480	174.00	26.00		
							13.00	10.00
1	2,100	210	37.40	3,740	187.00	23.00		
							17.50	10.00
1	2,200	220	40.90	4,090	204.50	15.50		

[a]The calculations serve to estimate best stocking and best growth rate.

Source: The source for columns (2) and (5) in the input section, from which all other figures are derived, is Duerr, William A., and Neils B. Christiansen, *Exercises in the Managerial Economics of Forestry,* O.S.U. Book Stores, Corvallis, Oreg., 1973, p. 80.

would change the output-to-input relationship. Or else this would have elements of a **long-run production function.**

> What do you think might be some innovations that would change the relation of growth to stock? Why would such innovations be adopted?

You will observe in column (5) that growth increases with stock, but at a diminishing rate. That is, the derivative of growth with respect to stock diminishes as stock increases. In economics terminology, this derivative is the stand's **marginal output,** or **marginal return.** Its negative trend signifies **diminishing** (marginal) **return.** Diminishing return is encountered in the **short run** because of the fixed input, which is more crowded when variable input is greater. There is more competition for growing space and other essentials.

In columns (3) and (6), the quantities in columns (2) and (5) are converted to dollar values (*stumpage,* or *liquidation* values, related to *conversion surplus* in Chap. 50) at the arbitrary rate of $100 per M board feet. The result is total value of stock (3) and **total revenue** from growth (6). Again diminishing return (diminishing **marginal revenue**), column (8), is in evidence. The figures are marginal revenue per 2,000 bd ft of stock input. Since they are the successive differences in column (6), they are printed between the lines in the table.

So much for revenue. Let's hear the cost side of the story. As with revenue, we must have a value per unit. We must know how much it costs the firm to hold 1,000 bd ft of stock for a year. Now presumably the firm has alternative investments for its capital. The best of these alternatives, other things equal, is an **alternative cost,** an **opportunity cost,** of leaving growing stock to grow rather than liquidating it. Let's assume that the firm's best alternative is 5 percent per year. This is the net rate of return from a money-lending enterprise available to the firm. Every $100 that the firm lends will return $5 annually after all costs. Any part of stock that will earn more than a 5-percent return is worth keeping. Any part that promises to earn less than 5 percent is best liquidated and put to the alternative use. Thus 5 percent is the firm's **guiding rate of return.**

The total annual costs of holding column (3) stocks are figured at 5 percent in column (4). The successive differences in total cost are uniformly $10. **Marginal cost** per 2,000 bd ft of stock input equals $10. The figure is recorded in column (9).

Comparing columns (8) and (9), you find that extra stock will pay its way up to a total stocking of about 32,000 bd ft per acre. This, then, is the best stocking, and **best combination** of these resources.

What about total net revenue? Does it peak at about 32,000 bd ft per acre? See column (7), which is column (6) minus column (4). Its story is consistent with that of columns (8) and (9).

BEST GROWTH OF THE DOUGLAS-FIR STAND

Table 17-1 (top half) derives marginal revenue and cost per unit of input (stock). But how about marginal cost and revenue per unit of output (growth)? Is there such a thing? Can we derive the figures? Will they give us the same reply about best stocking (best growth) as Table 17-1 (top half) gave us?

The answer is yes all around. The figures are in Table 17-1 (bottom half). The procedure is analogous in every respect. The independent variable, now, in column (2) is output (growth) rather than input (stock). Stock is made a dependent variable, so that the question asked is a bit unusual, but still basically the same as before: How much growth can the firm aim for and still have a growing stock that pays its way? This is the stand's "best growth."

Column (7) and columns (8) and (9) give us an optimum of about 1,900 bd ft of annual growth per acre, which is the equivalent of our previous optimum of 32,000 bd-ft stocking. See columns (2) and (5) in Table 17-1 (top half)—or columns (2) and (4) in Table 17-1 (bottom half).

> Marginal revenue per unit of output turns out to be the same ($10) as marginal cost per unit of input. Is this invariable or is it a coincidence?

In Table 17-1, we have a clear illustration of a principle that is universal in economics as a study of choice, the **principle of relevance:** What is variable among alternatives (stock and growth in this case) is relevant to the choice among them; what is fixed (land in this case) is irrelevant.

THE GEOMETRY

Finally, how about the geometry of the best stocking or best growth? The graphs are drawn in Figs. 17-1 and 17-2. Have a look at Fig. 17-1. The left-hand side of the figure is the story of Table 17-1 (top half). Figure 17-1(a) comes from columns (2) and (5). The solid curves in Fig. 17-1(b) come from columns (2), (4), and (6). The vertical distances between the curves are in column (7). Figure 17-1(c) is taken from columns (2), (8), and (9).

The right-hand side of Fig. 17-1 is taken from Table 17-1 (bottom half).

> In Figs. 17-1(d), (e), and (f), see if you can identify columns (2), (3), (4), (6), (7), (8), and (9) of Table 17-1 (bottom half).

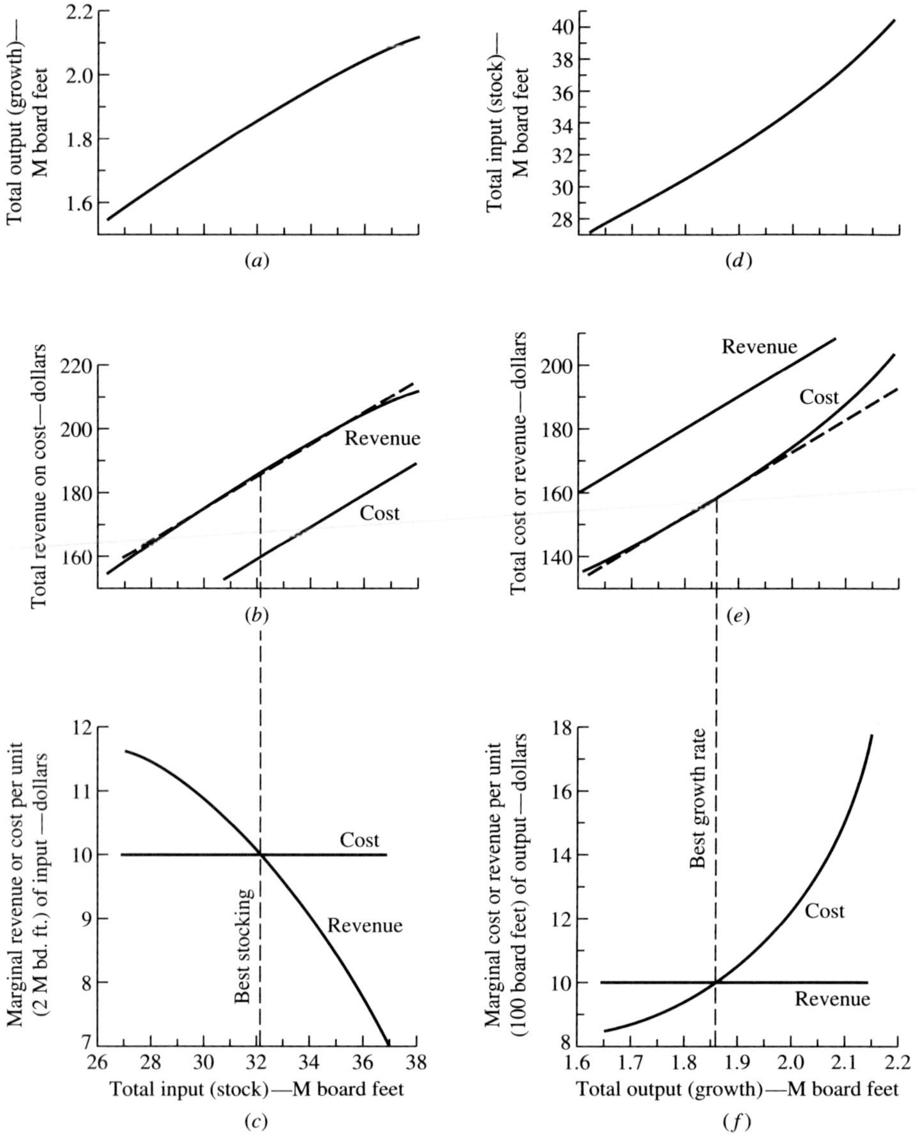

FIGURE 17-1
Annual per-acre production functions for a 50-year-old Douglas-fir stand on site I in western Oregon. (a), (b), and (c): Quantities and values in relation to input (stock), determining the best stocking. (d), (e), and (f): Quantities and values in relation to output (growth), determining the best growth rate. (*Source:* Table 17-1.)

We want to note two features of Fig. 17-1 that will give us valuable insight into **marginal analysis.** They both concern Figs. 17-1(b) and 17-1(e). Observe in Fig. 17-1(b) that total cost per unit of input is a straight line with a slope of 5 percent. (Convert the horizontal scale to dollars to see this.) This slope is repeated in the broken tangent to revenue, a concave function on account of diminishing return. Revenue and cost are diverging up to the point of tangency and converging beyond, so that the tangency point marks maximum total net revenue per unit of input per acre per year—that is, best stocking.

In Fig. 17-1(e) total revenue per unit of output is a straight line, also with a 5-percent slope, while cost is convex because of diminishing return. Again, there is a point of tangency where the cost function, too, has 5-percent slope and the two functions are farthest apart vertically. Here is the point of maximum net revenue per unit of output per acre per year—that is, best growth rate.

Now for the two features of Figs. 17-1(b) and 17-1(e).

First, we can see why marginal revenue and cost must be equated *in the region of diminishing return* in order for the equating to mark the optimum. Suppose we were in a region of *increasing* return. Then revenue in Fig. 17-1(b) would be convex; cost in Fig. 17-1(e), concave; and the points of tangency, where marginal revenue and cost would be equal, would be points of *minimum* net revenue.

Second, consider the slopes of the curves in Figs. 17-1(b) and 17-1(e), the derivatives of the functions with respect to input or output, as the case may be. We know that these slopes represent marginal values, as in Figs. 17-1(c) and 17-1(f). But we have seen that they can just as well be expressed in percent. The percentage slopes are graphed in Fig. 17-2.

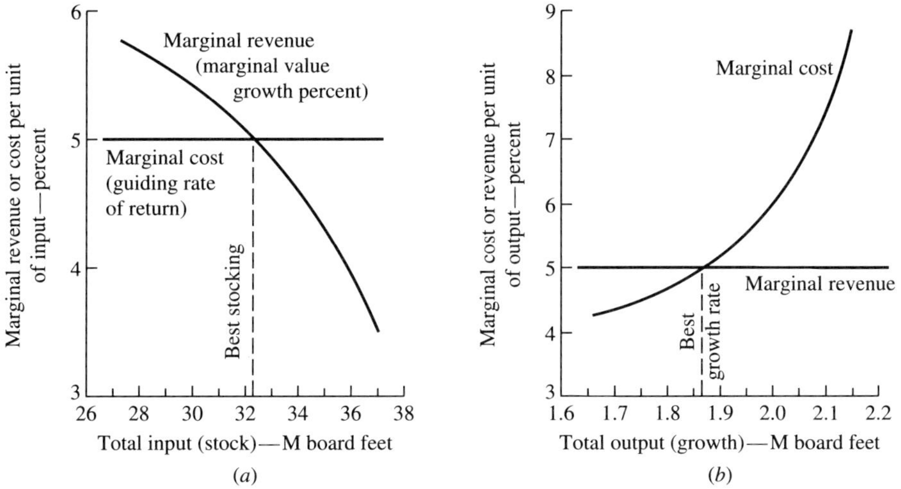

FIGURE 17-2
Best stocking and growth rate of a 50-year-old Douglas-fir stand on site I in western Oregon, as determined by the percentage marginal revenue and cost (a) per unit of input in relation to total input and (b) per unit of output in relation to total output. (*Source:* Fig. 17-1.)

Figure 17-2(a) is of special interest. The marginal revenue is **marginal value growth percent** of the stand. It tells what annual **internal rate of return** in the form of growth the firm can expect from successive additions to investment in growing stock. Another term for the percentage of marginal revenue is **marginal efficiency of investment.** The marginal cost is the firm's guiding rate of return. Investments that promise to pay more than the guiding rate are desirable investments. Those that promise less are not. Hence the optimum is found at the intersection of the curves. This is the model that we use in Chap. 18 to calculate a tree's financial maturity.

SUMMARY

The best stock of divisible capital such as a timber stand is found where the stand's marginal value growth percent equals the firm's guiding rate of return. In the case of divisible capital, marginal growth is the derivative of current growth with respect to stocking.

The stocking where the two marginal values are equated in terms of percentage is the same as that where the two are equated in terms of dollars, or where total net revenue is maximized.

The short-run optimum may be estimated either from total marginal values per unit of variable input or from values per unit of output.

REFERENCES

1 Duerr, William A., and Walter E. Bond: "Optimum Stocking of a Selection Forest," *Journal of Forestry,* January 1952, pp. 12–16.

CHAPTER 18

FINANCIAL MATURITY OF INDIVISIBLE TIMBER CAPITAL

TREE'S MARGINAL REVENUE AND MARGINAL COST

Now for a change of scenery. We move from western Oregon to eastern Virginia, from Douglas-fir to loblolly pine, and from the timber stand to the individual tree. The problem is this: What is the best time during the tree factory's life to harvest its product and thus liquidate the factory? What is the tree's optimal harvest age—the age when total gross revenue from liquidation most greatly exceeds the total cost of running the factory—when marginal revenue equals marginal cost? The model is the **financial-maturity** model. The financial-maturity question is opened in this chapter and closed, so far as the tree is concerned, in Chap. 21.

Let's view the tree as a candidate for thinning in any sort of stand or for harvest cutting in a many-aged stand. The tree, like the stand, is a capital investment that yields the firm a return in the form of growth.

Have a look at Figs. 18-1(a), (b), and (c), where some production functions for our tree are graphed. Figure 18-1(a) shows the tree's total merchantable wood content (growing stock) as a function of its age. The function is concave because of short-run diminishing return. Return diminishes for the same reason as that explained on page 118, the crowding of fixed inputs. Since we are dealing with the region of diminishing return, we should be able to find an optimal age, a financial maturity.

Diminishing return is discernible also from the negative slope in Fig. 18-1(b). The data are the successive differences in stock read from Fig. 18-1(a), and since they relate to the intervals between the points of time in Fig. 18-1(a), they are plotted between these points. They are the current annual growth of the tree, the derivative of stock with respect to age. They are the tree's marginal growth, or marginal return.

Take note of this: In Chap. 17, we figure the marginal growth of the stand as the

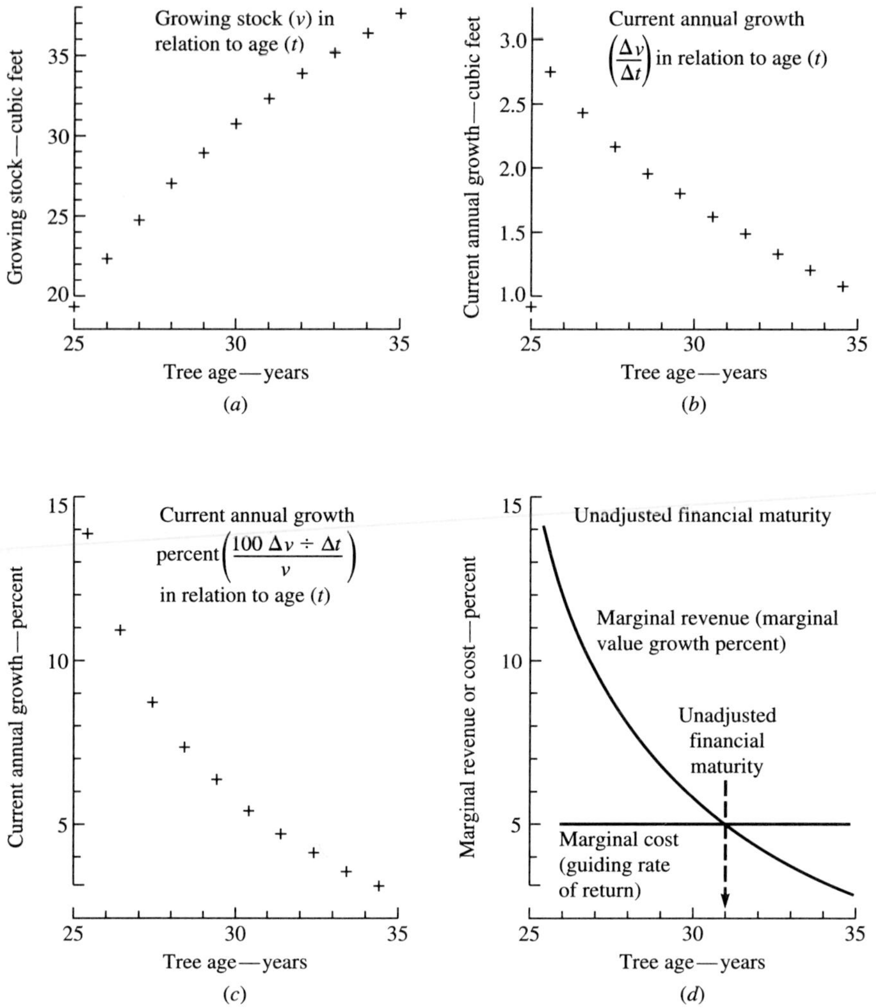

FIGURE 18-1
Some timber production functions [(a), (b), and (c); *Source:* Table 21-1] and derivation of unadjusted financial maturity (d) of a loblolly pine tree on excellent site in eastern Virginia.

derivative of its current growth with respect to stock, $\Delta G \div \Delta S$. Now in the present chapter, we figure the marginal growth of the tree as the derivative of stock with respect to age, $\Delta S \div \Delta A$, which is simply the current growth, G. Why the difference? It's the difference between divisible and indivisible capital. In Chap. 17, the question is, "Shall I hold stock at the given level, or shall I carry a little more?" And the reply is, "If an addition to stock will pay its way in the form of added growth (if $\Delta G \div \Delta S$ is greater than your guiding return), add to stock. If it won't, don't." But in the present chapter, looking at the individual tree, any question of manipulating stock is irrelevant.

At any given time, there are only two alternatives: Liquidate all the stock or liquidate none. The question, then, is, "Shall I cut the tree?" And the reply is, "If the tree is paying its way in the form of growth (if $\Delta S \div \Delta A$, equal to G, is greater than your guiding return), keep the tree. If not, don't." With the stand, any separable part of it is marginal. But with the tree, the whole thing is marginal; there is no smaller marginal quantity or value.

In Figs. 18-1(a) and 18-1(b), if the cubic feet on the vertical axes were multiplied by dollar values per cubic foot, the resulting function in Fig. 18-1(a) would be total revenue, and that in Fig. 18-1(b) would be marginal revenue. If you can suppose that all the wood is worth $1 per cubic foot, you can read the graphs directly as revenue graphs.

> Suppose that the tree's value is not $1 per cubic foot all during its life, but that its unit value increases with age because of higher wood quality or lower harvesting costs. How would you convert Figs. 18-1(a) and 18-1(b) to revenue terms?

In Fig. 18-1(c), growth during each year—Fig. 18-1(b)—is expressed as a percentage of stock at the beginning of the year—Fig. 18-1(a). If all wood has the same unit value—say, $1 per cubic foot—these current annual growth percents can be read as marginal value growth percents.

TREE'S PAST AND PROSPECTIVE PERFORMANCE

Imagine that you are the manager of a firm that has a very simple structure. You have just two enterprises. One is a young loblolly pine tree growing on excellent site in eastern Virginia [Fig. 18-1(c)]. The other is an investment opportunity that earns 5 percent per year, net. Every $10 that you invest yields 50 cents annually after all your costs are paid. Suppose that you can converse with the tree:

Q. Happy twentieth birthday, Tree! During the next few years, until I can get back to see you, how fast is your value going to increase?

A. I estimate that my value growth will be at least 15 percent per year.

Q. Well, that's splendid! The best I can do elsewhere is 5 percent. Carry on, I'll see you later.

After some years, you revisit Tree.

Q. Hello! I can tell by remeasuring you that your growth estimate was pretty good. Now, looking ahead again, what rate of return do you think you will pay me on my *present* investment in you?

A. It's going to be very close to 7 percent.

Q. Good tree! My other enterprise doesn't do that well. Success to you!

A third visit to Tree develops like this:

Q. Congratulations! Your rate of return was what you promised. Tell me how the further prospects look to you.

A. I'm sorry to say it, but I'm already 31 years old, and I'm running short of elbow room, too. The best I can look forward to is $4\frac{1}{2}$ percent.

Q. Too bad! I can't afford to keep my investment in you any longer. I have a better alternative. Off to the mill you go!

So you take the net proceeds from selling 31-year-old Tree and lend them out so as to earn a 5-percent net return. In doing so, you have improved your prospective net annual revenue by $\frac{1}{2}$ percent: 5 cents for every $10 of your capital.

> Notice that the estimates by Tree that are influential in the firm's decisions are *forecasts*. Data in hand, relating to the past, are valuable only as they help predict. Why can't a firm's decision be based on a sure thing?

The 5-percent yearly return on loans is your opportunity cost of leaving Tree to grow. It is an opportunity cost in the sense that it is revenue you forgo when you leave your capital invested in Tree rather than putting it into the best alternative project. Five percent is, at the same time, the marginal cost of holding Tree uncut: Each additional dollar invested in Tree costs you an additional 5 cents per year. So long as Tree grows in value faster than 5 percent annually, Tree is a satisfactory investment. Thus 5 percent is your guiding rate of return: the rate that guides your acceptance or rejection of investments alternative to the guiding one. Your use of this guiding rate gives you an investment policy consistent with the principle of the firm's unity: You are protected against the possibility of carrying any enterprise beyond the point where the firm as a whole benefits the most.

> Explain why the use of the guiding rate of return ensures consistency with the principle of unity of the firm.

On the revenue side of the combination is Tree's prospective current annual growth. Tree's current annual growth percent in value is its marginal value growth percent, or marginal revenue per year.

And so you determine Tree's optimal harvest age as the age when Tree's marginal value growth percent falls to equal your firm's guiding rate of return—that is, the age within the region of diminishing return when marginal revenue per unit of variable capital input equals marginal cost per unit of variable input. (Note that you can equally well say, "the age when marginal revenue per unit of output equals marginal cost per unit of output," since, with indivisibility and genetic identity, output and variable input are the same thing.) This best combination of capital is found in Fig. 18-1(d), which is derived from Fig. 18-1(c). It is assumed that all wood, both stock and growth, has the same unit value, so that the percentages in Fig. 18-1(c) can be interpreted as value growth percentages. Value growth percent in Fig. 18-1(d) is depicted as a continuous function of tree age.

UNADJUSTED FINANCIAL MATURITY

Financial maturity in Fig. 18-1(d) is labeled "unadjusted" for reasons now to be explained.

A person who was early to grasp the financial-maturity idea was Kenneth Boulding [1]. He applied it to the example of wine "laid away in a cave," pointing out that as wine ages in the barrel, it grows in value, but at a diminishing rate. When its rate of value increase has fallen to equal the "normal rate of interest," it is ready for bottling. To hold it in the barrel longer would be to retain an insufficiently productive investment.

> Suppose that wine continued to increase in value with age at a rate greater than the "normal rate of interest." What would lead vintners to release their wine to consumers? Suppose that Tree's value tended to increase indefinitely at a rate greater than the guiding rate. What analogous adjustments would keep open the channels of wood utilization?

But the winery that behaved as Boulding described would be aging its product too long! It would be overlooking the gain obtainable by emptying its barrels a bit earlier, thus releasing them (and cave space) for use with a new vintage. You can picture this gain as a reduction in the barrel and cave-space investment required for a given wine output. Surely the investment could not be drastically reduced without a more than offsetting loss in wine quality and value, but some reduction at a net gain is always possible.

The foregoing flaw in Boulding's financial maturity was pointed out by Albert Worrell [2]. He argued that the simple equating of value growth percent to guiding rate of return, as in this chapter, derives a faulty tree age—an unadjusted financial maturity, as it is termed here. The unadjusted calculation recognizes the cost only of retaining the present tree (cf. wine). It disregards the cost of delaying the use of the land (cf. barrel and cave space) by successive new generations of trees (cf. new vintages) that can follow the present one on the same ground. You can think of this disregard of cost as a failure to use the land more efficiently. You can think of it as using more than the optimal amount of land to achieve a given rate of timber output.

"Hold on a moment!" someone may say. "Suppose the firm doesn't plan to follow Tree with other trees. Suppose it plans to go out of timber forestry, using the land for farming or a factory or even a hunting ground. That is, suppose there *won't* be any 'new generations.' Why, then, speak of 'unadjusted' financial maturity?" The answer: So long as the firm plans to make *any* productive use of the land, it is postponing this use if it clings to Boulding's financial maturity. But our questioner is still unsatisfied. "Suppose the firm plans to *sell* the land after Tree is harvested and go out of business altogether." The answer: In such an event, the potential productivity of the "new generations" will be reflected in the value of the land for sale. Worrell's point is still valid.

The question to tackle, then, is how to go about adjusting the unadjusted estimate of financial maturity. To manage this question, it is necessary first to consider how a firm can use its guiding rate of return to change the time of reference of values and events such as the maturing of a tree and, beyond this, where the firm's guiding rate comes from. Chapter 19, then, takes up the use of the guiding rate, and Chap. 20, its origin. The question of adjusted financial maturity is held off until Chap. 21.

SUMMARY

This chapter opens the question of timing the tree harvest. As a wood factory, the tree is indivisible capital in the sense that if the firm takes any of the product, it liquidates the entire factory. As a first estimate, the financially optimal time for liquidation is the time of unadjusted financial maturity of the tree, when the tree's marginal value growth percent (marginal revenue per unit of capital input or output) equals the firm's guiding rate of return (marginal cost per unit of capital). The unadjusted financial-maturity calculation neglects the cost of waiting for generations of trees beyond this one. Adjusting the unadjusted estimate is the subject of Chap. 21.

REFERENCES

1 Boulding, Kenneth E.: *Economic Analysis,* Harper, New York, 1941, pp. 720–726.

2 Worrell, Albert: "Financial Maturity, A Questionable Concept in Forest Management," *Journal of Forestry,* October 1953, pp. 711–714.

CHAPTER 19

WORKING WITH COMPOUND INTEREST

This chapter explains procedures for changing the time of reference of values by allowing for compound interest. The chapter's subject is useful throughout forestry economics, management, and finance. And specifically, it will help make the transition from unadjusted (Chap. 18) to adjusted financial maturity (Chap. 21).

INTEREST ON SINGLE PAYMENTS

If you lend your friends some money to tide them over until payday, your friends will return the money, but ordinarily not with interest: It has been an **interest-free** transaction.

If, on the other hand, you lend money to the bank (as by opening a savings account) or borrow from the bank or other lending institution, interest is apt to be involved, and it's probably **compound interest.** Suppose you borrow \$100 at 2 percent per month. At the end of the first month, you will owe \$100 plus 2 percent of \$100—that is, \$102: $\$100 \times 1.02 = \$102$. In the multiplier 1.02, the 1 is the principal that you borrowed, and the .02 is the interest on this principal. If you let the debt run through a second month, you will then owe \$102 plus 2 percent of \$102—that is, $\$102 \times 1.02 = \$100 \times 1.02 \times 1.02 = \$100 \times (1.02)^2 = \$104.04$. One can say that V_2, the end value of the principal, is equal to V_0, its beginning value, **compounded** for 2 months—that is, multiplied by $(1 + i)^2$, one plus the **interest rate** in decimals raised to a power equal to the elapsed time, 2 months. In general, for any number of months, n,

$$V_n = V_O (1 + i)^n \tag{1}$$

If n is 12 months, $V_n = \$100 \times (1.02)^{12} = \$126.82$.

TABLE 19-1
SOME VALUES OF $(1 + i)^n$

	i									
n	0.03	0.04	0.05	0.06	0.07	0.08	0.09	0.10	0.11	0.12
1	1.0300	1.0400	1.0500	1.0600	1.0700	1.0800	1.0900	1.1000	1.1100	1.12
2	1.0609	1.0816	1.1025	1.1236	1.1449	1.1664	1.1881	1.2100	1.2321	1.25
3	1.0927	1.1249	1.1576	1.1910	1.2250	1.2597	1.2950	1.3310	1.3676	1.40
4	1.1255	1.1699	1.2155	1.2625	1.3108	1.3605	1.4116	1.4641	1.5181	1.57
5	1.1593	1.2167	1.2763	1.3382	1.4026	1.4693	1.5386	1.6105	1.6851	1.76
6	1.1941	1.2653	1.3401	1.4185	1.5007	1.5869	1.6771	1.7716	1.8704	1.97
7	1.2299	1.3159	1.4071	1.5036	1.6058	1.7138	1.8280	1.9487	2.0762	2.21
8	1.2668	1.3686	1.4775	1.5938	1.7182	1.8509	1.9926	2.1436	2.3045	2.47
9	1.3048	1.4233	1.5513	1.6895	1.8385	1.9990	2.1719	2.3579	2.5580	2.77
10	1.3439	1.4802	1.6289	1.7908	1.9672	2.1589	2.3674	2.5937	2.8394	3.10
12	1.4258	1.6010	1.7959	2.0122	2.2522	2.5182	2.8127	3.1384	3.4985	3.89
14	1.5126	1.7317	1.9799	2.2609	2.5785	2.9372	3.3417	3.7975	4.3104	4.88
16	1.6047	1.8730	2.1829	2.5404	2.9522	3.4259	3.9703	4.5950	5.3109	6.13
18	1.7024	2.0258	2.4066	2.8543	3.3799	3.9960	4.7171	5.5599	6.5436	7.69
20	1.8061	2.1911	2.6533	3.2071	3.8697	4.6610	5.6044	6.7275	8.0623	9.64
25	2.0938	2.6658	3.3864	4.2919	5.4274	6.8485	8.6231	10.834	13.585	17.00
30	2.4273	3.2434	4.3219	5.7435	7.6123	10.062	13.267	17.449	22.892	29.96
35	2.8139	3.9461	5.5160	7.6861	10.676	14.785	20.414	28.102	38.574	52.80
40	3.2620	4.8010	7.0400	10.286	14.975	21.725	31.409	45.259	65.000	93.05
45	3.7816	5.8412	8.9850	13.765	21.003	31.920	48.327	72.891	109.53	163.99
50	4.3839	7.1067	11.467	18.420	29.457	46.902	74.358	117.39	184.57	289.00
55	5.0821	8.6464	14.636	24.650	41.315	68.912	114.41	189.06	311.00	509.32
60	5.8916	10.520	18.679	32.988	57.946	101.26	176.03	304.49	524.06	897.60
65	6.8300	12.799	23.840	44.145	81.273	148.78	270.85	490.37	883.07	1581.9
70	7.9178	15.572	30.426	59.076	113.99	218.61	416.73	789.75	—	—

n	0.14	0.16	0.18	0.20	0.25	0.30	0.35	0.40	0.45	0.50
1	1.1400	1.1600	1.1800	1.2000	1.2500	1.3000	1.3500	1.4000	1.4500	1.50
2	1.2996	1.3456	1.3924	1.4400	1.5625	1.6900	1.8225	1.9600	2.1025	2.25
3	1.4815	1.5609	1.6430	1.7280	1.9531	2.1970	2.4604	2.7440	3.0486	3.37
4	1.6890	1.8106	1.9388	2.0736	2.4414	2.8561	3.3215	3.8416	4.4205	5.06
5	1.9254	2.1003	2.2878	2.4883	3.0518	3.7129	4.4840	5.3782	6.4097	7.59
6	2.1950	2.4364	2.6996	2.9860	3.8147	4.8268	6.0534	7.5295	9.2941	11.39
7	2.5023	2.8262	3.1855	3.5832	4.7684	6.2749	8.1722	10.541	13.477	17.08
8	2.8526	3.2784	3.7589	4.2998	5.9605	8.1573	11.032	14.758	19.541	25.62
9	3.2519	3.8030	4.4355	5.1598	7.4506	10.605	14.894	20.661	28.334	38.44
10	3.7072	4.4114	5.2338	6.1917	9.3132	13.786	20.107	28.925	41.085	57.66
12	4.8179	5.9360	7.2876	8.9161	14.552	23.298	36.644	56.694	86.381	129.75
14	6.2613	7.9875	10.147	12.839	22.737	39.374	66.784	111.12	181.62	291.93
16	8.1372	10.748	14.129	18.488	35.527	66.542	121.71	217.80	381.85	656.84
18	10.575	14.463	19.673	26.623	55.511	112.46	221.82	426.88	802.83	—
20	13.743	19.461	27.393	38.338	86.736	190.05	404.27	836.68	—	—

Source: Duerr, William A., and Neils B. Christiansen: *Exercises in the Managerial Economics of Forestry,* O.S.U. Book Stores, Corvallis, Oreg. 1973, 264–323.

Formula (1) applies to any compounding intervals, and *n,* then, is the number of these intervals if *i* is the rate of compound interest per interval. The amount of $126.82 can be thought of as the end value of $100 held for 12 years at 2-percent annual interest compounded annually. Or it can be thought of as the end value of $100 held for 1 year at 24-percent annual interest compounded monthly. The amount of interest in this last case is $26.82, not $24.00, because it is the result of 12 compoundings, not 1.

There are *two* phenomena to account for. First, the amount of principal has grown; interest has been charged. Long ago, interest was termed "usury," a condemnatory word applied also to the act of lending at interest. Nowadays, interest is recognized as payment for the use of a productive resource, capital. It is analogous to rent in the case of land and wages in the case of labor. The interest that capital can earn may helpfully be viewed as an opportunity cost of holding the capital (page 118). Interest can also be thought of as an expression of a fundamental human viewpoint: that a bird in hand is worth two in the bush; that a dollar today is worth the promise of more than a dollar any time in the future, because the future value is shrunk by deprivation and uncertainty.

Second is the phenomenon that growth has occurred, not only in the principal, but also in the past interest due on the principal. There has been interest upon interest: There has been compounding. This phenomenon recognizes that the lender has forgone the opportunity, not only to invest, but also to reinvest. The $102 that the lender left on loan at the end of the first period is treated as a new loan, with its own opportunity cost of 2 percent. If there were no reinvestment opportunity and thus no compounding, financial calculations would be made with **simple interest,** interest on only the original principal.

A distinction can be drawn between the rate of interest and the rate of inflation. The latter is the rate of increase in the current-dollar prices of things. Inflation affects current interest rates, just as it affects current wage rates and rents. But for present purposes, let's abstract from inflation and work with real (uninflated) interest rates, such as might hold if the dollar's value were unchanging and were not expected to change.

At the same time, bear in mind that any recurring compounding process, whether the payment takes the form of interest on capital or some other form, can be dealt with as is done here. If inflation proceeds at a rate of 2 percent each month, by the end of the year prices have risen 26.82 percent. If a tree's cubic-foot content increases by 26.82 percent in 12 years, its annual growth rate has averaged 2 percent, which is to say that

$$i = \sqrt[n]{V_n \div V_0} - 1 \qquad (2)$$

Such relationships can be solved with ease on a small calculator or with a **compound-interest table:** values of 1 compounded at successive interest rates for successive time periods. Table 19-1 is such a table.

Just as one compounds a current value to translate it to terms of a later point of time, so one can do the reverse: At 2-percent interest compounded annually, the promise of $126.82 in 12 years is worth $100 now. The process is **discounting.** In general, rearranging formula (1),

$$V_0 = V_n \div (1 + i)^n \tag{3}$$

> At 10 percent annual interest, what is the current value of a $10,000 tax that must be paid 2 years hence? What sum must be invested today at 10 percent in order to pay off a $10,000 tax in 2 years?

INTEREST ON PAYMENT SERIES

Now enlarge the problem, so as to consider, not single values, but *series* of values. Suppose you borrow $100 at 2 percent per month, but instead of letting the debt build up, you pay off the interest at the end of each month. What is the amount of your monthly payment? Obviously, $2. If you let r stand for the amount of payment, you have

$$r = V_0 \times i \tag{4}$$

and it follows that any capital instrument that promises to pay $2 per month and remain intact (that is, promises a **perpetual series** of $2 monthly payments) must, if the interest rate is 2 percent per month, be worth $100:

$$V_0 = r \div i \tag{5}$$

That is, the instrument has a **capitalized value** of $100, and formula (5) is for capitalizing a perpetual series of payments, r. Ordinarily, this is an **annual series,** and i is the rate of interest *per year,* compounded *annually.* When you capitalize r, you are using a quick and easy method of finding the sum of discounted values of an infinite series of r:

$$V_0 = [r \div (1 + i)] + [r \div (1 + i)^2] + [r \div (1 + i)^3] + \cdots = r \div i \tag{5a}$$

Notice that when you capitalize r by formula (5), you are being given the present value of a series of r that *commences one year hence*—one year later than the time of reference of V_0. If you want to capitalize a series that starts now, formula (5) gives you its value a year ago. Its value now is formula (5) compounded for 1 year. To compound $(r \div i)$ for a year, either multiply it by $(1 + i)$ or add r to it; the result is the same either way.

> At 5 percent annual interest, what is the current value of a bond that promises a perpetual net annual payment of $100 commencing in 2 years?

Suppose you wish to capitalize a perpetual series in which r promises to recur at some interval different from the interval at which i is compounded. You can still take the approach of formula (5), provided only that you first adjust the interest rate to represent the same interval as that of the recurrence of r. For example, a timber stand promises a perpetual net return of $10,000 *per 12 years, starting 12 years hence.* The interest rate is 2 percent per year, compounded annually. What is the capitalized value of the stand? As already observed, 2 percent compounded annually for 12 years comes to 26.82 percent. That is, $(1.02)^{12} - 1 = .2682$. The stand, then, has a capitalized timber value of $10,000 ÷ .2682 = $37,286. In general,

$$V_0 = r \div [(1 + i)^t - 1] \tag{6}$$

where t is the number of compoundings of i per recurrence of r. In most cases, i is annual, and therefore t is the number of years between periodic recurrences of r.

Notice that if r is annual, so that t is 1, formula (6) reduces to formula (5).

Notice, too, that formula (6) can be modified so as to capitalize a series that commences at any time other than t. Suppose t is 10 years. If the series starts now, multiply the right side of formula (6) by $(1 + i)^{10}$ or add r to it. If the series starts 4 years hence (and then continues at years 14, 24, and so on), multiply the right side by $(1 + i)^6$, since the series has a 6-year advantage; it is timed 6 years earlier than the unmodified formula (6) allows for.

> At 10 percent annual interest, what is the capitalized timber value per acre of an even-aged forest tract from which the stand, now 10 years old, will be clean-cut on a 30-year rotation with a net yield of $1,000?

Suppose that the **periodic series** of r is not perpetual, but **terminable:** The payments consist, not only of interest as in formula (6), but also part of the principal, which then becomes exhausted after n years, and the enterprise is liquidated. Think first about the *end value* of such a series: the sum of *rs* plus interest on them. Since the payments, r, in formula (6) may be viewed as interest on V_0, the total value of payments (including interest on them) after n years is the total interest $[(1 + i)^n - 1]$ accrued up to that time on the beginning value. Thus, for the end value of a terminable periodic series, you have

$$V_n = r[(1 + i)^n - 1] \div [(1 + i)^t - 1] \tag{7}$$

The beginning (capitalized) value of the series is the end value discounted for n years. [Note that discounting is introduced by putting $(1 + i)^n$ into the divisor, as in formula (3).]

$$V_0 = r[(1 + i)^n - 1] \div [(1 + i)^t - 1](1 + i)^n \qquad (8)$$

By setting t in formulas (7) and (8) equal to 1, you derive the end value of a terminable *annual* series

$$V_n = r[(1 + i)^n - 1] \div i \qquad (9)$$

and the beginning (capitalized) value of a terminable annual series [again with $(1 + i)^n$ in the divisor].

$$V_0 = r[(1 + i)^n - 1] \div i(1 + i)^n \qquad (10)$$

which says that if you want to buy an **annuity** for $\$r$ per year commencing 1 year hence and running for n years, and if the annual interest rate in decimals is i, you must pay $\$V_0$ today for the annuity.

Any of the formulas for the value of a series r can be modified just as formulas (5) and (6) can be modified. The general rule is that if the series commences y years earlier than year t (remember that t can equal 1 or more), the value of V derived from the formula is too low and is corrected by compounding it for y years: multiplying it by $(1 + i)^y$. Or if the series starts y years later than t, the value of V is too high and is corrected by discounting: dividing by $(1 + i)^y$.

We could elaborate further, but the ten formulas derived are largely sufficient for your purposes. Eight of them are summarized in Table 19-2. The Gunter-Haney key to selection is added as Table 19-3. More formulas may be found in specialized books on the subject of finance; for example, see Vichas [2] and Methur [1].

CHARACTERISTICS OF INTEREST CALCULATIONS

A little study of the relationships in Tables 19-1 and 19-2 brings out some principles about the working of compound interest. Since the numbers in the columns of Table 19-1 are geometric series with ratios of $1 + i$, (for $n = 1$ to 10), $(1 + i)^2$ (for $n = 10$ to 20), or $(1 + i)^5$ (for $n = 20$ to 70), the numbers in each of the three series plot as straight lines on semilogarithmic scale. The greater the value of i, the steeper the line. It follows that the end value of compounding for a given period or of accumulating a given terminable series is larger and larger as continually higher interest rates are used.

Or conversely, in the process of discounting of capitalizing, for given end values of series of payments, low interest rates lead to larger values at time zero than do high rates. Low rates emphasize the present value of the farther future; high rates, that of the nearer future. If \$100 annual net income is capitalized at 3-percent interest—\$3,333 by formula (5)—the first 10 years of returns—\$853 by formula (10)—represent about 25 percent of the total capitalized value. However, at 50-percent interest, \$100 annually capitalizes at \$200, and the first 10 years of returns—\$196.53—represents more than 98 percent of the total capitalized value. Low interest rates are like binoculars that you might use to look at future values; they bring close and enlarge

TABLE 19-2
SOME COMPOUND-INTEREST FORMULAS

Dependent variable	Formula[a]
Compounding and Discounting	
(1) Value at time n of payment at time zero	$V_n = V_0 (1 + i)^n$
(3) Value at time zero of payment at time n	$V_o = V_n \div (1 + i)^n$
Capitalizing Annual Series of Payments	
(5) Value at time zero of perpetual series that begins at time one	$V_o = r \div i$
(10) Value at time zero of terminable series that begins at time one	$V_o = \dfrac{r[(1 + i)^n - 1]}{i(1 + i)^n}$
Capitalizing Periodic Series of Payments	
(6) Value at time zero of perpetual series that begins at time t	$V_o = r \div [(1 + i)^t - 1]$
(8) Value at time zero of terminable series that begins at time t	$V_o = \dfrac{r[(1 + i)^n - 1]}{[(1 + i)^t - 1](1 + i)^n}$
End Value of Terminable Series of Payments	
(7) Value at time n of periodic series that begins at time t	$V_n = \dfrac{r[(1 + i)^n - 1]}{(1 + i)^t - 1}$
(9) Value at time n of annual series that begins at time one	$V_n = r[(1 + i)^n - 1] \div i$

[a] i = Annual rate of interest, compounded annually, expressed in decimals.
n = Number of years of compounding or discounting, or in a terminable series.
r = Payment recurring annually or at longer intervals ("periodically").
t = Number of years between periodic recurrences of r.
V_o = Value at time zero.
V_n = Value at time n.

what you see today. To understand high interest rates, reverse the binoculars and look through the large lenses. You will see today's value of future events withdrawn and minimized.

It is easy to see one reason forest property, as in parts of the west and southeast, has tended to move from the possession of smaller firms with high guiding rates into the hands of big firms with lower guiding rates: The big firms could afford to buy such

TABLE 19-3
GUNTER-HANEY KEY TO SELECTION OF COMPOUND-INTEREST FORMULAS

Key				Formula
Single sum			Future value	(1)
			Present value	(3)
Series	Terminating	Annual	Future value	(9)
			Present value	(10)
		Periodic	Future value	(7)
			Present value	(8)
	Perpetual	Annual	Present value	(5)
		Periodic	Present value	(6)

Source: Gunter, John E., and Harry L. Haney: *Essentials of Forestry Investment Analysis*, O.S.U. Book Stores, Corvallis, Oreg., 1984, p. 53.

property for a price greater than its value to the sellers. Both parties to the bargain may be presumed to have gained. It would be good to suppose that society, too, came out ahead, since low interest rates favor conservation and long-term planning, but the considerations aren't simple.

No wonder that foresters over long years have argued for low interest rates! There are even remnants of thinking in the profession that a zero rate of interest is appropriate for timber management. At this interest rate, a future decision, no matter how distant, has the same significance as today's decision. And any property capable of yielding a perpetual return has an incalculably high value. Or, to view the same proposition from a different angle, zero interest implies that forest capital has no alternative uses, no opportunity cost. To liquidate while it is still putting on any net growth is irrational.

A last thought—a caution—about compound interest: The calculations are perfectly clear, but, where they involve the future, their implications and even appropriateness often are not. Perhaps their most dangerous implication is that the future can be clearly foreseen. The longer the planning period and the lower the rate of interest, the greater the hazards of compound-interest calculations are. The formulas are referred to as devices for changing the time of reference of values. For these devices literally to perform this function is impossible, the future being unknown. Compound-interest formulas are models that yield replies. The manager of the firm, private or public, can use a reply, but the manager must translate it into the firm's language of the moment, with the right mixture of pessimism and optimism, confidence and misgiving, farsightedness and blindness to the future, in order to create a decision out of it. In other words, only by the addition of judgment are the formulas made usable in practice.

How should you respond to a forest owner who seeks financial advice and rejects the idea of interest, implying that the appropriate rate is zero?

SUMMARY

Compound interest is payment for the use of capital that has other productive uses, the proceeds from which can be reinvested. Formulas are available for shifting the time of reference of financial events by means of compound interest. If used with thoughtful judgment, the formulas can be helpful in solving economic problems in forestry. In capitalizing or in simple discounting, low interest rates magnify the farther future; high rates, the nearer.

REFERENCES

1 Methur, Iqbal: *Introduction to Financial Management,* Macmillan, New York, 1979, pp. 167–225.

2 Vichas, Robert P.: *Handbook of Financial Mathematics, Formulas and Tables,* Prentice-Hall, Englewood Cliffs, N.J., 1979, pp. 31–38, 71–73, 107–109.

CHAPTER **20**

FINDING A GUIDING RATE OF RETURN

THE PROBLEM: AVENUES FOR SOLUTION

In Chap. 19, we moved values from one time of reference to another—in fact, we used an arithmetical approach to this dialectical problem of weighing events at one point of time against those at another. Thus we worked with interest rates, guiding rates of return, or **hurdle rates,** as they are sometimes called. If you are studying a forestry question that uses compound interest, how will you decide on a rate? This is a problem now up for solving.

We will look for the determinants of the guiding rate among four circumstances: the national monetary context, the kind of firm, its financial alternatives, and its uncertainties about the question at hand.

MONETARY CONTEXT

The general context in which an interest rate is chosen is the national monetary system and policy. The monetary system provides markets in which buyers and sellers of loans exert their influences upon the rates that borrowers pay lenders.

> Are borrowers buyers or sellers of loans?

This complex market system is overseen and influenced by a host of major agencies whose collective activity represents U.S. monetary policy. The chief policymaker is the Federal Reserve Bank, controlled by a board of governors headed by a chairman. The bank has regulatory devices. Charged with stabilizing prices and steadying national economic growth, the bank can tighten money and encourage rising interest rates when the economy threatens to overheat and inflate, and it can loosen credit when activity is lagging.

Over the decades since 1940, the monetary system has seen dramatic changes in interest rates. During World War II, the federal Treasury was offering $\frac{3}{8}$ of 1-percent annual interest on its short-term bills. In the 1940s, the **prime rate** (charged by large commercial banks to their preferred corporate customers) was as low as $2\frac{1}{2}$ to 3 percent. Home buyers were paying 4 percent on mortgages. As the 1980s decade opened, all these rates had moved up toward or above 20 percent per annum, and later in the decade, they fell again.

Inflation is a big determinant of interest rates. Inflation means a rise in the average price of all the current output included in GNP.

Suppose the market is such that the **real interest rate** (the rate in the event of stable prices) would be 6 percent for a given type of loan. And suppose the inflation rate (that is, the *anticipated* inflation rate; this is what counts) is 12 percent. Then the **current interest rate** (the actual rate) is bound to be at least $18\frac{3}{4}$ percent ($1.06 \times 1.12 = 1.1872$). And if there's uncertainty about inflation (which there always is!) and a tendency to add a margin of safety, the rate can easily go above 20 percent. If it doesn't, lenders may find themselves ending the year with resources slimmer than at the beginning, since borrowers can pay off their loans with cheapened dollars.

In general, a given anticipated rate of inflation, having had its effect on the interest rate, does not subsequently affect it further. The anticipated inflation rate must *change* for inflation's effect to change.

The prospect of inflation works in two major ways to raise interest rates. First, it tends to raise rates through the negotiations between borrowers and lenders, as described. Second, it encourages the central bank to fight inflation by tightening the money supply—that is, by raising rates.

Whether the guiding rate of return used in a given analysis should be a real or a current rate depends on the monetary values that the rate will operate on—whether they are real or current values.

How does one estimate the real interest rate when inflation is in prospect? One starts with the current rate. If this is an annual **discount rate,** d (the charge for a loan that is paid by the borrower at the *start*—not, as with interest, at the end—of a period), one recognizes that

$$d = 1 - \frac{V_0}{V_1} \tag{1}$$

whereas [from Chap. 19, formula (2), taking $n = 1$]

$$i = \frac{V_1}{V_0} - 1 \tag{2}$$

and therefore

$$i = \frac{1}{1-d} - 1 \tag{3}$$

Suppose that the annual current discount rate is 10 percent. Then the annual current interest rate is 11.11 percent. If the anticipated annual inflation rate is 5 percent, the 11.11 figure is reduced accordingly:

$$\frac{1.1111}{1.05} \cong 1.058 \tag{4}$$

and the real interest rate is about 5.8 percent per year.

TYPE OF FIRM: CAPITAL BUDGETING

Next, turn your attention to the type of firm and its financial alternatives. Let's see how these matters influence the firm's interest rate for forestry and related planning. Remember that the question *What are the alternatives?* is one of the most useful questions in forestry economics.

If a firm has no productive uses for its capital, its interest rate (opportunity cost of holding this capital) is zero. Now if a productive alternative appears, the interest rate becomes positive. How so? Because now the present use of the capital has a competitor, and the firm has a choice to make, just as it had in Chaps. 17 and 18. The competing opportunity provides a standard, a guiding rate of return, against which to judge whether the present use of capital is satisfactory.

The essence of the relation between the firm's alternatives and its guiding rate of return is given in the **capital-budgeting** model, with which the name of Joel Dean [3] is associated. You are asked to picture a firm that has a number of investment opportunities, and for simplicity to assume that these are independent: The return to any project is unaffected by the inclusion or omission of any other. (Christiansen [2] discusses less simple cases.) The investment projects promise various rates of return. You are asked to imagine that the projects are arrayed in descending order of promise, as in Fig. 20-1, where each bar is a project. The bar's width represents the size of the investment; its height, the promised net rate of return on an investment of this size. The costs that are netted out of the return include income taxes—that is, this is an **after-tax** rate of return.

Converting the stair-stepped bar tops into a smooth curve, one gets the down-sloping function labeled "return on capital" in Fig. 20-2. This is the firm's short-run demand for capital: the "subjective values" that the firm places on the successive projects in accordance with their promised percentage rates of return, and thus what the firm "stands ready to pay" for the projects.

The smooth down-sloping curve in Fig. 20-2 describes the firm's prospective marginal net revenue from successive increments of investment, plotted in descending order of percentage return. Strictly speaking, these increments are much smaller than the

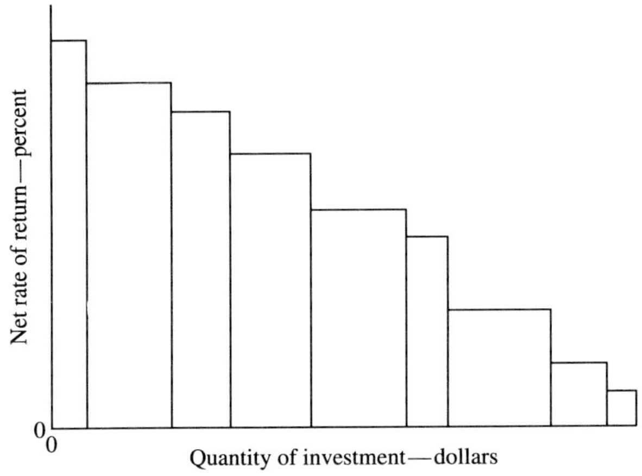

FIGURE 20-1
Net annual rate of return on investment in relation to its quantity.

whole project investments of Fig. 20-1. The increments are in amounts that recognize the working of diminishing marginal return *within* projects [4]. We have already seen, in Chap. 18, how the marginal return from Project "Tree" diminishes with increasing total investment in Tree. Under the same principle, increasing investment in Project R (reforesting an acre) or Project S (stand improvement on an acre) or any other such project is subject to diminishing marginal return. Strictly speaking, then, one must think of the down-sloping curve in Fig. 20-2 as representing marginal return from a bit of Project R, the best bit, and then bits of Projects S and so on, and then more of S and of R, always in descending order of percentage return.

Next, in our development of the capital-budgeting model, suppose that the firm has various sources of capital that it can draw upon for investment. Although it's not an essential assumption, you can picture that the price (percent interest) the firm pays for additional capital increases with the total quantity borrowed, as in the up-sloping func-

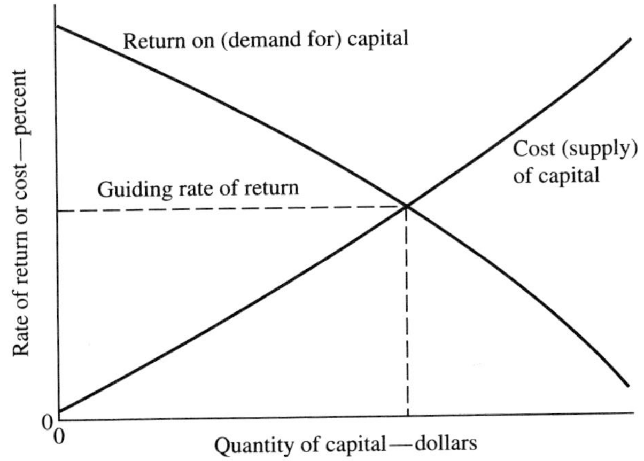

FIGURE 20-2
A firm's capital supply and demand and guiding annual rate of return: the capital-budgeting model.

tion labeled "cost of capital" in Fig. 20-2. This is the marginal unit cost of capital: the short-run supply of capital to the firm.

> Do you think the up-sloping cost curve in Fig. 20-2 is realistic? Why would a firm's unit cost of capital increase with quantity?

Clearly, the investment projects to the left of the demand-supply intersection in Fig. 20-2 are acceptable: They promise to return more than they cost, down to the very last increment that just breaks even. But projects to the right are unacceptable. The rate of return that guides the firm in these judgments is the rate at the intersection, marked with a broken horizontal line in Fig. 20-2.

By taking its guiding rate of return as its prospective marginal unit revenue from capital where this equals its marginal unit cost of capital, the firm limits its total interest charge to the area of the rectangle beneath the broken horizontal line in Fig. 20-2 and maximizes its profit, represented by the area of the triangle above the line. Or perhaps you should think of the firm, for a moment, as a consumer. Then the rectangular area is what the firm pays for the projects: their market value. And the triangular area is the excess of subjective value over market value: consumer surplus.

How does the firm know what rates of return its investment projects will pay? How does it even know what its cost of capital will be? It doesn't. It must forecast. The firm is uncertain about the future. How, then, does it know what guiding rate of return is best in a given case? It doesn't. Figure 20-2, as a model, represents a reply. The firm's answer must be based on the estimates that it chooses to make, tempered by such other considerations as you are glimpsing as you go along.

The important thing for you to grasp is that the guiding rate of return is a reality, but its appropriate numerical value is not subject to verification—except, perhaps, too late. As for Fig. 20-2, it is highly valuable as a simple, sharply defined model, even though a few ifs, ands, and buts have to be hung on it.

If the marginal unit cost of capital to the firm is independent of the quantity borrowed, the supply curve is horizontal, representing both marginal and average unit costs (refer to page 169). Supply in this case determines the guiding rate of return independently of demand, while demand influences only the *quantity* of capital invested. Such a redrafting of Fig. 20-2 would illustrate the method used by many, if not most, corporations to decide upon a guiding rate: They calculate it as the *average* cost of all capital, both the cost of equity capital (dividends on common stock) and that of debt capital (interest on medium- and long-term borrowing [7]).

TRANSFER COST

For many uses of the guiding rate of return, one determinant of the rate is **transfer cost.** This is the cost of transferring capital out of Project A into Project B. It may take the form of a tax or a brokerage fee or any other expense that the firm incurs only if it carries out the transfer of capital.

Take for illustration a transfer cost consisting of a 20-percent income tax. If the firm liquidates Project A, the proceeds are income and, after the tax is paid, the firm retains only 80 percent of these proceeds. The guiding rate of return is 5 percent. What bearing does the transfer cost have on this guiding rate? Consider two contrasting cases:

CASE 1

Project A is a Christmas tree enterprise that pays a $4\frac{1}{2}$-percent annual after-tax return on its liquidation value. Project B pays 5-percent after tax, the supposed guiding rate of return. Should the firm liquidate Project A and put the proceeds into B? No, because the proceeds would be taxable at a 20-percent rate, which means that for every \$100 taken out of A, the firm could transfer only \$80 to B. This \$80 would earn, in B, 5 percent, or \$4, which is less than the \$4.50 being earned in A. In this case, the effective guiding rate is not 5 percent, but is reduced by the 20-percent transfer cost to 4 percent.

CASE 2

Project A is Tree in Chap. 18, which promises $4\frac{1}{2}$-percent after tax. Project B pays 5 percent after tax, the supposed guiding rate of return. Should Tree be liquidated? Is it financially mature (overlook the question of "adjustment"; it won't affect the conclusion)? Yes, Tree *is* ready for liquidation. Each \$100 of the proceeds will (after transfer cost) furnish \$80 for Project B, where it will grow to \$84 at the end of a year. If Tree's liquidation is postponed for a year, the \$100 invested in it will have grown to \$104.50. These proceeds, after transfer cost, will provide only \$83.60 (= \$104.50 × .8) for Project B. The guiding rate *is* 5 percent!

How does the second case differ from the first? As always, you can ask *What are the alternatives?* In the first case, the alternatives are to incur a transfer cost or not to incur it. The cost is variable with the choice and therefore relevant to it. In the second case, the alternatives are to incur a transfer cost now or to incur it at some other time. The cost is fixed with respect to the choice and therefore irrelevant to it.

> Explain why a quantity or value that varies among alternatives is relevant to the choice among alternatives, whereas one that does not vary is irrelevant to the choice.

INDIVIDUALS VERSUS CORPORATE FIRMS

Up to this point, firms and their alternatives have been considered from the viewpoint more of corporations than of individuals. As one moves from corporate forestry and wood industry toward the realm of nonindustrial private forestry, one finds two things happening to the guiding rate of return. First, it becomes less and less a manager's tool and more and more an index that *we,* as students of forestry, may use to understand and explain how forest owners behave. Second, the guiding rate is determined less by financial considerations and more by owners' subjective judgment. The rate arises in part from what owners have been taught to believe and in part from the thoughts and behavior that they have come to expect of themselves.

Do you remember the pirates in Stevenson's *Treasure Island*? After they had mutinied, rum had the upper hand with them day in and out, and their greed suppressed anxiety about tomorrow. Two of them who had been left as watchmen aboard ship were brawling while their anchor line was being cut and their mates' plans destroyed. By what rate of interest were these fellows guided? If they valued today at 10 times tomorrow, which one can easily believe, their guiding rate per annum was $(1 + 10)^{365} - 1$, or approximately 10^{382} percent.

> What sort of forest management would such persons be expected to practice?

At the other extreme are small-scale private forest owners in western Europe who, through a sense of obligation to the community, manage their timber so as to achieve a perpetual yield of maximum value, a marginal revenue of zero percent. Their guiding rate, apparently, is zero.

The multitude of forest owners who fall between the extremes are swayed by their own highly varied circumstances. The family that is financially strapped acquires spending alternatives that are far greater in earning power, perhaps in the hundreds of percent, than anything the living forest could contribute for them, and liquidation is the obvious move. But the environmentalists who can afford to practice what they advocate may perform at a 1- or 2-percent guiding rate. There is no general rule, except to observe managerial behavior and interpret it in percentage terms.

Or look at it this way: Forest owners who have high guiding rates for financial reasons include individuals whose wealth and income are chronically low. But also included may be persons of substantial income who are suffering from an adverse **cash flow.** Consider the case of the family whose **disposable** (after-tax) **income** is $10,000 per month. The family is capable of living comfortably within its means, at a low guiding rate of return. But it has committed itself to such heavy investment and spending outlays during the next few years that it is, at least for the time being, financially embarrassed. The family seeks cash and is prepared to borrow at high interest rates. If it owns forest, it may liquidate timber that is growing fast in value.

Not only individuals, but also corporations, have cash flows that affect their guiding rates of return.

A prominent feature of individuals' guiding rates is that they fluctuate. Even the poor, and certainly the wealthy, operate at low rates sometimes and in certain types of choice. But inevitably the time and circumstance arrive when an individual has inescapable pressing alternatives—perhaps illness, death, or other family emergency—and must perform at a high guiding rate of return.

GOVERNMENTS AS FIRMS

And what about governments? Consider the federal government. If it were to view itself as corporations view themselves, its guiding rates of return might well be higher than the corporate rates. This is because its alternatives (that is, the alternatives of its taxpayers) include a large fraction of high-rate personal-spending opportunities. Only in exceptional times, such as during World War II, might the government's borrowing rates be low enough to depress its cost of all capital significantly.

Look back through the years, especially through the decades preceding 1950. You find government rates being set, not primarily on financial grounds, but on the basis of a general philosophy: Society does, and should, take a longer-term view of things than individuals do. It has, and should have, a more distant **planning horizon.** Therefore government, as the representative of society, operates at low rates of interest as a matter of morality. For forestry projects, a guiding rate of 2 percent, 1 percent, or even zero, is appropriate.

As time has gone on, a financial philosophy has, little by little, replaced the moral. Today, it is generally conceded that the guiding rate of return for appraising a government investment should equal the rate of return that the required funds would have earned in the use that was deprived of them [1]. It has been argued, not necessarily in disagreement, that a rate should be set equal to the maximum internal rate of return available from unfunded government projects. This implies a rate of 10 percent or more [5].

Since 1969, the federal Office of Management and Budget has set a 10-percent guiding rate of return for federal projects, except where some other rate is derived by law. The office explicitly identifies this as a *real* rate: Prospective costs and benefits to be capitalized at 10 percent are to be expressed in dollars of constant value. In practice, an allowance for inflation of costs and benefits may be permitted to creep into the calculation. The effect of such an allowance is the same as the effect of a low interest rate, mentioned earlier.

Federal rates other than 10 percent that are derived by law include the Water Resources Council's real rate for water projects, which is based upon the yield of 15-year or longer-term Treasury securities. Further, the National Forest Management Act of 1976 directs the Forest Service to recommend a real rate for its own use in long-term planning. The Service's recommendation, approved by the Office of Management and Budget, is 4 percent [6]. Such percentages may be referred to as **exception rates** of interest.

To the foregoing neat concepts of governments' guiding rates, we owe it to ourselves to add that much government forestry spending and investment is guided in ways that resemble those of the nonindustrial private managers and, indeed, of some managers of all sorts: The decisions are rational in the accepted sense, but the values that are maximized or minimized are arrived at by judgment, compromise, or emotion, or, at any rate, are not readily quantified. Just as for the little holdings, the guiding rate of return is a useful tool for the analyst if not always for the manager. It is, nevertheless, a real parameter in the scheme of forestry.

UNCERTAINTY AND THE GUIDING RATE

Finally, there is one more interest-rate determinant to recognize: the firm's uncertainty about the question at hand (the question that called for the use of interest) or about the method employed for arriving at an interest rate. The uncertainty involved is that which arises in the firm's mind when its view of the future is clouded, either because this future is remote or because, though not remote, it is thought to contain a wide scatter of possibilities—of unexpected outcomes (see Chap. 31). In either case, the greater the firm's uncertainty, the higher its guiding rate of return: It demands a higher probable rate of return from its projects to allow for the possibility that some of the return may not materialize at all.

Incorporating a risk allowance in the interest rate may involve a technical fault. Suppose that the firm's guiding rate of return is 4 percent, which it raises to 5 percent in order to get capitalized values that allow for risk. If the revenue series to be capitalized is annual, the risk allowance is 20 percent—$(r \div .05) = .8(r \div .04)$.

But if r occurs at 20-year intervals, the denominators become 1.6533 and 1.1911, respectively (Table 19-1, $n = 20$), and the risk allowance is 28 percent—$(r \div 1.6533) = .72(r \div 1.1911)$. And if a firm with an 8-percent guiding rate raises it to 10 percent for a supposed 20-percent risk allowance, the denominators ($n = 20$) are 5.725 and 3.661 and the actual risk allowance is 36 percent.

It is better to allow for risk by reducing estimates of net benefit or increasing estimates of net cost, leaving the guiding rate untouched. Nevertheless, incorporating a risk allowance in the guiding rate is common practice, and one must recognize the fact.

> Why do you suppose that risk is so commonly allowed for by increasing the interest rate, even though this practice may be technically at fault?

SUMMARY

To put the foregoing ideas in a nutshell, the rate of interest for firms to use in their planning is a guiding rate of return. This rate may be thought of, and in many cases derived, as a function of the supply and demand for capital for the firm's use. The rate depends on the national monetary context, the nature of the firm and its alternatives, and its uncertainties about the future.

You have encountered the first illustration of a useful **principle of relevance** in economics and management: What is variable among alternatives is relevant to the choice among them; what is fixed is irrelevant.

REFERENCES

1 Baumol, William J.: "On the Social Rate of Discount," *The American Economic Review,* September 1968, pp. 788–802.

2 Christiansen, Neils B.: "Models for Time Adjustments of Costs and Revenues," In Duerr, William A., Dennis E. Teeguarden, Neils B. Christiansen, and Sam Guttenberg, eds.: *Forest Resource Management, Decision-Making Principles and Cases.* O.S.U. Book Stores, Corvallis, Oreg. 1982, pp. 131–146.

3 Dean, Joel: *Capital Budgeting,* Columbia University Press, New York, 1951.

4 Lundgren, Allen L.: "Banking Investment Alternatives, a New Look," *Journal of Forestry,* September 1971, pp. 568–573.

5 Quirk, James, and Katsuaki Terasawa: "Choosing a Government Discount Rate, an Alternative Approach," *Journal of Environmental Economics and Management,* January 1991, pp. 16–28.

6 Row, Clark, H. Fred Kaiser, and John Sessions: "Discount Rate for Long-Term Forest Service Investments," *Journal of Forestry,* June 1981, pp. 367–369, 376.

7 Segur, John A.: "Dealing with Financial Agents," *In* Duerr et al.: Op. cit., pp. 403–417.

CHAPTER 21

ADJUSTED FINANCIAL MATURITY OF INDIVISIBLE CAPITAL

THE PROBLEM

Chapter 18 derived for Tree a financial maturity that turned out to be "unadjusted." Now, having in the meantime reviewed the use of compound-interest formulas and the process of finding an interest rate, we are ready to adjust the unadjusted procedure. The data are those of Chap. 18. The models are from Chaps. 19 and 20.

REVIEW OF UNADJUSTED MATURITY

Let's begin by again calculating Tree's unadjusted financial maturity, using the same age, stock, and growth data and the same guiding rate of return, 5 percent, that went into Fig. 18-1.

In Table 21-1, column (1) is Tree's age, t. Column (2) is Tree's stumpage, or liquidation, value, v or r, at age t. Column (3) is Tree's current annual value growth (marginal value growth), which in column (4) is expressed as a percentage of the stock at the beginning of the year. This marginal value growth percent falls to 5.0 in year $t = 31$, the same age of unadjusted financial maturity that we estimated in Chap. 18. We can verify this estimate by bringing each of the column (2) values to a single point of time, in this case the year $t =$ zero. The verification is carried out in column (5).

Why do the column (5) values peak at age $t = 31$, where the marginal value growth percent equals the firm's guiding rate? Note that column (2) is a geometric series with a declining ratio, and the problem is to find the point where the ratio is 1.05. Column (4) identifies this point directly as $t = 31$. Column (5) is derived by discounting each value in (2) to $t =$ zero, using 5-percent interest. If (2) were a series with a perfectly constant ratio of 1.05, all the values in (5) would be the same. But the ratios are not

TABLE 21-1
FINANCIAL MATURITY OF TREE

Tree age, t, in years (1)	Growing stock value, v or r, in dollars (2)	Marginal annual value growth, g, in dollars (3)	Marginal annual value growth, p, % (4)	Capitalized single-stock value, $\frac{v}{(1.05)^t}$ in dollars (5)	Capitalized series of stock values, $\frac{r}{(1.05)^t - 1}$, in dollars (6)	Adjustment factor $f = \frac{(1.05)^t - 1}{(1.05)^t}$ (7)	Adjusted marginal annual value growth, pf, % (8)
26	22.45			6.31	8.78	.7188	
		2.44	10.87				7.81
27	24.89			6.67	9.11	.7322	
		2.17	8.72				6.38
28	27.06			6.90	9.27	.7449	
		1.98	7.32				5.45
29[a]	29.04			7.06	9.32[a]	.7571	
		1.81	6.23				4.72[a]
30	30.85			7.14	9.29	.7686	
		1.65	5.35				4.11
31[b]	32.50			7.16[b]	9.19	.7796	
		1.50	4.62[b]				3.60
32	34.00			7.14	9.03	.7901	
		1.36	4.00				3.16
33	35.36			7.07	8.83	.8001	
		1.23	3.48				2.78
34	36.59			6.97	8.60	.8096	

[a]Adjusted financial maturity.
[b]Unadjusted financial maturity.

constant. Column (4) says that they are above 1.05 in the early years; therefore, the values in column (5) must be increasing. In the last years, the ratios are below 1.05, and so column (5) values are decreasing. Where column (4) signals a 1.05 ratio, column (5) values are going neither up nor down: They are at their maximum.

ADJUSTING THE UNADJUSTED

Now the route to an adjusted estimate of financial maturity opens up to us. Pages 125 and 126 explain that a 31-year financial maturity is at fault because it fails to take into account the opportunity cost of postponing other productive uses of the ground after Tree is felled. You can see that 31 years has this fault because we were able to derive it in column (5) by using the denominator $(1.05)^t$, which merely brings each of the values in column (2) to the year t = zero. We "capitalized" each of the values by discounting it as a unique event, not to be repeated. In effect, we disregarded the use of the land after Tree is felled.

But what we want is to capitalize a perpetual *series* of events, a series of periodic returns, r. Formula (6) in Chap. 19 tells us how to do it: Use the denominator $(1.05)^t - 1$ beneath the successive values of r. The calculation is performed in column (6) of Table 21-1. Adjusted financial maturity, denoted by the maximum in column (6), occurs at age $t = 29$.

> You understand from the explanation at the end of the last section why the column (5) figures peak at (unadjusted) financial maturity. But why do the column (6) figures peak at financial maturity?

The design of columns (5) and (6) in Table 21-1 reveals an *adjustment factor* usable for estimating adjusted financial maturity by the simple method of equating Tree's marginal value growth percent to the firm's guiding rate of return. This adjustment factor is the ratio of $(1+i)^t - 1$ to $(1+i)^t$. The ratio is calculated in column (7). When the marginal value growth percent is multiplied by this ratio, the result, shown in column (8), is an *adjusted* marginal value growth percent. The adjusted percent is seen to equal the guiding rate, 5.0, at age $t = 29$, which verifies the finding in column (6).

> Instead of adjusting the marginal value growth percent by using the ratio of $(1+i)^t - 1$ to $(1+i)^t$ as a multiplier, we could as well adjust the *guiding rate of return* by using the ratio as a divisor. The result would be the same. Can you see a reason for preferring one approach over the other?

PHILOSOPHY ABOUT METHOD

Evidently financial-maturity adjustments measured in years may be small. Certainly the adjustment for Tree is small. If the annual decrements in marginal value growth percent were less or the firm's guiding rate were lower, the adjustment would be larger. Even so, when you consider the firm's uncertainties about tree value and growth rate and about the tree's influence on other trees and even about the firm's own guiding rate of return, you may well conclude that the gain from adjustment may be negligible.

However, it behooves us as students of the subject to concern ourselves with principles and seek logically correct replies to our questions. We want to know what we are doing before we dismiss a procedure as unnecessary fine-tuning.

Financial-maturity guides have been developed for a number of tree species in various localities [1–10]. These help the timber manager or marker to recognize financially mature trees from such characteristics as vigor, size, and quality. They allow for alternative assumptions about the firm's guiding rate of return.

SUMMARY

The nub of the financial-maturity adjustment is this: It takes account of the opportunity cost of postponing new returns. It estimates the tree age that maximizes the capitalized value of a continuing series of harvests, $r \div [(1+i)^t - 1]$, not the capitalized value of a single harvest, $v \div (1+i)^t$. The ratio of the two denominators can be used as a multiplier for estimating an adjusted marginal annual value growth percent to be equated to the firm's guiding rate of return. The adjustment, in terms of tree age, may be small.

REFERENCES

1 Chappelle, Daniel E.: *Financial Maturity of Eastern White Pine in New York State,* New York State College of Forestry, Syracuse, Tech. Pub. 91, 1966.

2 DeBald, Paul S., and Joseph J. Mendel: "Determining the Rate of Value Increase for Oaks." *In* USDA, Forest Service, Northeastern Forest Experiment Station: *Oak Symposium Proceedings,* West Virginia University, Morgantown, August 1971, pp. 142–151.

3 Dobie, J.: *Lumber Values and Financial Maturity of Young-Growth, Good-Site Douglas-Fir in British Columbia,* British Columbia Forest Products Laboratory, Vancouver, Inf. Rep. VP-X-37, 1968.

4 Duerr, William A., John Fedkiw, and Sam Guttenberg: *Financial Maturity, A Guide to Profitable Timber Growing,* USDA Tech. Bull. 1146, 1956.

5 Guttenberg, Sam, and J. A. Putnam: *Financial Maturity of Bottomland Red Oaks and Sweetgum,* USDA, Forest Service, Southern Forest Experiment Station Occasional Pap. 117, 1951.

6 Guttenberg, Sam, and R. R. Reynolds: *Cutting Financially Mature Loblolly Pine and Shortleaf Pine,* USDA, Forest Service, Southern Forest Experiment Station Occasional Pap. 129, 1953.

7 Leak, William B., Stanley M. Filip, and Dale S. Solomon: *Rates of Value Increase for Yellow Birch in New England,* USDA, Forest Service Res. Pap. NE-120, 1968.

8 Mendel, Joseph J.: "Financial Maturity of Paper Birch," *In* USDA, Forest Service, Northeastern Forest Experiment Station and cooperators: *Birch Symposium Proceedings,* University of New Hampshire, Durham, August 1969, pp. 135–144.

9 Naughton, Gary G.: *Evaluating Economic Maturity of Individual Trees, In* USDA, Forest Service General Tech. Rep. NC-4 *(Black Walnut as a Crop),* 1973, pp. 97–100.

10 Smith, H. Clay, George R. Trimble, Jr., and Paul S. DeBald: *Raise Cutting Diameters for Increased Returns,* USDA, Forest Service Res. Pap. NE-445, 1975.

CHAPTER 22

FINANCIAL MATURITY OF DIVISIBLE TIMBER CAPITAL

PROBLEM TO BE SOLVED

This chapter aims to explore the (adjusted) financial maturity of *forests,* just as Chap. 21 explored the question for trees. What is financially the best age for cutting a stand within a forest: the best rotation age for the forest? The model is to find the age of the oldest stand within the forest when, in the region of diminishing return, the *forest's* marginal revenue per unit of variable input (marginal value growth percent) equals the firm's marginal cost per unit of variable input (guiding rate of return). Here, *value* means the dollar value of the standing timber.

The forest is taken to be divisible capital, because part of it is harvested and parts are left to grow some more. And the firm has a plan, or prospect of some sort, for managing the parts left and the part cut over.

If forest stands are many-aged, so that the parts cut and left are intermingled on the same acres, then the focus is on the individual trees or groups, and the principles of Chap. 21 hold. In the present chapter, we look at forests of even-aged stands.

A SPECIAL CASE

To be sure, in theory at least, some forests for which we may want to plan the rotation are indivisible. We should at least glance at such forests. Suppose that you manage a forest that is all of one age. You plan, for some reason, to hold the entire forest untouched until it is financially mature as a whole, and then to clear-cut all of it and start over again. This is an indivisible forest. The model for its maturity is that of Chap. 21—specifically, column (8) in Table 21-1. The marginal value of growth percent is the current *(total)* annual growth percent of column (4) adjusted for perpetuity by the factor in column (7). Marginal and total growth are the same because the capital is indivisible. See pages 124 and 125.

Referring again to Table 21-1, notice that adjusted maturity is found in column (6) by the model

$$\text{Max.: } V = r \div [(1+i)^t - 1] \tag{1}$$

Maximization occurs at the point of maturity for the reason explained on pages 148 and 150.

Thus, for indivisible capital, expression (1) is an alternative form of

$$(\text{Adjusted marginal value growth percent at year } t) = (\text{Guiding rate of return}) \tag{2}$$

If there are thinnings, their value can be compounded to year t and added to r in expression (1) or taken into account in computing growth percent in expression (2).

The identity of expressions (1) and (2) holds for one-aged forests. But let an element of divisibility be introduced into the forest regime—let the cutting cycle be some fraction of the rotation—and expression (1) is invalidated, because marginal and total are no longer the same. Refer again to pages 124 and 125. Note that the model offered as valid for the thinning problem of Chap. 17 is not expression (1), but expression (2)—with the first word omitted, since no adjustment is necessary.

> Can you explain why adjustment of the marginal value growth percent is unnecessary in Chap. 17 and in the present chapter? Note that adjustment is necessary in Chap. 21.

Expression (1) is a simplified form of the **Koenig-Faustmann formula,** devised in Europe in forestry's early days, for estimating the bare-land expectation value of a forthcoming forest. The best rotation, it was said and has been said many times since, is the one that maximizes *V.* We are beginning to learn that this is true only for some special cases [1].

MARGINAL ANALYSIS: FIRST THREE STEPS

Enough of special cases. Let us focus our interest on an ordinary managed forest in which the timber capital is divisible. Divisibility consists of a thinning regime or a change in the number or arrangement of age classes in relation to rotation length or of *planning* for any such measures. These are measures that call for change in forest structure, so that the forest's stands interact. The optimal forest rotation can no longer be learned from the stand in isolation, because its optimum depends on the optima for other stands. Here is the forest counterpart of the principle of unity of the firm: The manager is obliged to look at all enterprises together, not one at a time. The manager is obliged to look for the best *forest* combination, to look at the individual stand as only part of a system.

Our example is a Douglas-fir forest comprising stands like the one studied in Chap. 17. Suppose that we are making a management plan for the forest. We want an optimal thinning regime and rotation length. As in Chap. 17, we assume that all wood—and therefore growth, too—has the same unit value. Consequently, it does not matter whether our plan is phrased in board feet or in dollars.

The model is marginal analysis and runs like this: Learn the firm's marginal cost of timber capital, or guiding rate of return; then estimate marginal revenue from thinnings and final harvest by calculating how much extra yield can be realized from a unit increase in growing stock (rotation length). Lastly, find where marginal cost and revenue are equal, where consequently both rotation and thinning are optimized.

Step 1 Learn the firm's guiding rate of return. It is 4 percent after tax. Note that this is different from the 5-percent rate of Chap. 17.

Step 2 Estimate timber growth in relation to stocking in stands of each age class and find the best stocking for each class. Figure 22-1 shows the prospective annual board-foot growth per acre during the forthcoming decade, for stands currently 30, 40, 50, and 60 years old. The 50-year curve is the same as the one drawn in Fig. 17-1. Allowing for the shift from board feet to dollars, the curve is the same as the revenue curve in Fig. 17-1(b). The broken line in Fig. 22-1, with a 4-percent slope, is analogous to the broken line in Fig. 17-1(b), with its 5-percent slope. Best stocking at 50

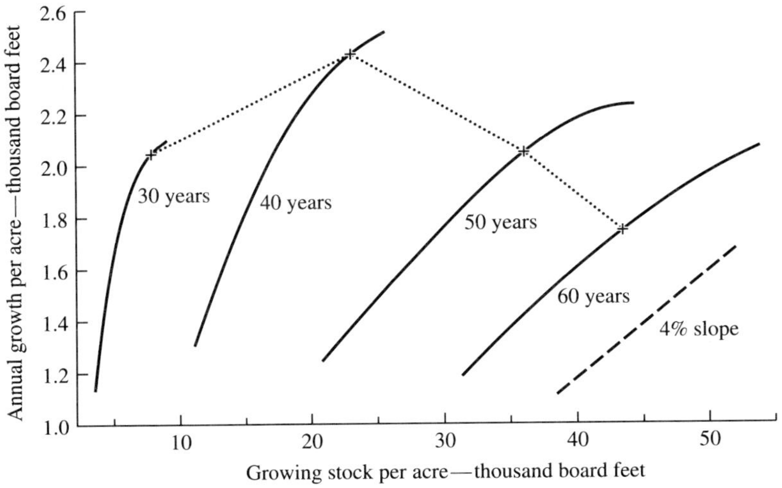

FIGURE 22-1
Best stocking of site-I Douglas-fir in western Oregon at 30, 40, 50, and 60 years of age. (*Source:* Duerr, William A., and Neils B. Christiansen: *Exercises in the Managerial Economics of Forestry,* O.S.U. Book Stores, Corvallis, Oreg., 1973, p. 80.)

years in Fig. 22-1, the point where the growth function has a 4-percent slope, is 36,000 bd ft per acre.

Best levels of stocking at 30, 40, and 60 years are similarly estimated and marked in Fig. 22-1. The line of dots linking the four marked points traces approximately the optimal development of the stand from age 30 to age 60—or to final harvest if it comes before age 60.

Step 3 From the best-stocking data, deduce how much timber can be harvested as thinnings. For this purpose, the four best stocks from Fig. 22-1 are recorded on line 1 of Table 22-1. These show the amount of the final harvest per acre if the rotation ends at any of the four stand ages. Then the prospective annual growth from Fig. 22-1 on the first three quantities of stock is entered on line 2(a), between the columns. The quantities are multiplied by 10 to account for all the new wood available during the decade. The manager will want to leave some of this new wood, line 2(b), to build up the stock to the desired level shown on line 1. The rest is available as thinnings during the decade, line 2(c). Ideally, from the financial viewpoint, the estimated quantity of thinnings equals the volume of financially mature trees in the stand, trees identifiable with the help of financial-maturity guides (Chap. 21).

MARGINAL ANALYSIS: FOUR FINAL STEPS

Step 4 Add final harvest cut (step 2) and thinning harvest (step 3) for each age class to estimate total harvest revenue for each rotation length. In this step, we shift our attention from one-aged timber stands to a many-aged forest made up of such stands. Assume that timber stands become merchantable at age 25 and that thereafter they acquire growing stock along the path traced on line 1 in Table 22-1 (the broken curve in Fig. 22-2). Or, to make the calculations simpler without changing the figures much, assume that stands carry the growing stocks represented under the stairsteps in Fig. 22-2: 8,000 bd ft per acre between ages 25 and 35, and so on. Assume that the line 2(c) quantities are available as thinnings during the decades 30–40, 40–50, and 50–60.

One forest that you can envisage is **regulated** for annual yields on a 30-year rotation. It comprises stands of 30 ages from 1 to 30. Let t = rotation length in years = number of age classes in regulated forest = number of acres in a cross section of the regulated forest containing one acre in each age class.

> How many age classes are there immediately after the harvest cutting of 30-year-old timber?

Now set down in column (2) of Table 22-1 the total annual yield from a 30-acre forest cross section, line 3(a). The yield in the current year, as already estimated, is 8,000 ft, all from the 30-year-old acre. Next year there will be a similar yield, from the

TABLE 22-1
FINANCIALLY BEST STOCKING AND GROWTH IN RELATION TO STAND AGE, AND THINNING AND FINAL HARVEST YIELDS AND MARGINAL YIELD PERCENT IN RELATION TO ROTATION AGE, OF A REGULATED DOUGLAS-FIR FOREST ON SITE I IN WESTERN OREGON

	Stand age or forest rotation, years						
Item (1)	30 (2)		40 (3)		50 (4)		60 (5)
1 Optimal stand yield (best stocking) per acre—board feet	8,000[a]		23,000[a]		36,000		43,500
2 Decade growth per acre—board feet:							
(a) Total		20,400[a]		24,200		20,450	
(b) Leave		15,000[a]		13,000		7,500	
(c) Available as thinnings		5,400[a]		11,200		12,950	
3 Total annual yield—board feet:							
(a) Per *t* acres	8,000		28,400		52,600		73,050
(b) Per managed acres	267		710		1,052		1,217
4 Growing stock—board feet:							
(a) Total per *t* acres	40,000		195,000		490,000		887,500
(b) Total per managed acre	1,333		4,875		9,800		14,792
(c) Marginal per managed acre		3,542		4,925		4,992	
5 Marginal annual yield per managed acre:							
(a) Board feet		443		342		165	
(b) Percent		12.5		6.9		3.3	

[a]Where these numbers come from: 8,000 is stocking at 30 years; 23,000 is desired at 40 years; desired growth, then, is the difference: 15,000; but actual growth (Fig. 22-1) is 2.040 times 10 = 20,400. This leaves the difference, 5,400, available for thinning.
Source: See Fig. 22-1.

stand now 29 years old. And so on. The average annual yield per acre of the entire forest is $\frac{1}{30}$ of 8,000, or 267 bd ft, line 3(b). This is the *total output per acre* on the *forest* production function. It is the **allowable cut,** or **allowable annual harvest** per acre of a forest managed under a 30-year rotation.

Similarly, numbers are derived for lines 3(a) and 3(b) in each of the other three columns. Line 3(a) is the sum of the yield in line 1 plus all previous thinning in line 2(c). For example, the yield of 52,600 bd ft in column (4) is the sum of 36,000 plus 11,200 plus 5,400. Line 3(b) is 52,600 divided by 50, the number of age classes for a 50-year rotation. Does it appear that the annual thinning yield is overstated, since each annual amount has, in effect, been multiplied by 10 in order to derive lines 2(a), 2(b), and 2(c)? What you need to remember is that there are 10 *acres* available for thinning in each of the age intervals in Table 22-1. You can think of 10 *years'* worth of thinning being done on these 10 acres each year.

Step 5 So much for outputs; what about inputs? We must calculate the total growing stock (capital input) necessary to produce each of the step 4 yields.

Take the 30-year column, for example. The total growing stock on the entire forest, line 4(a), is 8,000 ft on each of 5 acres: 40,000 ft, the area under the left half of the first stairstep in Fig. 22-2. Only 5 acres are involved because we are assuming that stands have no merchantable growing stock until they reach age 25, just 5 years (acres) from age 30. And so line 4(b), the average stock per acre, is $\frac{1}{30}$ of 40,000: 1,333 bd ft. This is the *total variable input per acre* on the *forest* production function. It is the input on which the output of 267 ft is obtainable.

Take the 50-year column as another example. Line 4(a) shows the area under the stairsteps up to age 50: 10 acres at 8,000 ft each, plus 10 acres at 23,000 ft each, plus 5 acres at 36,000 ft each. Then line 4(b) is $\frac{1}{50}$ of line 4(a).

Step 6 From step 5, figure the marginal growing-stock investment required at each rotation length. Marginal investments are figured as differences in line 4(b) and are recorded in line 4(c). These are the *marginal inputs per acre of forest.*

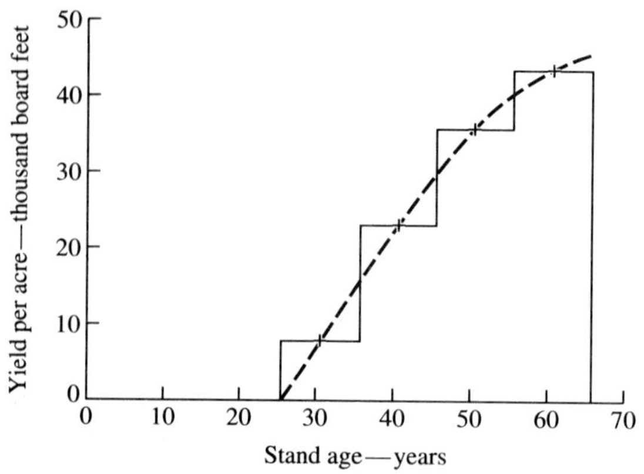

FIGURE 22-2
Optimal stand yield (best stocking) in relation to age. (*Source:* Table 22-1, line 1.)

Then from step 4, figure the corresponding annual revenues, or outputs. These are calculated as differences in line 3(b) and recorded in 5(a).

The marginal amounts respond to this question: What would happen if the manager should restructure the forest, changing the rotation length, *t?* The marginal input says what difference the restructuring would make in average growing-stock investment per acre. The marginal output says what difference the restructuring would make in average per-acre annual return on the growing-stock investment.

Step 7 From an analysis of the marginal amounts in step 6 comes insight into the question of the best rotation—and thinning—regime. The marginal yield percents of line 5(b) are derived by expressing line 5(a) values in percent of line 4(c). By virtue of the assumption that the unit value of all wood is the same, these are marginal value growth percents analogous to those in Fig. 17-2(a) and can be compared with the firm's guiding rate of return, 4 percent. Best rotation (Fig. 22-3) turns out to be 53 years.

> So far as you can on the basis of Table 22-1, describe the best thinning regime.

CHARACTERISTICS OF THE MARGINAL ANALYSIS

What are some of the characteristics of the marginal-analysis model just used for responding to the financial-maturity question for divisible timber capital?

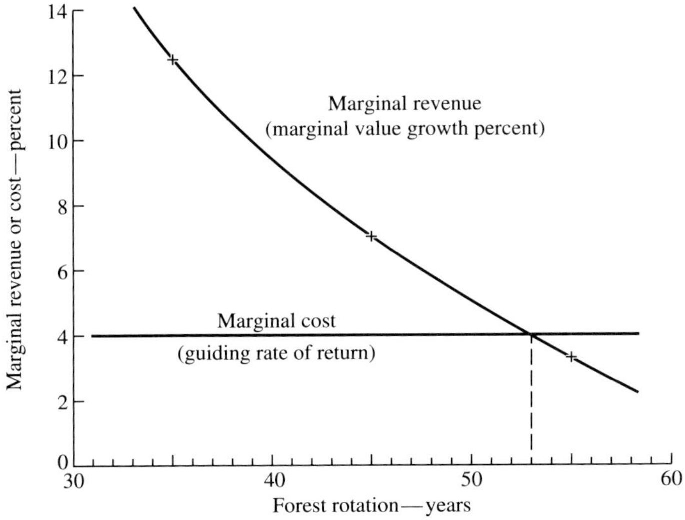

FIGURE 22-3
Best rotation of a regulated Douglas-fir forest on site I in western Oregon. (*Source:* Table 22-1.)

One characteristic of the particular model we used is that the forest cutting cycle is one year. This is not a necessary characteristic of such models. We saw on page 153 that, as a special case, the cycle can be as long as the rotation. But in the general case that concerns us here, all that's required is that the cycle be shorter than the rotation, and thus that the age-class distribution of the forest be a function of t.

For insight into another characteristic of our model, observe that in the Koenig-Faustmann model, marginal revenue percent is the current annual growth percent of the stand corrected by the multiplier $[(1+i)^t - 1] \div (1+i)^t$ [column (7) in Table 21-1]. In this multiplier, t has a single value, equal to the number of years in the rotation being studied. But in our marginal analysis, the revenue percent is derived with t as a variable divisor. For example, see lines 3(b) and 4(b) in Table 22-1. This use of t recognizes the penalty of carrying heavier (more costly) forest growing stocks as the rotation is lengthened, whereas the Koenig-Faustmann model has nothing to say about growing stocks in any but one-aged forests (pages 153 and 154). As a result, our marginal-analysis model derives a lower revenue percent at any given t—that is, a shorter rotation—than the Koenig–Faustmann model.

If the cost of achieving regulation varies with rotation length, it is to be taken into account as a marginal cost. And if, as is likely, this cost increases with rotation length, it becomes another cause of a shorter marginal-analysis rotation.

Still another characteristic of our model is that it maximizes the current net income from the forest and therefore the capitalized value of the income. The model limits investment in growing stock to the amount that just equates the marginal return on investment to the marginal cost, the firm's guiding rate of return. It follows that any greater investment would yield a lower marginal return, involving a lower net revenue. And any lesser investment would forgo satisfactory net revenues, also involving lower revenue.

> Early in this chapter, the firm's guiding rate of return was said to be 4 percent *after tax.* Study the revenue and cost estimates in the chapter. Nothing is said about taxes. Explain how it is that we can use 4 percent as a guiding rate in our calculations.

But the marginal-analysis model has at least one weakness, the amount of data and effort that may be demanded to apply it. If the manager follows a thinning or other cultural regime and wants to find the best combination, then there may be no way that is more attractive unless local practice provides reliable guides. If cultural regimes are not in the picture, but only some stab at regulation, and if simplicity is to govern choice of a model, then the manager can use the Koenig-Faustmann formula and reduce its finding by 10 percent or so.

In any case, there is this to hold in view: A financially optimal regime can still lose money. The cost of regulation may contribute to loss. The question of profit or loss in forestry is raised in Chap. 25.

SUMMARY

In a few words, just as one can work out the best stocking of a single-aged stand, so one can analyze stands of successive ages and picture them as the components of a regulated forest. And just as one can calculate the stand's marginal revenue and cost (per unit of variable input), so one can make analogous calculations for the regulated forest. Forest revenue and cost lead to estimates of the financially best *forest* stocking, thinning regime, rotation, and allowable timber harvest. The benefit of the present chapter is the insight it gives into the general dimensions of the problem. You are learning how to discover what these general dimensions are.

REFERENCES

1 Oderwald, Richard G., and William A. Duerr: "Koenig–Faustmannism, a Critique." Comments by Theodore E. Howard, Sun Joseph Chang, Jay O'Laughlin, and the Authors. *Forest Science,* March 1990, pp. 169–185.

CHAPTER 23

SOME FINANCIAL INVESTMENT CRITERIA

THE QUESTION

The purpose of this chapter is to return to a question that was touched on at the beginning of Part Three: How does one judge whether an enterprise or project is financially attractive? How does one identify the most attractive alternative? A study of the question in greater depth will help prepare for chapters that follow.

A project's attractiveness is a matter of its promise. What can the firm and the community look forward to? The reply depends—*always*—on predictions, and the future holds few sure things. One needs to remind oneself of this at every step in this chapter. There are two chapters on forecasting methods at the start of Part Four.

A NOTE ON EQUITY

Before getting down to the specifics of investment criteria, it behooves us to look around, if only briefly, at the context of the matter. What considerations—financial or, if you like, economic—are *socially* relevant when a choice among alternative investments is at hand? One is the consideration to which this chapter is mainly devoted: How *efficient* is each alternative? How great is its benefit in relation to the sacrifice it requires in the form of cost? But there is another consideration: *Who*—what individuals and groups—bears a project's costs; and who enjoys its benefits? Largely the same persons or largely different ones? If different, do they differ in affluence? Perhaps the wealthy benefit at the expense of the poor, or perhaps net benefits are shared. Does the incidence of benefits and costs differ with location? The whole world can enjoy forest products derived from a small, heavily exploited region. Do benefits and costs occur at different times? Perhaps a project benefits the current generation, leaving costs to be borne by descendants. Or perhaps the other way around: a forest established today, to be enjoyed later. All these questions about *who* are questions of **equity.**

Each culture has its standards respecting efficiency and respecting equity. These standards, in effect, define efficiency and equity by stating or implying what relationships of benefit to cost and what incidence of benefits and costs are socially acceptable. They may draw distinctions between private and public projects or among private projects undertaken by different groups. Chapter 20, you may recall, in the section on Governments as Firms, traces the distinction in the U.S., once wide, now narrowing, between private and public guiding rates of return. Other sections of the same chapter point out differences within the private sector in this country and abroad.

By and large, prescriptions for economic efficiency are much more clearly written in most nations of the world than are prescriptions for economic equity. The great difficulty of understanding the forces involved in equity and of giving clear voice to policies respecting equity is implied in Chap. 3, under the topic, "Relation to Income Distribution," pages 16 and 17.

It can be, then, with some relief that you turn to the remaining sections of the present chapter, which are devoted entirely to the topic of efficiency. But don't forget that this is only one-half of the story, and unquestionably the less human half.

> How do you explain the fact that equity is less clearly defined than efficiency?

ANNUAL NET REVENUE AS A CRITERION

How does one judge whether an enterprise or project is financially attractive? One must have two pieces of information. First, what are the alternatives to this investment and how do they compare in terms of risk? Second, what is a suitable criterion for weighing the alternatives? The second question depends on the first.

The simplest financial criterion is net revenue (benefit) per unit of time—for example, per year. By this criterion, an opportunity that promises an annual revenue greater than its annual cost is attractive, and the one that promises, at equal risk, the greatest excess of revenue over cost is the most attractive. Here is a criterion used frequently in Part Three. To maximize total annual net revenue is to equate marginal revenue and cost per unit of input or per unit of output within the region of diminishing return. The point is elaborated in Chap. 17. The most recent example is Fig. 22-3.

Recall that to equate marginal revenue and cost is not necessarily to earn any net revenue. If fixed costs are high, it may be simply to minimize net loss.

The criterion of annual net revenue is valid provided two conditions hold: First, the alternatives' efficiency measures must differ only in respect to promised annual net revenue. If they differ in any other respect, then this respect, too, is relevant and net revenue is an insufficient criterion. Second, each alternative must promise an annual net revenue that is unchanging over time. If change is in prospect, then the analyst is obliged to use some device for transferring values in the time dimension: one of the formulas, say, in Chap. 19. The most recent example is the use of the Koenig-Faustmann formula to set a rotation, page 160. Now for some further examples.

INTERNAL RATE OF RETURN

A firm invests $100 in Project B. This project promises to pay (net of taxes and any other coincident expenses, if any) $10 at the end of the first year and of every following year. One can say, then, that Project B promises an internal rate of return (marginal efficiency of investment) of 10 percent per annum. This is the same sort of internal rate as was introduced on page 122. And it's the same sort of rate as that promised by Tree in Chap. 18.

Notice that 10 percent in the present case is not merely $10 as a fraction of $100. It is also the interest rate at which the total revenue from the project ($10 yearly) and the total investment cost of the project ($100 today) are equated at any given point of time. Thus total cost as of now is $100; total revenue as of now is $10 ÷ .10 [Chap. 19, formula (5)], or $100. No other rate of interest will equate these two.

To estimate the internal rate as that which equates all revenues to all costs at a point of time is a useful procedure where revenues or costs are changeable in amount or timing interval, as they are in the following illustration of Project A. In this project, the firm invests $50 now, $25 in a year, and $25 in 3 years. The firm's revenues are $10 (net, as before) in 2 years and every 2 years thereafter for a total of four payments, and then $10 every year. What is Project A's internal rate of return? Let's find it by trial and error. First, try 10 percent.

Today's value of revenues:	
First four $10 payments:	
$10 \div (1.10)^2 + 10 \div (1.10)^4 + 10 \div (1.10)^6 + 10 \div (1.10)^8 =$	$25.40
Subsequent payments:	
$(10 \div .10) \div (1.10)^8 =$	46.65
All payments	$72.05
Today's value of costs:	
$50 + 25 \div (1.10) + 25 \div (1.10)^3 =$	$91.51

As you might guess, 10 percent is too high, for revenue is going to rise faster than cost as interest rate falls. So try some successively lower rates. The results are plotted and curved in Fig. 23-1(a). The internal rate of return proves to be about 8.2 percent for Project A. (Disregard "current net worth" for the moment.)

> Does it matter what point of time is chosen for estimating the internal rate of return? That is, does it matter whether prospective revenues and costs are brought back to the present or to 10 years ago or to 10 years hence—will the estimated internal rate be the same?

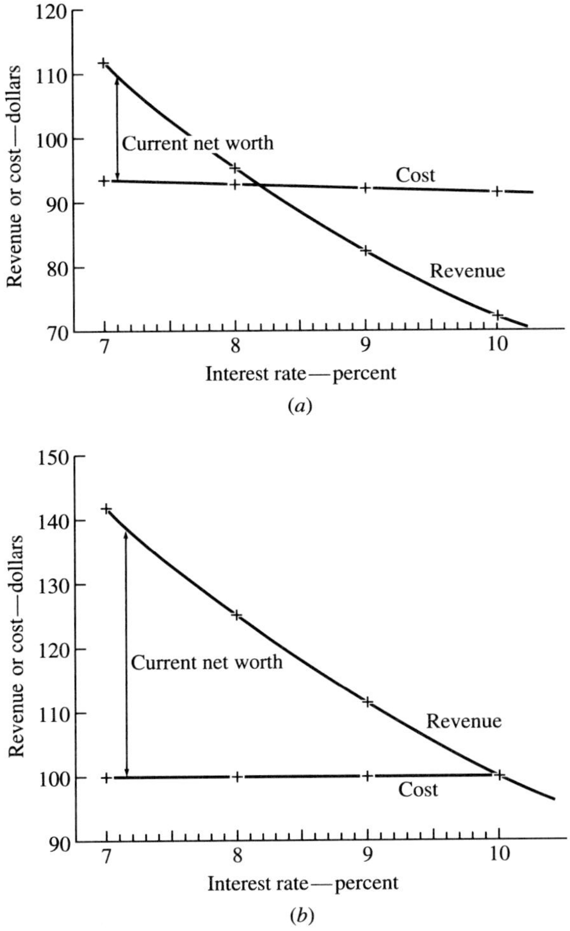

FIGURE 23-1
Today's capitalized value of the anticipated costs and revenues of (a) Project A and (b) Project B over a range of interest rates.

Does Project A "pay" in a financial sense? Does Project B "pay"? Surely, if the firm's guiding rate of return is 8 percent or less, they both pay. Furthermore, other things equal, the firm will invest in Project B, with its higher internal rate, before turning to Project A.

In the foregoing comments and in those that follow, assume that all rates of return and all dollar amounts underlying them are comparable: Assume that they are after-tax rates, that the amounts are expressed in dollars of constant value, and that revenues are net of taxes and other coincident outlays, if any, not separately recorded as costs.

CURRENT NET WORTH

An investment project's internal rate of return can be calculated (though not interpreted) without knowledge of the firm's guiding rate. A reliable estimate of the internal rate is usable by any firm. Now consider two financial investment criteria the magnitudes of which depend on the firm—that is, on the firm's guiding rate. These criteria are (1) current net worth and (2) benefit-cost ratio.

The **current net worth** of an investment project is today's capitalized value of the project's anticipated revenues minus today's value of anticipated costs. In other words, it is the capitalized promised flow of net revenue. Consider Project B, and take a 7-percent guiding rate for illustration:

Capitalized revenues:	10 ÷ .07 =	$142.86
Capitalized costs:		100.00
Current net worth		$ 42.86

At 8 percent, the current net worth is $125 minus $100. At 9 percent, it is $111.11 minus $100; at 10 percent, $100 minus $100, or zero. The revenues and cost just calculated are plotted and curved in Fig. 23-1(b). The current net worth at any interest rate from 7 to 10 percent is the reading on the revenue curve minus the reading on the cost curve.

For Project A, revenue and cost data are already in hand. Project A's current net worth appears in Fig. 23-1(a) just as Project B's appears in Fig. 23-1(b). A rule comes into view for relating the internal and guiding rates of return and the current net worth: *At the interest rate that equals the internal rate of return, the current net worth is zero.* The current net worth is a manifestation of profit: It is the capitalized profit. In order for there to be profit on a capital investment, the return on the capital (internal rate of return) must be greater than the cost of the capital (guiding rate of return).

> The guiding rate of return has been dubbed by many, including your present author, the "alternative rate of return." This is a poor term because it focuses on capital demand and neglects supply. Explain.

There is a first cousin of the current net worth that is widely useful in forestry. It is calculated by capitalizing all the promised revenues from forest land and then subtracting the capitalized value of all the anticipated costs except the cost of the property itself. The result is an estimate of what the property is worth to the firm—say, the firm that is contemplating acquisition. This first cousin is termed the **forest expectation value**—or, if the property is cutover land awaiting reforestation and management, the **land expectation value** (sometimes called "soil expectation value").

Go back to Project A. At 7 percent interest, it has a current net worth of $18.22 per $100 total outlay, signifying that a firm with a 7-percent guiding rate of return is promised a capitalized net profit of $18.22 from Project A. Now suppose that Project A is a forestry project and that the first of the three cost items is the estimated outlay for the acre of bare land where the project is to be installed. At 7 percent, the land expectation value is

$$\begin{aligned}&[10 \div (1.07)^2 + 10 \div (1.07)^4 + 10 \div (1.07)^6 + 10 \div (1.07)^8)\\&+ (10 \div .07) \div (1.07)^8] - [25 \div 1.07 + 25 \div (1.07)^3]\\&= \$68.22\end{aligned}$$

Thus \$68.22 is the most that the firm can afford to pay per acre of land and just break even. If the firm can get the land for \$50 per acre, it stands to enjoy an \$18.22 capitalized profit, the same conclusion that was reached just above.

For the firm with an 8.2-percent guiding rate of return, the expectation value of the land is \$50 per acre, and if this is the price of the land, the firm's profit will be nil. This, too, confirms the earlier finding.

BENEFIT-COST RATIO

The **benefit-cost ratio,** in financial terms, is today's value of all prospective revenues divided by today's value of all prospective costs. Note the relation of the benefit-cost ratio to the current net worth: The latter is the benefit-cost *difference.* For Project B, at 7-percent interest [Fig. 23-1(b)], the benefit-cost ratio is \$142.86 ÷ \$100.00 = 1.43. At 8.5 percent, the ratio is 1.18; at 10 percent, 1.00; and at higher interest rates, less than 1. *At the interest rate that equals the internal rate of return, where current net worth is zero and the expectation value of property equals its price, the benefit-cost ratio is 1.*

To tie things together a little more firmly at this point, recognize that the foregoing discussion of Project A and Project B implies that the projects have a fixed size (\$100) and fixed internal rates of return (8.2 and 10 percent, respectively). Either that or the projects can absorb additional investments at the same rate of return, so that the marginal rate and the average are equal: 8.2 percent for Project A, 10 percent for Project B.

> Can you cite examples of forestry projects that are not subject to diminishing return—projects that can produce a constant rate of return on successive inputs of investment, at least over a significant range of investment? What technological characteristic do all your examples have in common?

Now consider a contrasting case, Project C, which is subject to diminishing return, as are all the timber cases in earlier chapters. The firm's guiding rate is 10 percent. If it invests \$100 [capitalized cost, column (1) in the tabulation below] in Project C, the prospect is a \$13 annual return, column (3); a \$130 capitalized return, or benefit (5); a 13 percent internal rate, (7); a \$30 current net worth, (9); and a 1.3 benefit-cost ratio, (11). Different initial investments promise different results:

Initial investment		Annual return		Capitalized return		Internal return		Current net worth		Benefit-cost ratio	
Total (1)	Marg. (2)	Total (3)	Marg. (4)	Total (5)	Marg. (6)	Avg. (7)	Marg. (8)	Total (9)	Marg. (10)	Avg. (11)	Marg. (12)
$ 0		$ 0		$ 0		—		$ 0		—	
	$100		$13		$130		13%		$ 30		1.3
$100		$13		$130		13%		$30		1.3	
	$100		$12		$120		12%		$ 20		1.2
$200		$25		$250		12½%		$50		1.25	
	$100		$11		$110		11%		$ 10		1.1
$300		$36		$360		12%		$60		1.2	
	$100		$10		$100		10%		$ 0		1.0
$400		$46		$460		11½%		$60		1.15	
	$100		$ 9		$ 90		9%		$−10		0.9
$500		$55		$550		11%		$50		1.1	

Clearly, the scale at which to stop investing is $300 or so, where the marginal internal rate (4) has fallen to 10 percent, the marginal current worth (10) is zero, the marginal benefit-cost ratio (12) is 1.0, and total current net worth (9) is as large as possible. In terms of capitalized values—marginal revenue, or benefit (6), and marginal cost (2) are equated. Marginal value growth percent (8) equals the guiding rate of return, and since this is the region of diminishing return, total net revenue is maximized.

Thus, with Project C and its diminishing return, the rules that are valid for Projects A and B are called off. At the interest rate that equals the (marginal) internal rate, where (marginal) current worth is zero and the (marginal) benefit-cost ratio is 1, the project doesn't just barely break even; it is at its profit peak! Diminishing return unhitches marginal value from average (total) value, just as a timber-size premium (page 174) unhitches growth value from stock value.

RELATION OF AVERAGE VALUE TO MARGINAL VALUE

The foregoing tabulation for Project C raises the question of whether there is some general rule governing the relation of average to marginal quantity or value.

Indeed, there is such a rule, and we can bring it out with the help of an experiment. You have a bucket containing two stones that weigh a total of 11 ounces: 5.5 ounces on the average. Add a third (marginal) stone with a weight of 7 ounces. The total weight rises to 18 and the average to 6. A fourth stone weighing 6 marginal ounces puts the total at 24 and holds the average at 6. Further stones lighter than the average draw the average down; heavier stones raise it:

Number of stones	Total ounces	Average ounces	Marginal ounces
2	11	5.5	
			7
3	18	6.0	
			6
4	24	6.0	
			5
5	29	5.8	
			4
6	33	5.5	
			2
7	35	5.0	
			5
8	40	5.0	
			6
9	46	5.1	
			6
10	52	5.2	

Hence the rule: *The marginal attracts the average, irrespective of how (or whether) the marginal is moving.*

PAYOUT PERIOD

There is just one other investment criterion to be recognized here: the **payout period.** This is the number of years from the start of a project until the accumulated sum of its revenues has come to equal the sum of its costs. Project B, at $10 per year, pays out its cost in 10 years. For Project A, the payout period is 14 years. Both projects have $100 to pay out; no account is taken of the difference in timing of this investment. No compound interest charge is exacted for late timing, as it is with other criteria, and this rule applies to the revenues as well as the costs. Thus, for even the most complex case, figuring the payout period is simply a matter of keeping track of the **cash flows** from year to year. Using Project A as an example gives us the following:

Year	In	Out	Balance	Year	In	Out	Balance
0	—	$50	$−50	8	$10	—	$−60
1	—	25	−75	9	10	—	−50
2	$10	—	−65	10	10	—	−40
3	—	25	−90	11	10	—	−30
4	10	—	−80	12	10	—	−20
5	—	—	−80	13	10	—	−10
6	10	—	−70	14	10	—	Payout
7	—	—	−70				

In terms of payout, Project B ranks ahead of Project A because, disregarding any compounding or discounting, the firm gets its money back sooner from Project B. The investment in B is at risk less long than the investment in A. The allowance for risk in the payout-period criterion, and the simplicity of this criterion, often outweigh its naiveté in the minds of analysts and make for its popularity.

Does it strike you that the criterion of the payout period is merely the criterion of simple interest? This *is* the case! Project B's internal return in simple interest is $100 ÷ 10 years = $10 per year = 10 percent, the same as the compound rate because the whole investment is made at time zero and the return is constant. Project A pays about 7-percent simple interest—that is, it pays out $100 in 14 years. Here the simple rate of return is lower than the compound rate because the simple rates take no credit for the delay in part of the $100 investment and reflect only the 4 years (1, 3, 5, 7) when there is no revenue. You and I are obliged to argue that so long as the returns from an investment can be reinvested—that is, so long as there can be interest on interest—we have the fact of compound interest and there is no escaping it.

Show that if I is the initial investment and P is the payout period in years, the rate of simple interest (in decimals) returned on the investment is $1 \div P$.

SUMMARY

To review, this chapter opens by recognizing that equity is an essential social criterion for choosing among alternatives. The chapter then turns to efficiency as a criterion and defines six models for weighing the financial attractiveness of alternative investment projects, or resource combinations: annual net revenue, internal rate of return, current net worth, land expectation value, benefit-cost ratio, and payout period.

The models may not all rank a set of alternatives in the same order. In this event, it is for the firm to decide which modeled criteria are best suited to the circumstances [1].

Whatever model is chosen for use, it cannot guarantee a reliable weighing of alternatives. Reliability depends on the forecasts put into the model.

REFERENCES

1 Christiansen, Neils B.: "Models for Time Adjustment of Costs and Revenues," *In* Duerr, William A., Dennis E. Teeguarden, Neils B. Christiansen, and Sam Guttenberg, eds.: *Forest Resource Management, Decision-Making Principles and Cases,* O.S.U. Book Stores, Corvallis, Oreg. 1982, pp. 131–146.

CHAPTER **24**

WHAT AFFECTS THE BEST COMBINATION?

THE PROBLEM

In our study of financial maturity, optimal stocking, and other instances of the best combination in genetic production, we make a good many simplifying assumptions—about product values and other revenue matters, about costs, about forest behavior, and about the firm's behavior. What difference does it make if any of these assumptions is relaxed? Here is a good question to ask at this point, because the replies serve all four chapters in Part Three that concern financial maturity and best stocking. These are the chapters that seek a best combination where forest capital is the variable input to genetic production. Furthermore, the replies serve to clarify the story in many a chapter in parts beyond Part Three.

The problem-solving model used throughout this chapter is Fig. 24-1, which is generalized from Fig. 17-2(a), Table 21-1, and Fig. 22-3. You are to view both revenue and cost in Fig. 24-1 as *real* percentages, applicable to dollars of constant value. The question then is, "Does a difference in variable x_1 change the picture in Fig. 24-1? How about a difference in x_2? And so on. As each variable is studied, the others are held constant.

Why should it be necessary to hold the others constant?

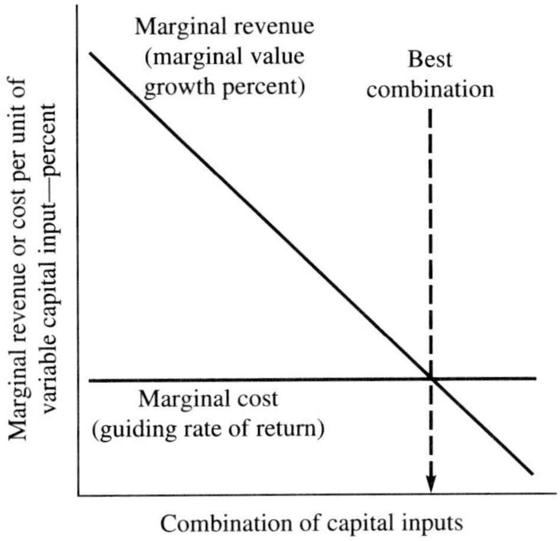

FIGURE 24-1
Best combination of capital inputs in genetic production.

DIVISIBILITY OF FOREST CAPITAL

The variable to take first is the character of the forest capital, whether the capital is divisible or indivisible. If the capital is divisible, marginal value growth percent is used without adjustment. But if the capital is indivisible, the growth percent is adjusted downward with the multiplier $[(1 + i)^t - 1] \div (1 + i)^t$. This adjustment recognizes the opportunity cost of postponing the liquidation of capital input and reduces the input represented in the best combination. The rotation is shortened; stocking is reduced; thinning is intensified.

TIMBER-SIZE VALUE PREMIUM

A second variable is price, or unit liquidation value of the tree. Retaining the assumption that all wood has the same unit value, regardless of tree size or quality, assume a new unit value—say, 50 cents per cubic foot instead of one dollar. Does it matter? No; neither the guiding rate nor the value growth percent is affected. The point to note in this case is as follows: First, this is genetic production in which genetic identity holds, so that factory and product are the same substance. Second, forest capital is the variable input to production. Third, all forest capital has the same value per unit of measure. Therefore, any difference in product price level (representing revenue) is accompanied by an equal difference in the capital value of the factory on which interest must be charged (representing cost). And so cost and revenue are locked together. A difference in one means an equal difference in the other: a standoff. The principle: In production where genetic identity holds and forest capital is the variable input, product price *level* is irrelevant to the best combination.

Now abandon the assumption that all wood has the same unit value, regardless of tree size or quality. To be sure, the assumption makes some sense for trees to be used solely as pulpwood, firewood, or particles for composition board, products little sensitive to knottiness or other sound defect. But when it comes to sawlogs and veneer logs, unit value generally increases with diameter. And surely the same relationship holds where little trees are usable only for firewood, but the trees become merchantable for higher-value products as they grow bigger. Furthermore, if logging cost per unit of measure diminishes with increasing tree size, the result is a rising tree value even if end-product value is constant.

Let's say a 10-in d.b.h. (diameter at breast height) pine is worth 50 cents per cubic foot. When the tree grows to 12 in, it is worth 75 cents per cubic foot; and at 14 in the unit value is one dollar. What is the unit value of the tree's *growth?*

	Tree d.b.h. (in)				
	10		12		14
(1) Tree volume (cubic feet)	15.0		24.0		33.0
(2) Tree value per cu. ft. (dollars)	.50		.75		1.00
(3) Total tree value (dollars)	7.50		18.00		33.00
(4) Growth in value (dollars)		10.50		15.00	
(5) Growth in volume (cubic feet)		9.0		9.0	
(6) Value of growth per cu. ft. (dollars)		1.17		1.67	

In this particular case, growth has a unit value, line (6), approaching twice that of growing stock, line (2), and it is the rise in unit value with tree diameter in line (2) that produces this result. If all three figures in line (2) were the same, then these figures would be repeated in line (6). In other words, a **timber-size premium** that takes the form of a rise in unit value with tree diameter lifts the unit value of growth above that of stock. Thereby it raises the revenue function of Fig. 24-1, determining an older and greater capital tree or stand than otherwise.

The point to stress about a timber-size premium is that it unfastens the unit value of growth from that of stock: It unfastens the source of revenue from the source of cost, the two entities that are inseparable in the absence of a premium. The premium accomplishes the separation because stock exists at a point in time, whereas growth is a flow during a subsequent interval of time.

What is termed a "timber-size premium" here has been termed a "quality premium" elsewhere. The latter term slights at least one cause of separating the unit value of growth from that of stock. Explain.

EXTERNALLY CAUSED TIMBER-VALUE CHANGE

Now leave the subject of value differences resulting from a timber-size premium and turn to changes in value arising externally to the tree.

Suppose the firm anticipates a rising or falling real price. For instance, imagine that the manager learns of two new wood-using plants to be built in the vicinity and expects fresh competition for wood raw material to drive up the price. Does this affect the firm's best combination?

Yes; because a promised change in real price is a change in the tree's real value growth percent. If the manager thinks that prices will rise, the firm can respond by postponing the harvest to take advantage of higher returns. In terms of Fig. 24-1, the manager sees an upward shift in marginal revenue and a new best combination at an older age and greater stocking, the same as when a timber-size premium is introduced. Observe that it's what the manager *expects* will happen that decides the move. Whether the anticipation materializes is irrelevant.

Furthermore, observe that if the change *does* materialize and the manager thinks that there will be no further change, the best combination drops back to where it was before. Or if change *doesn't* materialize and the manager then concludes that it will not, the best combination drops back. For the best combination to stay at the new point, the expectation that brought it there must remain in force year after year: The manager must expect successive changes. The rule: Any change in real-price-level expectation produces a change in the best combination.

What is the effect of inflation on the best combination? Suppose that inflation is under way, so that the value of the dollar is deteriorating. Thus the **current price** of timber is rising, even if its real price is unchanging. Or suppose there's a rise in the inflation rate or an expected rise. Is the picture in Fig. 24-1 changed in any of these cases? No; not if one assumes that inflation has no bearing on the real guiding rate; and not if inflation deals evenhandedly with tree values and other values.

But what if Fig. 24-1 were viewed as depicting *current* revenue and cost percentages? Wouldn't a rise, say, in the anticipated inflation rate affect the best combination? The reply is still no: Conceptually, the curves in Fig. 24-1 would shift upward equally, leaving their intersection at the same combination of inputs as before.

SILVICULTURAL MEASURES

The tree's value growth percent can be affected by silvicultural measures and related procedures and by other circumstances over which the forest manager has some control.

One such measure, not common in American forest practice, is artificial pruning. If pruning is financially attractive, its promise lies in raising the quality of the tree and thus tree value. It is a means for heightening the timber-size premium. What is its effect on the best combination? In many cases, the optimal pruning program is begun early enough in the life of the tree so that the achievement of quality goals—and, thereafter, a dropoff of value growth percent and the arrival of the best combination—comes at younger ages than otherwise. Thus the firm's waiting period is shortened while its reward is enhanced [2].

A more common measure is fertilization. Here the effect on the best combination is much the same as that of pruning. Fertilization is a means of raising the quality of the forest site. Its rewards are much the same as those available on naturally high site: the earlier arrival of each stage in forest stand and tree development, including the stage of maturity. But the productivity of the site is high enough to produce more valuable trees at this early maturity than could have been grown economically in a longer period on lower site. In Chap. 26, where forest fertilization is analyzed pointedly, this conclusion is verified.

Another influential silvicultural measure is tree release, as by thinning out the trees' competitors so as to give the trees more elbow room [3]. What is the effect on the best combination? If one imagines a forest manager who keeps an eye on the trees until they are about to become mature and optimally stocked and then runs out and releases them, one can easily believe that the optimum will thereby be postponed. Indeed, when one stops to think of it, the same postponement *could* be achieved by a late timing of pruning or of fertilizing.

But the rational manager will want to avoid postponement, for it is a source of extra cost. As with pruning or fertilizing, the manager ideally aims at a thinning that will commence early enough and be repeated often enough to produce a result better than otherwise in a time shorter than otherwise. The manager's goal is to maintain a timber stand of maximum value, with every tree growing in value at a rate at least equal to the guiding rate, and to declare all remaining trees mature when they are better replaced.

> Granted that postponement raises cost, may it not raise revenue, too, and in an even greater amount?

But, of course, the manager operates under financial constraints that are wider than just the performance of the tree or stand. There may, for example, be a wood-using plant to supply with raw material, and the plant's requirements may fluctuate. At times, trees otherwise ready for harvest will be held back; at other times, some overcutting may be necessary.

And then there is the influence of forest values other than those of timber alone. Is scenery a consideration? If so, its effect may be to lift the trees' revenue function and postpone silvicultural operations. Is the value of deer a consideration? If so, its effect may be to depress the revenue function.

In general, any tree revenue that is variable with the timing or amount of the harvest affects the trees' marginal value growth percent and best combination of capital inputs. And any variable cost chargeable against tree revenue is likewise influential. The property tax may be a case in point, as explained in Part Six.

But there are many costs, and revenues, too, that are fixed with respect to harvest time and quantity and are thus irrelevant. You may have run across cases in point. Perhaps you have heard some forestry-minded landowner explain his program in this way: "I paid such a high price for my land that I have got to manage it intensively if I am to come out ahead." To such a declaration, the sensible response (whether aloud or to oneself) is, "How so? If you are concerned with financial aspects, then you should try to plan out your financially optimal forest program. Whatever the program is, maybe it will pay off your land investment and maybe it won't, but you can't improve your position by departing from the optimum. What you paid for the land is irrelevant."

GUIDING RATE OF RETURN

Finally, have a look at influences on the best combination that work through the firm's guiding rate of return. Here we can fall back on the analysis in Chap. 20: The guiding rate depends on the national monetary context, the nature of the firm and its alternatives, including those transfer costs that are relevant, and the firm's uncertainties about the future. Anything that makes for a higher guiding rate makes for lower optimal capital inputs. And the lower the guiding rate, other things equal, the higher the inputs.

Thus, in general, the guiding rate is an index of the firm's economic conservatism: the firm's tendencies toward **conservation,** or saving. The firm with a very low guiding rate of return is the firm with very modest, undemanding alternatives. Such a firm is not pressed for cash. Its annual budget for consumption may be high, with only its most trivial wants forgone. In its eyes, the marginal utility of money is low. The firm's marginal propensity to consume is low; to save, high. Here is the ideal conservationist. If the firm owns forest land, the firm may be expected to retain its trees to a comparatively old age and to carry heavy growing stocks. It may be expected to practice capital-intensive management and to enjoy a high allowable annual harvest.

Concerning the firm with a very high guiding rate of return, all the opposite of the foregoing can be said. Here is a firm with its back to the wall financially. Its tendencies toward conservation are negligible. Refer back to the *Treasure Island* illustration, page 144.

An interesting and common case is the firm whose fortune fluctuates over time, and along with fortune, the guiding rate of return and the tendency to conserve. Most small-tract forest is owned by such firms. When the owning family suffers financial losses, illness, death, or other emergency—which it inevitably does, sooner or later—it turns to the woods as a source of cash. Tree's financial maturity suddenly arrives. The highest guiding rate in the fluctuating series turns out to be the effective rate, for Tree, once cut, cannot be managed at any other guiding rate. Here is a way of interpreting one of the nation's outstanding forestry phenomena: vast acreages which in the aggregate are simultaneously the target for derogatory comment and the source of inexhaustible wood supplies.

> It has been said that the "problem" of the nonindustrial private forest holdings is a cash-flow problem. Do you agree? Explain.

SUMMARY

To make a long story short, the best combination of forest capital inputs in genetic production is affected by the trees' marginal value growth percent and the firm's guiding rate of return. Value growth is a function of certain forms of price change, but not of price level. It is a function of some of the nontimber values that the firm places on the forest. It is a function also of the trees' physical growth, as influenced in part by silvicultural measures. The guiding rate is affected by the monetary context and the firm's circumstances, as sketched in Chap. 20.

In general, the best combination is influenced by any cost or revenue that is variable with capital input. This principle has been enunciated by the Forest Service [1], using West Coast Douglas-fir as the illustration.

REFERENCES

1 Division of Forest Economics Research. *Timber Trends in Western Oregon and Western Washington,* U.S. Department of Agriculture, Forest Service Res. Pap. PNW-5, 1963, pp. 11–16.

2 Fight, Roger D., James M. Cahill, Thomas D. Fahey, and Thomas A. Snellgrove: *Financial Analysis of Pruning Coast Douglas-Fir,* U.S. Department of Agriculture, Forest Service Res. Pap. PNW-RP-390, 1987, 17 pages.

3 Miller, Richard E., Gary W. Clendenen, and David Bruce: *Volume Growth and Response to Thinning and Fertilizing of Douglas-Fir Stands in Southwestern Oregon,* U.S. Department of Agriculture, Forest Service Gen. Tech. Rep. PNW-GTR-221, 1988, 38 pages.

CHAPTER 25

DOES FORESTRY PAY?

FINANCIAL PERSPECTIVE

This chapter takes a glimpse into the question of whether the practice of forestry—the deliberate continuing production of forest resources—makes sense economically. Are its results a sufficient reward for the sacrifices it exacts?

To be sure, society takes an affirmative view of forestry practice and has done so as far back in the past as one can reach. Through public ownership and the regulation of private activity, society administers a forestry policy aimed at material abundance, financial efficiency and equity, spiritual well-being, and other complex and partly conflicting goals. This chapter avoids the question of whether forestry pays in a social sense. It emphasizes the private forest owner who is oriented largely to self-betterment.

To be concerned with private forestry is not, of course, to interpret *pay* as purely monetary. We can't afford to overlook the nonmonetary criteria that strongly influence much private practice. However, the chapter stresses financial analysis and, indeed, begins with it, using models defined in Chap. 23.

> How do you distinguish between financial analysis and economic analysis? Is one a subset of the other? Explain. What elements are found in the set that are absent from the subset?

A LOBLOLLY PINE CASE

For a first case, consider a loblolly pine forest in southeastern United States, starting with cutover land contemplated for acquisition by a nonindustrial corporation. In this case study, it's necessary to make a number of assumptions about outputs and about revenue and cost rates [5], and here's the approach taken: Most of the assumptions are made with the 1990s in mind. For some of the data, single assumptions are made; but for others, a range of assumptions is adopted. The data are real values and quantities from the loblolly pine region, but the forest is hypothetical.

I begin with a set of assumptions that gives timber management the brightest prospect. The site quality is high, index 70 ft at 25 years. The condition of the land is such that a low regeneration investment of $70 per acre to cover the cost of broadcast burning and machine planting is sufficient to get a timber stand started. Property taxes are low, $1 per acre per year. Annual costs of management are $5 per acre. The rotation is 30 years, with thinnings at ages 15 and 20 or so, each yielding about 10 cords of pulpwood, chip-and-saw logs, sawlogs, and veneer logs per acre. The final harvest is 35 cords per acre. These yields are sold as stumpage, the only source of revenue, at prices rated high for the reference period [12]. The firm's guiding rate of return is 4-percent after tax. The final harvest is followed by a fresh regeneration investment and successive rotations like the first, except that regeneration costs then total $160 per acre. This last set of assumptions involves other assumptions: (1) Real values per unit of inputs and outputs will not change. (2) No cost will be incurred, or benefit gained, from achieving a regulated forest. However, in the income-tax calculations, revenues are assumed to occur as an annual flow equal in capitalized value to the actual periodic series. An allowance is made for uncertainty by reducing the gross-income estimate 30 percent. The result of estimating this "brightest prospect" is as follows:

Today's discounted or capitalized value of revenues per acre	
From yield at year 15, $550 ÷ 1.8009	$ 305.40
20 or so, $550 ÷ 2.1911	251.02
30, $1925 ÷ 3.2434	593.51
From subsequent yields, $1,149.93 ÷ 2.2434	519.58
All anticipated revenues	$ 1,662.51
Today's capitalized value of costs per acre:	
Regeneration investment, $160 ÷ 2.2434 + $70	$
141.32	
Property taxes	25.00
Management costs	125.00
Risk; 30% of $1,662.51	498.75
Subtotal costs	790.07
Net revenue before income taxes, $1,662.51 − $790.07	872.44
Income taxes, 39%	340.25
Net revenue after income taxes	$ 532.19

Thus the land expectation value is $532.19 per acre. If the firm can buy the land for $350, it will enjoy a current net worth per acre of $182.19, which is the capitalized value of an anticipated profit of $7.29 per acre per year. The internal rate of return, including the land as a $350 cost, is a tad more than 5 percent. The benefit-cost ratio, counting the land as a cost, is 1.124. The period required to pay out initial costs of land, regeneration, property taxes, and a charge for risk is 20 years or so, a period clearly justifying the 30-percent charge. The firm's judgment—its degree of optimism and its concern for other than timber finance—determines the approach to the risk allowance and the magnitude involved.

> Exercise: See if you can verify or contradict the foregoing estimates of internal rate of return, benefit-cost ratio, and payout period. 1.05^{15} is 2.0789; you have all the other necessary data.

So far, we have estimated the expectation value of only the "brightest prospect" loblolly pine cutovers. But what about the alternative situations? Table 25-1 lists expectation values for 54 combinations of site quality, property tax, stumpage price, and regeneration investment, each calculated just as in the first case study. Our $532 value appears at the upper-left corner of the matrix. The values descend more or less evenly from this point downward and to the right. The forestry situations that allow the firm to invest $350 or more per acre in the land are found at the top and left side of the

TABLE 25-1
EXPECTATION VALUE OF CUTOVER LOBLOLLY PINE LAND—MAXIMUM PRICE IN DOLLARS PER ACRE THAT A NONINDUSTRIAL FIRM CAN PAY FOR THE LAND AND JUST BREAK EVEN FINANCIALLY—IN RELATION TO STUMPAGE-PRICE LEVEL, INITIAL REGENERATION INVESTMENT, TIMBER SITE QUALITY, AND PROPERTY-TAX LEVEL

		Stumpage-price level								
		High			Moderate			Low		
		Initial regeneration investment								
Site quality	Property tax	Low	Mod	High	Low	Mod	High	Low	Mod	High
High	Low	532	518	504	403	388	375	274	259	248
High	Moderate	516	502	488	386	373	359	257	244	230
High	High	500	486	471	371	355	342	240	226	213
Medium	Low	412	397	384	304	290	276	197	184	169
Medium	Moderate	395	382	366	287	274	258	182	167	152
Medium	High	380	364	349	272	256	243	165	148	136

matrix. The promising section of the matrix is separated from the unpromising section by a stair-stepped line.

All the foregoing relates to the firm with a 4-percent guiding rate of return. What effect does the interest rate have on the estimates? From the fact that the case study internal rate of return is barely above 5 percent, one can deduce that at 5 percent, only the very best land category has a $350 value, and that at 6 percent no land has such a value. The 6-percent matrix corresponding to Table 25-1 shows a land-expectation value of $256 in the upper-left corner.

One note on loblolly pine forestry before we move along. We have been studying the timber-growing firm that sells stumpage. What can we say about the integrated firm that grows stumpage, processes wood products, and even produces nonforest commodities? Such a firm may have many good reasons for paying more than stumpage-expectation value for land. Benefits of assured raw-material supply are perhaps the most common case. The federal income-tax law used to give such a firm a dramatic advantage in land acquisition. But no longer. The tax advantage today is slight.

One more note on this. Have a look at Bill McKee's Alabama story in the *Forest Farmer* [10].

NEW ZEALAND RADIATA PINE

From southeastern U.S., we move to a region at about the same latitude on the opposite side of the equator: the islands of New Zealand. Though so remote geographically, the two regions are much alike culturally and in forest productivity. The timber species to be studied is Monterey pine, *Pinus radiata.* A tree of little glamour in its native California, this pine, introduced into New Zealand, blossomed into one of the world's most spectacular producers of wood.

There are no managed radiata pine forests well fitted to present wants, and so, just as for loblolly pine, we synthesize a firm and its forest. Robert Fenton [4] provides data and analysis. He describes a government-owned brush-covered tract of easy topography, to be cleared and burned over before planting. The plantation is not to be thinned, but the trees are to be pruned, to successive heights up to 36 ft. In a 26-year rotation, timber will be grown that can be processed into high-quality lumber and veneer at the plants operated in conjunction with the forest enterprise. On this good site (index 95 ft at 20 years), crop trees can be expected to average 24 in d.b.h.. Their yield will be about 8,250 cubic feet per acre, so that mean growth will be 317 cubic feet per acre per year in the course of the 26-year rotation.

The radiata pine assumptions lend the forest some advantages: integration with processing facilities, freedom from taxes, no necessity for any special risk allowances. However, one constraint is imposed that was absent in the loblolly pine case. A regulatory program is started at the outset: Only $\frac{1}{26}$ of the property is planted each year. In the absence of thinnings, the forest is not ready to yield revenues until the twenty-sixth year; but then revenues arrive in an annual series.

The internal rate of return to the *integrated business* is about 10 percent. If the firm's guiding rate is the one first used with loblolly pine, 4 percent, then the land expectation value is $560 (New Zealand dollars) per acre; the current worth with land at $20 [9] is $540 per acre; and the benefit-cost ratio is 3.0.

However, we apparently cannot conclude that radiata pine plantations are generally self-supporting. One of New Zealand's consulting economists [9] wrote as follows at about the time to which the radiata pine estimates apply:

> Tree growing, as distinct from utilization, is not a business proposition at current stumpage rates. If it was, new forest planting companies would be springing up again [as] they did in the 1920s. Only large existing forest enterprises can finance new plantations, because they must ensure their own raw materials for the future, and not because there is profit in this side of their operations.

In the consultant's comment and in the loblolly pine example, you get your first glimpse into a principle that will thrust itself upon you time and again: Timber forestry, if required to stand on its own as a separate enterprise independent of other sources of benefits, is, for the most part, a financially unattractive venture.

> The consultant appears to blame stumpage price for the pine's low profitability. Our Chap. 24 makes the point that price level does not affect the best combination. Is there a contradiction?

SCOTTISH DEDICATED FORESTS

From 1919 until at least recently, the British Forestry Commission administered a series of laws that authorized owners of landed estates to "dedicate" their forest land. The owners made management plans, had them approved, carried them out, kept records, and opened their books and woods to Commission representatives. In return, they received public grants. The dedication program was originally meant to build a national emergency timber reserve. Recently it emphasized wise land use and the cultivation of aesthetic values.

As part of the dedication program, forestry schools in England and Scotland surveyed each year a sample of dedicated estates. From the sample, the revenues and costs of all dedicated forests were inferred, an inference that has been questioned in some quarters, but which we can regard as valid for present purposes. The Scottish data from the program in its original form are available for the years 1961 through 1985 [6]. They are suitable for analysis because the forests run more heavily to conifers than elsewhere in Britain and thus extend the conifer theme of this chapter. At the same time, the Scottish forest sites are not so lush as those studied until now and thus should give some contrasts. And private forest management is more commercially oriented in Scotland than in England and Wales.

In 1985, there were about 2,400 dedicated estates in Scotland, with a total of nearly 960,000 acres of forest, an average of some 400 acres apiece. Individual forests ranged from less than 100 acres to 10,000 acres. Revenues and costs on the average acre of all dedicated forests are estimated for 1985 in Table 25-2. The estimates show costs overrunning revenues, including grants, by nearly $20 per acre.

The revenue–cost relationship for 1985 shows some marked differences among estate size classes. The biggest holdings, where revenues came closest to equaling costs, look best. The littlest holdings look least good:

Forest size, acres	Revenue–cost ratio
Less than 250	.30
250–1,249	.51
1,250–2,499	.60
2,500–4,999	.85
5,000 or more	.92

Perhaps private forestry can pay its own way on tracts that are large enough to enjoy economies of scale—efficiency in the purchase of equipment and supplies and in their use and the use of labor. But one cannot be sure, because information is wanting.

1 What happened in years other than 1985? What effect did fluctuations in the forestry economy have on the relation of revenues to costs in those other years?

TABLE 25-2
REVENUES AND COSTS IN U.S. DOLLARS PER ACRE ON THE DEDICATED PRIVATE FORESTS OF SCOTLAND, 1985

Revenues:	
Timber sales	$ 9.66
Sales of other forest products	1.00
Government grants	5.01
Other revenues	1.00
All revenues	$ 16.67
Costs:	
Laborers	$ 7.90
Managers and supervisors	5.75
Contractors	7.55
Vehicles, machines, transport, plant, tools	7.18
Materials	5.04
Buildings, land acquisition	.71
Other costs	1.79
All costs	$ 35.92
Net revenue (loss)	$ (19.25)

Source: Holmes, W. D., and J. D. Todd: *Economic Survey of Private Forestry, Income and Expenditure, Scotland, 1985,* University of Aberdeen Department of Forestry, February 1987. One British pound is taken to equal $1.35.

2 What happened to timber-growing stocks in 1985 and other years? When there was net revenue, did it result simply from a drawing-down of stocks? When there was net loss, was it the result only of stock accumulation?

3 What about taxes as an element of cost? This is a hard question, because taxes on the forest enterprise cannot be separated meaningfully from taxes on other estate enterprise. From the mid-1960s through the mid-1980s, one clear effect of taxes was to encourage afforestation, because the cost of it could be charged against any estate income, however earned.

4 What is the value (cost) of the investment in the forest property?

The first of the four questions can be answered in part. This is done in Fig. 25-1, which covers 25 years. In two of these years, revenues other than grants exceeded costs on the average acre of all Scottish dedicated forests. Bear in mind the purpose of the Scottish surveys: to help the Forestry Commission set the amounts of the grants. If the Fig. 25-1 data were used in a way that was never intended, as a time series of cash flows for a forestry investment project, the internal rate of return would prove to be negative, the forest expectation value negative at any nonnegative rate of interest, and the payout period indefinite and conceivably infinite.

> If you had responsibility for setting the amounts of the grants, how would you use the results of the surveys reported here?

Thinking of the payout period reminds one of a major consideration not yet recognized. The British forestry policy was set up after World War I and renewed after World War II as a response to forest depletion and land denudation. Of all the forestry practices for which grants were given, tree planting was first by a wide margin. Hence much of the forest represented in Fig. 25-1 is young. The figure tells only the first part of what may be a long and more affirmative story.

To remark on the limited scope of the Scottish data is to remind oneself that the Scottish case study responds to a question fundamentally different from the loblolly and Monterey pine questions. The pine questions concern the long-term survival of a whole business enterprise. The Scottish question picks out only specific *investment projects* during a limited time and asks about their return. Remember, some fruitful investments can be made by an *enterprise* that is failing. And a profitable enterprise can make some losing investments.

PRIVATE HOLDINGS IN GENERAL

The foregoing three case studies give, not a clear response, but some insight to the question of whether forestry pays in a financial sense. They give information about some of the variables in the question. Now we broaden the question so as to consider, at least briefly, whether forestry pays in a general sense.

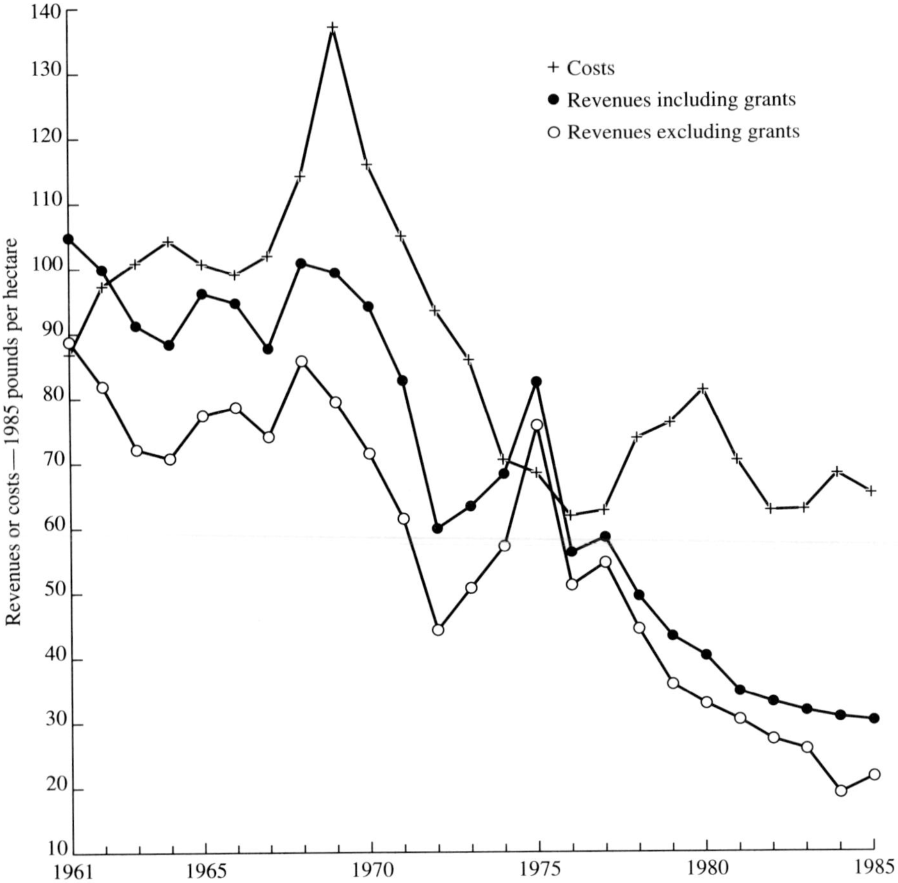

FIGURE 25-1
Annual revenues, including and excluding government grants, and annual costs, in 1985 pounds per hectare, on the dedicated private forests of Scotland, 1961–1985. (*Source:* Holmes, W. D., and J. D. Todd: *Economic Survey of Private Forestry, Income and Expenditure, Scotland, 1985,* University of Aberdeen, Department of Forestry, 1987, p. 27.)

The Scottish survey analysts make a strong point that their forest estate owners derive large benefits beyond those expressed in money. The same can be said of private owners almost anywhere in the world. Certainly in eastern and central U.S., the preponderant nonindustrial private holdings offer a prime example of the wide variety of benefits that accrue to owners. Does forestry, or at least the holding of forest property, pay these owners? Evidently it must in some sense, else this vast, scattered, dog-eared, productive resource would not flourish as it has.

One of the pioneer students of nonindustrial private forest owners was Solon Barraclough [1], who surveyed such owners in New England in the 1940s. He found that most owners were not primarily or even appreciably interested in their timber, a discovery that was greeted with shock and skepticism by the forestry fraternity of the day. How greatly the professional folklore has become modified in the meantime is illustrated by the opening remark in a 1981 *Journal of Forestry* article [8]: "It is generally thought that timber production is not a primary objective for the majority of nonindustrial private forest owners."

> How do you explain the radical change in foresters' attitudes between the 1940s and the 1980s?

Barraclough, indeed, learned that owners valued their woods as shelter for livestock and crops or as scenery or as a pleasant place to enjoy recreation such as hunting. One interviewee said that his woods were "a place where I can cuss at the top of my lungs and the neighbors won't hear me." And most research workers who have followed Barraclough have recognized the wide range of values that owners as a whole place upon their woodlands.

William Kurtz and Bernard Lewis [9], working in the Missouri Ozarks, identify four types of owners, based upon motives and objectives, that are most common in the region. The "timber agriculturist" values timber growing as a business. The "timber conservationist" values timber for its future promise to consumers and for its association with wildlife and aesthetic amenities. The "forest environmentalist" seeks "a host of benefits—beauty, privacy of living, wildlife—that contribute to the satisfaction of ownership." And the "range pragmatist" looks upon land as a source of grazing income, a long-term investment, and a habitat for wildlife.

Paul Ellefson, Sally Palm, and David Lothner [3], in a survey of small-tract buyers of tax-forfeited forest in Itasca County, Minnesota, find that current primary uses of the tracts rank as follows:

Primary use	Owners, %
Recreation	26
Timber growing	14
Permanent residence	13
Farming	12
Investment	11
Satisfaction of owning land	9
Forest preservation	7
Second home	5
Mineral extraction	1
Other	2
All uses	100

The diversity of the forest benefits is emphasized in the fact that an owner's average period of tenure is $8\frac{1}{2}$ years, and some tracts have had six or more owners in 18 years.

Archibald Jones [7], writing about a case-study farm woodland near Montreal, Quebec, emphasizes the use of the woods for maple products and firewood. Wallace Christensen [2] describes a case-study forested property in northern Vermont, whose purchasers, an absentee urban couple, look forward to enjoying benefits such as these: a rural retreat; a place for the children to experience new ways and challenges; an opportunity to hike, ride horseback, ski; a secluded den for business conferences; a wildlife sanctuary for hunting, fishing, and watching; and a property of great value simply as a possession.

And so it goes. One is confronted, time and again, by the suggestion that forest ownership or the practice of forestry pays the private owner only if the tree-growing enterprise is linked to some other use. Here is what John Segur [11], a business consultant with long experience in forestry and forest industry, has to say on the subject:

> At current prices for land and timber, the long-term growing of timber is not sufficiently profitable to carry its own financing and debt. To [my] knowledge, there is presently no forest-resource unit anywhere that exists as an independent, self-sufficient unit. If the forest is held by individuals, its income, taxes, and financing are intertwined with the other affairs of these individuals. If it is held by the government, its financial benefits and costs are difficult to isolate—and seldom determine policy anyway. If the forest is held by a corporation, it must usually be subsidized by the firm's other operations such as processing plants.

> Can you offer a simple, convincing explanation of the fact that timber growing as such is not an attractive investment?

SUMMARY

The question in this chapter is whether private forestry pays. The models of Chap. 23 are applied to cases in southeastern U.S., New Zealand, and Scotland. Some nonfinancial models are used, too—studies of forest owners in central U.S., Lake States, eastern Canada, and New England. Consistently, this response to the question is encountered: It is difficult to make a case for timber growing by itself as a money earner. But the general rule is that private *forestry* does *pay* if one accepts both terms in a broad enough sense.

REFERENCES

1 Barraclough, Solon L.: "Forest Land Ownership in New England with Reference to Forest Holdings of Less than Five Thousand Acres," Ph.D. thesis, Harvard University, Cambridge, Mass., 1949.

2 Christensen, Wallace W.: "Owner-Consumer Decisions on an Amenity Forest," *In* Duerr, William A., Dennis E. Teeguarden, Neils B. Christiansen, and Sam Guttenberg, eds.: *Forest*

Resource Management, Decision-Making Principles and Cases, O.S.U. Book Stores, Corvallis, Oreg. 1982, pp. 474–486.
3 Ellcfson, Paul V., Sally L. Palm, and David C. Lothner: "From Public Land to Nonindustrial Private Forest, a Minnesota Case Study," *Journal of Forestry,* April 1982, pp. 219–222, 234.
4 Fenton, R.: "Economics of Radiata Pine for Sawlog Production," *New Zealand Journal of Forestry Science,* October 1972, pp. 313–347.
5 Haney, Harry L., professor in the Department of Forestry at Virginia Tech, Blacksburg, on the basis of his experience in working with forest owners and his familiarity with the published literature, has provided me with most of my loblolly pine estimates.
6 Holmes, W. D., and J. D. Todd: *Economic Survey of Private Forestry, Income and Expenditure, Scotland, 1985,* University of Aberdeen, Department of Forestry, February 1987.
7 Jones, Archibald R. C.: "Integrating Farm and Forest Decisions," *In* Duerr et al.: op. cit., Item 2, p. 474.
8 Kurtz, William B., and Bernard J. Lewis: "Decision-Making Framework for Nonindustrial Private Forest Owners, An Application in the Missouri Ozarks," *Journal of Forestry,* May 1981, pp. 285–288.
9 Larsen, C. R.: "Returns Expected from Forestry as a Business," *New Zealand Journal of Forestry,* 1968, pp. 146–148.
10 McKee, Bill: "Economic Tradeoffs, Timber and Game," *Forest Farmer,* November–December 1986, pp. 18–20.
11 Segur, John A.: "Dealing with Financial Agents," In Duerr et al.: op. cit., Item 2, p. 410.
12 Timber Mart South, Inc. *Timber Mart—South.* vol. 16, Highlands, N.C., 1991.

CHAPTER **26**

FOREST FERTILIZATION

SIGNIFICANCE OF THE FERTILIZATION TOPIC

In previous chapters that have to do with the best combination in timber growing, the firm's guiding rate shows up as a principal determinant. The guiding rate is dominant because both the variable input (growing stock) and the output (growth) are wood. Consequently, the question is always about the rate of return of wood on wood: whether it is sufficient in view of the firm's guiding rate. To be sure, timber-*size* premiums and prospective price *changes* affect the best combination. But the *level* of wood prices in itself is immaterial. If wood is the source of both revenue and cost, the relation between the two is always the same, regardless of price level. This is the gist of the argument in Chap. 24, where determinants of the best combination are explored.

However, in some timber-growing operations, the inputs are other than wood. In this case, cost is unfastened from revenue, and timber price level becomes material to the best combination. One such operation is forest fertilization, the subject of this chapter. Another is forest protection, the subject of the next. In both instances, the prime inputs are labor, equipment, and materials.

Since 1960, forest fertilization has become an increasingly common silvicultural practice in the two principal timber regions of the United States. By 1986, forest owners in the Pacific Northwest had fertilized a total of 2.5 million acres; those in the Southeast, nearly 2 million [1]. Since fertilization practices in the Pacific Northwest are comparatively straightforward, we'll choose this region for study. Let's concentrate on a Douglas-fir stand.

DOUGLAS-FIR TECHNOLOGY AND DATA

On most Douglas-fir sites, the limiting element in soil fertility is nitrogen. In the natural forest, where growth and decay are in some balance, the supply of nitrogen is cycled in the ecosystem and thus maintained. But in the managed forest, as an even-aged stand develops toward maturity, the trees draw more and more nitrogen from the soil and hold it in storage. Thus a deficit develops in the area where the root systems are at work. Researchers have found that fertilizing at the time of planting or early in the life of the stand elicits very little growth response from the trees. It's later on, as maturity approaches, that fertilizing brings the largest reward in added timber growth.

And so the typical Douglas-fir fertilizer regime is to apply nitrogen some 10 or more years before harvest. The nitrogen is ordinarily in the form of urea pellets, which are broadcast by helicopter.

The determinants of financial success in Douglas-fir fertilization [2] are listed below. This is a simplified list. It says nothing about tax incentives or deterrents to fertilizing. It says nothing about prospective changes in the unit cost of fertilizer, a product of natural gas. Thinning and its timing are overlooked. So is rotation length. Furthermore, the possibility of replacing or supplementing commercial fertilizer with that produced biologically—as by red alder, a nitrogen-fixation agent that can be grown in rotation with fir [3]—is laid to one side.

1 Quantity of fertilizer per acre; assume 435 pounds of urea (46 percent nitrogen).
2 Current price of fertilizer per pound, 10 cents.
3 Per-acre current cost of applying fertilizer, $22.50
4 Length of waiting period until harvest, 10 years.
5 Real guiding rate of return, 4 percent.
6 Per-acre extra timber yield attributable to fertilizer, 1,250 bd ft.
7 Probability of obtaining the extra yield, .9.
8 Current net per-M value of extra yield, $180.
9 Prospective annual increase in real per M net value, 2 percent.

The list of determinants is extended to include the assumptions made in this chapter about the magnitude of each determinant. Note that only the quantities and values that are variable with the manager's choice—to fertilize as described or not to fertilize at all—are recognized. The rest are viewed as fixed and thus irrelevant to the choice. The result is a marginal analysis of fertilizing—a survey of the approach to such analysis. It is not a study of Douglas-fir management profitability, either with fertilizer or without it.

The assumptions about resource values relate to a time some years ago. Let us assume that the relation of unit revenues to unit costs holds about the same today as it did then.

Note that provision is made for expressing all revenues and costs as of the same point in time, the time of fertilizer application, referred to as "current." Thus you can view the analysis as though it were being made in response to the manager's question, "Is it a good idea to apply fertilizer this year?"

The per-acre cost of applying fertilizer, *c*, is a function of, among other things, the total acreage fertilized and the season of the year when the work is done. Apparently helicopter operators have more slack time in spring than in other suitable seasons. Timber-growth response to fertilizer, *f*, is influenced by quantity of fertilizer, *a*, length of waiting period, *d*, season of application (some of the urea is lost in warm, dry weather), and site quality. Fertilization is most rewarding on medium sites. Very high sites have limited room for improvement. Very low sites, on the other hand, have deficiencies besides those of the soil.

Despite the fact that very high sites are not usually the best bet for a fertilizer program, let's turn to the site-I forest of Chaps. 17 and 22 for case material.

MARGINAL ANALYSIS FOR DOUGLAS-FIR

Speaking of Chap. 22, let's recall that it is concerned with the rotation. Even though the rotation isn't listed as a determinant of financial success in fertilization, there must be some connection: Surely fertilizing affects the choice of a rotation.

Think of a timber stand's rotation in simplest terms, as coincident with the *stand's* adjusted financial maturity in the sense used in Chap. 21. Look at the stand as an independent enterprise, not as part of a regulated forest. This won't affect the general reasoning.

Ordinarily the stand responds to fertilizer within a year or so. Annual rings widen. Current annual growth percent increases, first at an increasing rate, then at a diminishing rate. Then growth percent peaks, perhaps in the fifth year. By about the tenth year, the fertilizer's influence has played out. The annual rings have narrowed to their prefertilizer width. But the trees are much larger now. Growth percent is depressed. The rotation has ended.

Apparently, then, the firm can control its best rotation through its fertilizer program. Probably it sets a shorter rotation than otherwise, since stand conditions associated with harvest time arrive sooner with fertilization than without. The firm, say, first plans a new rotation age. Then it plans an interval for fertilizer response appropriate to all the determinants in the list except *d*, the length of the waiting period itself. Ideally, the firm solves simultaneously for *a*, *d*, and *f*, taking the other determinants as given.

> If you had good information for your Douglas-fir stand on the effect of *a* (quantity of fertilizer) and *d* (length of waiting period) upon *f* (extra timber yield attributable to fertilizer), what would be your model for estimating the most efficient magnitudes of *a* or *d?*

As a practical matter, the production functions are not well enough understood for this purpose. The firm must follow the regional experience with Douglas-fir. It can make adjustments here and there to suit its own guiding rate of return and other prospective rates of cost and revenue. Some back-of-the-envelope trial and error may be helpful. Here are the estimates of financial promise that flow from the assumptions listed earlier.

The per-acre cost of fertilizing, as of the time of fertilizing, is 435 pounds of urea at 10 cents per pound, plus $22.50 for helicopter application. Total: $66.

Per-acre revenue from fertilizing, as of harvest time, is, tentatively, the anticipated extra yield of 1,250 bd ft times .9 probability times $180 per M, or $202.50. Since the $180 is in dollars as of the time of fertilizing, the $202.50 must be compounded for the 10-year waiting period at the forecast 2-percent annual increase in timber's real value per unit. Result: about $247.

The per-acre revenue figured as of the time of fertilizing so as to be comparable with the cost estimate: $247 discounted for 10 years at the 4-percent guiding rate of interest: about $167. The estimated cost is $66. Clearly, if the firm's estimates are anywhere near correct, fertilizing is financially worthwhile.

Now for a small **sensitivity test.** How sensitive is the finding to the firm's assumptions about unit revenues and costs? Here are the figures, which you can verify for yourself: Fertilizer could cost as much as 33 cents per pound, and the operation would still break even. Or the guiding rate of return could be as high as 14 percent. Or the value of the extra timber yield could be as low as $72 per M.

Further, there is at least one gain for fertilizer that hasn't been mentioned. This is that the program, by shortening the rotation, allows the firm to make more efficient use of its fixed inputs, notably land, personnel, and plant. The firm's evaluation of the added yield does not include this gain. Since data deficiencies make it difficult for the firm to evaluate the gain, the firm can just think of it as a safety factor supporting any fertilizer regime that otherwise looks reasonably attractive.

The resource combination described isn't necessarily the best. The firm could, given good production-function data, search for other combinations of fertilizer input, waiting period, timber output, and, indeed, the determinants omitted from the list that would widen the margin between the extra revenue and costs of *fertilization* (holding aside the general question of Douglas-fir management, as mentioned previously).

SUMMARY

In some regions of the U.S., forest site improvement through fertilization can be financially attractive under favorable conditions. Parts of the Douglas-fir region are a case in point. There, the application of nitrogen a few years before final harvest can increase timber growth enough to pay for fertilizer and application costs carried at interest, perhaps with something to spare.

REFERENCES

1 Allen, H. Lee: "Forest Fertilizers," *Journal of Forestry,* February 1987, pp. 37–46.

2 Miller, Richard E., and Roger D. Fight: *Fertilizing Douglas-Fir Forests,* U.S. Department of Agriculture, Forest Service, General Tech. Rep. PNW-83, January 1979.

3 Tarrant, Robert F., Bernard T. Bormann, Dean S. DeBell, and William A. Atkinson: "Managing Red Alder in the Douglas-Fir Region, Some Possibilities," *Journal of Forestry,* December 1983, pp. 787–792.

CHAPTER 27

WEIGHING ALTERNATIVES IN WILDFIRE MANAGEMENT

PROBLEM AT HAND

The problem at hand is to identify the wildfire-management program that is economically optimal. By **wildfire** is meant the nonprescribed burning-over of forest or range land, most commonly as a result of human mishap, incendiarism, or lightning.

In fact, the problem at hand is twofold. The first part is to describe a model that accounts for the benefits and costs of dealing with wildfire and that identifies the level and mix of costs that is sufficient but not excessive, benefits considered. In other words, the first part of the problem is to define economic efficiency in fire management. The second part is to be aware of the operational barriers that block precise use of the model and to recognize how the optimum is roughly approximated in practice.

Fire-control agencies in the U.S. have long been searching for guides to economic efficiency. Some of the earliest analyses of wildfire economics appeared in the professional journals in the 1920s [2, 5]. The reason for long-standing concern has been heavy losses from wildfire and high costs of combating it. In the nation as a whole, the annual cost of controlling wildfires on all forest and range lands rose from perhaps 250 million 1985 dollars in the early 1950s to the vicinity of $550 million in the mid-1980s [6].

BEHAVIOR OF FIRE

To begin the story at the beginning, consider how wildfire behaves. The occurrence and severity of wildfire in a given region and year depend not only on the people who start or prevent or stop fires, but also on the natural environment. *Weather* is a major environmental element: Fire danger is greatest during the fire seasons of the year, and within these seasons, when wind and temperatures are high and humidity low. *Topography* is another element: For example, other things equal, fire burns and spreads most vigorously uphill on steep slopes and on the comparatively hot and dry southerly and westerly exposures. And then there is the character of the *fuel:* size of pieces, moisture content, continuity (governing the ability of the fire to spread), and compactness (governing its access to oxygen).

One thinks of wildfire as a destructive agent, and so, by and large, it is. It destroys or damages vegetation, soil, and other property and even human life. It leads to water and air pollution. It deteriorates the scenery. It gives rise to rehabilitation costs. For all such negative results, wildfire is unwanted. But wildfire may have positive results, too. It may lessen the danger of subsequent fires; it may control vegetation beneficially; it may improve the range or the wildlife habitat. On the whole, however, wildfire's negative results outweigh the positive, and to manage it is to control it. The point to emphasize is that wildfire management is an integral part of resource management. It must, therefore, be evaluated and planned—its alternatives weighed—in consideration of resource management as a whole.

FIRE-MANAGEMENT TECHNOLOGY

Before we discuss the weighing of wildfire management alternatives, we must review very briefly the technology of fire control so as to understand what the alternatives are.

Fire-control technology is ordinarily thought of as a three-part procedure: (1) prevention, (2) presuppression, and (3) suppression. As in the case of fire control in residential, commercial, and industrial areas, all three parts of the procedure are mainly a public responsibility. In wildfire control, the responsibility rests with state forest services and other land agencies and with the federal land agencies such as Forest Service and Bureau of Land Management. Some private agencies, too, are involved, notably the landowning corporations and their associations. In all this, the key word is *cooperation.* Agencies try to organize the system in such a way that each participant can join with others for greatest effectiveness in fire prevention, presuppression, and suppression.

> Why, in their forest management, do private owners lean upon government for activities such as protection, but rely on private initiative for acquisition, silviculture, harvesting, and the like? What are the benefits of this division of responsibility? the penalties?

1 Prevention includes advertising—for example, with the assistance of Smokey Bear and friends. It makes use of displays and lectures. It involves publicizing and enforcing fire laws. High-danger areas are patrolled. The closing of areas is enforced. Visitors are cautioned and instructed. Indeed, prevention includes all activities directed to the general public and aimed at reducing the probability of fire starting through carelessness or incendiarism. Prevention also includes efforts to lessen the ease of ignition and the difficulty of suppression. Fuel management is the principal such effort: replacing higher-hazard fuels with those of lower hazard. Prescribed burning is a key measure.

2 Presuppression consists of all other fire-control activities preparatory to attack upon the flames. These activities come down to planning, organizing, staffing, equipping, supplying, and administering the systems for fire detection and suppression. Included here are personnel training, arranging for mutual assistance among agencies, and setting up a pool of workers who can be called upon in emergencies. Among specific activities are mapmaking, constructing fuel breaks and firebreaks, and running the fire danger rating system. Then there is building and staffing lookout stations, acquiring communications equipment, and arranging for the use of aircraft of many types, from big tankers and transport planes to light patrol craft.

Taken together, prevention and presuppression are the heart and soul of wildfire management. The level, or intensity, of the management program can be measured in units of the effort—expenditure—that is put into prevention + presuppression. If this expenditure is nil, the program is nil: Fires go unprevented, unprepared for, and unsuppressed. If the expenditure is moderate and funds are made available for extinguishing the fires that start, these funds are drawn upon heavily. But if the prevention + presuppression expenditure is high, the cost of suppression tends to be lower than otherwise: Prevention + presuppression dollars are substituted for suppression dollars. The general relationships are sketched in Fig. 27-1 (the axes are calibrated to emphasize that the horizontal and vertical scales differ; *n* is a constant of unspecified value).

In the U.S. as a whole, our wildfire suppression over the years has tended to cost about the same as our prevention + presuppression. But suppression costs vary from year to year, depending on the number and severity of the year's fires and on the operational goals of suppression agencies.

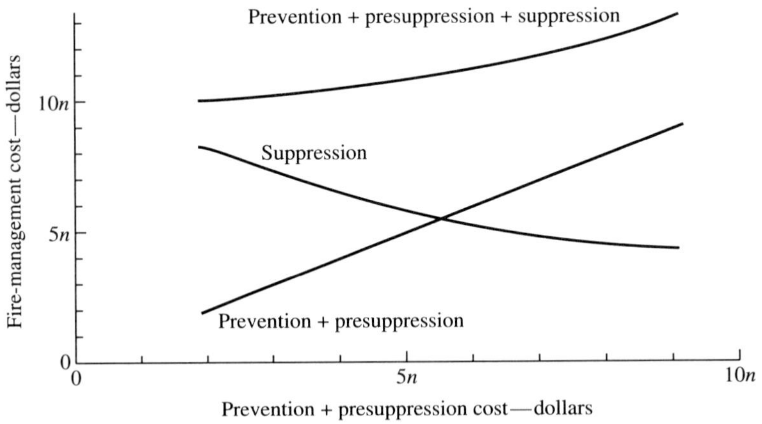

FIGURE 27-1
Hypothetical relationship of wildfire-management costs and loss to total prevention + presuppression cost in a given region and year.

3 Suppression begins when someone discovers a fire and tries to extinguish it or, in a more formal sense, when a fire is detected and the dispatcher is notified. A common suppression policy is to hold down the costs and losses of each fire by reaching it swiftly and stopping it while it is still small. Thus, in remote areas, use may be made of helicopter crews and smokejumpers and airborne fire retardants in the initial attack. Wherever possible, mechanized ground equipment is put into action. Even so, fire fighting is a labor-intensive operation. Although a well-trained crew of two may successfully handle a small surface fire on a low-danger day, thousands of workers may be called out on the line at a large fire in a spell of high-danger weather.

THE OPTIMAL PROGRAM

Now what is the optimal fire-management program? Look again at the total-cost function (prevention + presuppression + suppression) in Fig. 27-1. The optimal program can be approximated with the help of this function. But first, *in theory,* the function itself must be optimal: Every total cost read from the function must be the minimum for achieving a given benefit. Or, to view the matter from the opposite direction, each total cost must be the one that leads to maximum total net benefit, so that the marginal benefit-cost ratio is 1.0 (Chap. 23). This occurs when a difference, ever so slight, in cost produces an equal difference in benefit:

$$\frac{dB}{dC} = 1 \tag{1}$$

What is the benefit from fire management, and how is benefit to be measured? Benefit is the reduction of loss from fire, as defined in the next section. It is measured in the same units as cost. If cost is the dollar value of the inputs to fire management, then benefit is the dollar value of the outputs.

To be sure, it's impossible to measure the total benefit of fire management, for one cannot know what the resource value would be in the complete absence of fire management. Fortunately, this doesn't matter. All one needs to visualize is that for each point on the total-cost function, there is a total after-fire resource value that's as great as possible: a total loss from fire that's as small as possible. In Fig. 27-2, such an optimal total-loss function is drawn on the same scales as an optimal total-cost function derived as in Fig. 27-1. In Fig. 27-2, total loss is as low as it can be for the given cost; total cost is as low as possible for the given loss. Again, we are speaking of the theoretical meaning of these functions: the conceptual ideal.

> Why is it essential for us, as students of the subject, to understand the theory, the conceptual ideal? Indeed, why is it necessary for the "practical" fire-management officer to have the same understanding?

While it is fairly easy to define optimal cost and loss functions for fire management, it is hard to visualize in detail what they involve, because the variables are so numerous. Think, for a moment, what the optimum is. Total cost at each level is apportioned among the three major activities—prevention, presuppression, and suppression—in such a way as to minimize fire loss. For this to be possible, outlays *within* each activity are optimized. This means, for example, an optimal prevention budget [3]. It means an advertising budget of the right size, correctly divided among billboards, periodicals, flyers, and other media. It means having the right number of trucks, Pulaski tools, and fire guards, and having them in the right places at the right times. Think further about presuppression. The optimal program involves resource-management activities that are only indirectly linked to fire control: reforestation and range-reseeding plans, manipulation of vegetative cover, transportation systems for access and as firebreaks, and so on and on. You begin to see the significance of the thought that fire management cannot be analyzed independently of other resource-management functions.

And then there is the losses side of the optimization problem. Three difficulties, particularly, present themselves. First is the difficulty of evaluation: What, for example, is the value of a human life lost in a fire? Second is the great variability of fire losses from year to year. To produce a single loss function, one must estimate probabilities. Third is the practical difficulty of just measuring and recording losses.

One concludes that, as ideas subject to rough approximation in terms of numbers, fire cost and loss functions such as those sketched in Fig. 27-2 are useful—indeed, essential—in wise resource management. At the same time, one recognizes the complexity and elusiveness of the idea of the optimum. In concept, it is the outcome of the simultaneous solution of a multitude of production functions, most of which are unknown. Some *can* be accounted for, mostly through the indefinable processes of "judgment."

A prime device for guiding fire-control managers through their manifold uncertainties is the fire-management plan [1]. Depending on the degree of attention given to principles and philosophy on the one hand and specifics of operational procedure on the other, the planning unit varies all the way from the nation to the county or protection district. The ultimate plan recognizes three goals:

1 To hold down the number of fires that start.
2 To control fires that do start, before they become large.
3 To limit the size and destructiveness of fires that start and become large despite foregoing aims.

You should note that the degree of success in meeting goal *1* determines the size of the problem in *2* and *3*. And the degree of success with *2* controls the effort given to *3*. Although there are no easily measurable criteria for success, comparison between planning units with similar problems helps identify the parts of a plan that call for redrafting. A good plan may always be in process of redrafting.

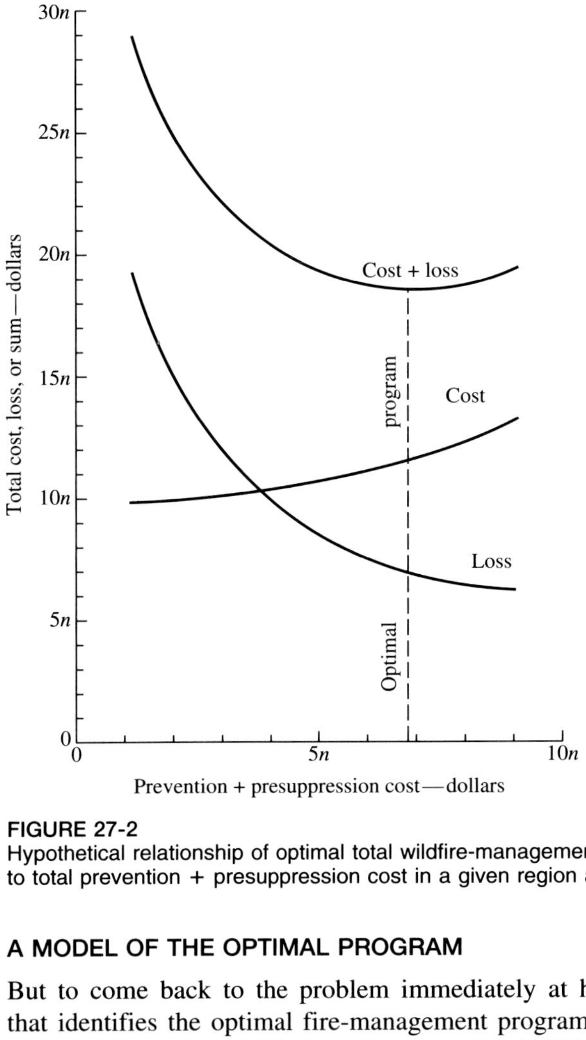

FIGURE 27-2
Hypothetical relationship of optimal total wildfire-management cost and loss to total prevention + presuppression cost in a given region and year.

A MODEL OF THE OPTIMAL PROGRAM

But to come back to the problem immediately at hand, we have to build a model that identifies the optimal fire-management program. There is one more preliminary step to take before a solution can be ventured. This is to generate a careful definition of **loss** (and of benefit as the avoidance of loss) and a passing thought about its estimation.

Loss from fire is the total value of the resource if it had not been burned minus the total value of the burned resource. If the resource is worth less when burned than if it had not burned, the loss is positive. Total loss in specific cases can be negative, as implied earlier, under Behavior of Fire. In general, though, total loss is positive, of a magnitude reduced to allow for a few beneficial effects of the fire.

"Loss" is the original and still persistent term in fire lore. Some writers have used the term "net value change"; but what is referred to is *not* a change: It is a *difference*, Δ, between two resource values at a point in time. Let's use the simpler and more accurate term, loss.

> *Difference* and *change* are not synonyms. Explain. Is there an analogy between these terms on thc onc hand and the terms *stock* and *flow* on the other? Glance at page 216.

Some types of loss can be expressed in dollars as restoration cost. Burned buildings are an example. Other types can be measured with market values. Loss of merchantable timber about to be sold as stumpage can be stated as the stumpage value of green timber minus the salvage value of burned timber. Consumers' subjective valuations (Chap. 12) may be a guide to losses in recreation and aesthetic opportunities and the like. In general, for forest and range resources that are usable as a flow over time, fire loss is the expectation value of unburned property minus the expectation value of burned property.

Just for an idea of what the dollar losses per unit of measure may amount to for some illustrative resources, see the list that follows. These are unit values used by the federal Forest Service in a fire study [4] on the Huron-Manistee National Forest in Michigan's Lower Peninsula. Remember that these are values for a specific area. Values in other areas are different.

Resource	Unit of measure	Unit value, 1990 dollars
Commercial timber:		
Mature hardwood	Thousand board feet	129.00
Mature softwood	Thousand board feet	108.00
Pole stands	Acre	2,169.00
Seedlings-saplings	Acre	3,981.00
Commercial range	Animal-unit-month	9.79
Water use or storage	Acre-foot	2.92
Fish habitat:		
Warm water	User-day	12.40
Cold water	User-day	18.30
Anadromous	User-day	57.00
Wildlife habitat:		
Big game	User-day	30.70
Other game	User-day	23.40
Recreation resources:		
Developed	User-day	8.77
Dispersed	User-day	16.10
Wilderness	User-day	40.90

The value of pole and of seedling-sapling stands is expressed per acre at age of maturity, to be discounted back to their age at the time of the fire. Seedling-sapling stands have the higher value because, if not burned, they would have yielded thinnings that pole stands have already produced. An animal-unit-month is the amount of range forage that will sustain for a month one cow or the equivalent in other species of stock.

AN OPTIMUM OF ALL OPTIMA

Having at hand optimal cost and loss functions after the style of Fig. 27-2, one can now specify the optimal scale of wildfire management program—the optimum of all optima: the best point on the horizontal axis.

We turn to our old friend among models, the equating of marginal unit cost and marginal unit revenue, which is to maximize total net revenue per unit of time. This is an appropriate criterion for the best combination, because all other things are held equal (page 163).

> *Before reading any further,* test yourself. Explain why the fire-management program that minimizes cost plus loss (Fig. 27-2) is the optimal program that equates marginal cost and marginal revenue per dollar of prevention + presuppression cost.

Reflect that the slope of the total cost curve in Fig. 27-2 is the marginal cost per unit of input. And that the (negative) slope of the total loss curve is the marginal (negative) loss—that is, the marginal revenue per unit of input. The optimal fire-management program, then, is found at the point where the two slopes are equal in absolute value. This is the point where the sum of total cost and total loss is a minimum, as pictured in Fig. 27-2. Where loss is falling more steeply than cost is rising, as on the left side of the graph, their sum must be falling. And where, as on the right, cost is rising more steeply than loss is falling, their sum is rising. Where the sum is neither falling nor rising, but is a minimum, the cost and loss slopes must be equal in absolute value. This is the **least-cost-plus-loss** rule for the fire-management program.

At the scales used in Fig. 27-2, the minimum cost plus loss is not sharply defined. The investment in the program, as measured on the horizontal axis, can depart as much as 30 percent from the marked optimum without changing cost plus loss by more than 5 percent. Such relative insensitivity to program investment may have the ring of reality, especially in view of data deficiencies. The agency that is planning for the period ahead is bound to be greatly uncertain about the losses that will actually be associated with given suppression costs. And it will be uncertain about the sum of losses and suppression costs that will accompany any given prevention-presuppression effort. What the agency must do is to be guided in general by past analysis and experience. It can use past program levels and results as a first step in reasoning about the future.

SUMMARY

This chapter addresses the problem of identifying the best wildfire-management program for a given region and year. Conceptually, the best program is arrived at in two steps. First, one visualizes optimal functions of (1) total fire-management cost (prevention + presuppression + suppression) and (2) total loss from fire, in relation to program level (prevention + presuppression cost). These functions are optimal in the sense that they are as efficient as possible: Their sum at every program level is as small as possible. Second, one identifies the optimal program level as the one where this sum of cost and loss is the least of all. The marginal revenue and cost of management are equated at this level.

As a practical matter, the foregoing least of all least-cost-plus-loss program levels cannot be determined. But it can be envisaged and approximated and is a vital concept in wildfire management.

REFERENCES

1 Brown, Arthur A., and Kenneth P. Davis: *Forest Fire Control and Use.* McGraw-Hill, New York, 1973, pp. 457–488.

2 Flint, Howard R.: "Adequate Fire Control," *Journal of Forestry,* May 1928, pp. 624–638.

3 Paananen, Donna M., Larry Doolittle, and Linda R. Donoghue: Evaluating wildfire prevention programs. *Fire Management Notes,* **51,** (3), pp. 5–8, 1990.

4 Schweitzer, Dennis L., Ernest V. Andersen, and Thomas J. Mills: "Economic Efficiency of Fire Management Programs at Six National Forests," U.S. Department of Agriculture, Forest Service Res. Pap. PSW-157, 1982.

5 Sparhawk, W. N.: "The Use of Liability Rating in Planning Forest Fire Protection," *Journal of Agricultural Research,* August 1925, pp. 693–762.

6 U.S. Department of Agriculture, Forest Service: "An Analysis of the Timber Situation in the United States, 1952–2030," Forest Res. Rep. 23, p. 233. Also later annual reports of the Forest Service.

CHAPTER 28

MULTIPLE USE OF FOREST RESOURCES

STATUS OF MULTIPLE USE

Multiple use is the forester's term for creating two or more joint products from forest land. The possible categories of products include timber, wildlife, forage and shelter for livestock, food crops, recreation, scenery, watershed regulation, and space for human residence. Multiple use can result from deliberate planning by the land-management firm or it can be inadvertent. After all, the firm need only leave trees alone in order for them to yield wood, bark, leaves, shade, and shelter; act as water pumps and soil stabilizers; and offer scenery for people and nesting sites for birds. Since our interest here is in the economics of the matter, let's think mainly about deliberate multiple use.

Multiple use has been an increasingly lively topic in forestry circles. There are at least two causes. One is the world's mounting consumption of all the forest outputs; the world's pressures to cut, graze, use. Here, multiple use is thought of as a response to world wants. The second cause is a U.S. phenomenon, the drive by wilderness enthusiasts and other environmentalists to withdraw lands from timber use. Here, multiple use is seen as the opposing philosophy.

In production economics where the effort is to identify a best combination of resources, two questions are to be answered: (1) What are the production functions? (2) What are the unit values of inputs and outputs? Both questions involve special difficulties where multiple use is concerned. This chapter will recognize both questions, but, for the question about values, it will lean on Part Five of the book, concerned with marketing and valuation.

The chapter goes at the multiple-use problem in two steps. First, it touches on the theory of joint products. Second, it surveys some of the approaches being taken to multiple-use management.

INTERPRODUCT RELATIONSHIPS

Think about production functions. Any two joint products are related to each other in one of three ways.

1 Two products are **rival** if an increase in the rate of output of one raises the cost of producing the other. Example: timber and forage. Picture a stand of ponderosa pine with grass beneath, on which cows are grazing. If the firm wants an enduring increase in timber output, it raises the stocking of pine. This cuts back on forage output unless the firm resorts to costly measures such as fertilization. Similarly, the rivalry of the two products becomes evident if the firm wants to increase the rate of forage output. Rivalry arises from the fact that both products draw upon the same fixed inputs, notably land: space, nutrients, water, sunlight, and so on. Thus rivalry is a common relationship among genetic products. Indeed, rivalry is the only relationship among joint products in which diminishing return obtains and a best combination can be found (page 121). More on this point as we go along.

2 Two products are **cooperative** if an increase in the rate of output of one lowers the cost of producing the other. An illustration is oak timber and eastern grey squirrels. If, with higher timber output in view, the firm raises the stocking, the resulting more abundant shelter and acorn crops support a higher squirrel population at no additional cost. Or if, with no initial change in the timber enterprise, the firm manages to enlarge the squirrel colony, more acorns will be "planted"—stashed away in the ground and then forgotten—and the tendency will be toward a heavier stand of trees at no extra cost in human effort. Two forest products are cooperative when each provides an input for the other, while the use that each makes of the land is essentially independent of the other's use.

Often cooperation is not reciprocal as with squirrels and oaks, but works in only one direction. Good scenery brings recreationists, but not vice versa. Scenery furnishes inputs to recreation, but receives none in return.

Cooperation is necessarily a transitory relationship among joint products. The forest calls a limit to such an increasing return, beyond which rivalry sets in and a best joint combination is possible.

3 Two products are **independent** if a change in the rate of output of one has no bearing on the cost of the other. Since you have had it dinned into you that everything depends on everything else, you may reject the notion of independence. And well you may, for it implies that a firm can have two land-based enterprises with totally different, and even unrelated, inputs and outputs. One must work with the idea of *virtual* independence. Strictly speaking, independence is not a joint-product relationship and so not multiple use. Each single product encounters its own diminishing return and best combination.

Can you describe a situation in which two forest products are virtually independent?

Whether products are rival, cooperative, or independent is a function of the circumstances. Let's recognize four kinds of circumstance.

First, the relationship of two products depends partly on the *intensity of resource management,* as measured, say, by outlays per acre of land. Under very extensive management, timber and forage may verge on independence; off-road cycling and horseback riding may be cooperative. But as the firm tries to step up the output of either product in either of these pairs, the relationship moves toward rivalry.

Second, the relationship of two products depends in part on their *inherent characteristics.* Hardwood timber and deer are often cooperative products; recreational hiking and range forage can be seen as virtually independent—up to a point. Pine and hardwood timber may be rivals.

Third, the manner in which two products are related is a function of the *pattern of their production* in either space or time. As an example of *geographic pattern,* consider the deliberate management of a large forest holding for home sites and timber. If the home sites are established in even a moderately close pattern, their occupancy and the effort to grow trees for harvest will conflict. But as the sites are more widely spaced, even though each is used just as intensively as before, the intervening ground can be given over to timber with less and less interference in either direction. The ultimate arrangement is a boundary line within the forest holding, separating homesites-but-no-timber on one side from timber-but-no-homesites on the other. One may term this ultimate arrangement **segregated multiple use,** in contrast to the closely interspersed pattern of **integrated multiple use.**

Is the segregated pattern properly termed "multiple use"?

Just as multiple use may be segregated in space, so it may be segregated in *time.* A tract of timber can be used as recreation ground until harvest age, when the recreation activity is moved and the timber cut and regenerated. Later, the recreation use is brought back and sustained until time for another harvest. This is crop rotation in an agricultural sense.

Fourth, interproduct relationships depend in part on the *culture.* In the U.S., with our tradition of lavish resource use, many of us are led to believe that clean-cutting in the forest is at odds with the output of aesthetic and recreation values. But in nations less well able to be fastidious, people think that the cleared patch of woods is as acceptable aesthetically as the cleared patch of corn.

CONCEPTUAL BEST COMBINATION

It's easy enough to think about multiple-use production functions in general qualitative terms, as in foregoing paragraphs. But to be explicit about product relationships to the point of attaching numbers to them is quite another matter. A first difficulty is that one often doesn't know how to measure forest *outputs.* You saw this in the case of aesthetic resourccs in Chap. 11. Finding a unit of measure amounts to finding a general definition of beauty, and this is impossible. Chapter 27, in measuring the output of fish, wildlife, and recreation resources, fell back on the "user-day," surely a weak proxy for what was sought. A second difficulty is that information about the *inputs* to production is poor. Even for timber, the best-understood forest product, production functions are built on such measures as forest type, site index, and stand density and age, indirect expressions for the climatic and edaphic inputs.

And if one doesn't know what the functions are for individual products, one can scarcely be expected to express product relationships explicitly. Turning back to the oft-repeated central question, What are the alternatives? One is abashed to recognize that in multiple use, one really doesn't know.

To be sure, one does have a fairly clear idea of what multiple-use functions might look like. Take the hypothetical case of two rival products, A and B. Assume that their outputs can be measured and that unit values in dollars can be attached. The joint production function is illustrated in Table 28-1. Dollars of total variable input (variable cost) are given in the body of the table. These are expenditures per acre per year for such things as interest, labor, plant materials, fertilizer, and other supplies. Physical outputs of A are shown along the top of the table; outputs of B, along the left side. Thus Table 28-1 is a story of **production possibilities.** It describes 121 discrete combinations, but you can picture the function as continuous, assuming that units of A and B are finely divisible.

That products A and B are rivals is demonstrated by the fact that as the output of either product increases, the (marginal) cost of the other also increases. For example, observe any two columns of figures in Table 28-1. There are 11 pairs of total costs in the two columns. The difference between the figures in any pair is the marginal cost of A at that level of output. Observe that this marginal cost rises with the output of B. Then study any two *rows* in the cost matrix and note that the marginal cost of B rises with the output of A.

The whole cost matrix of Table 28-1 lies within the region of diminishing return. You can recognize this by noting that the marginal cost per unit of output of A or of B or of the two together rises with total output. That is, the cost differences along any straight line are larger and larger as you run your eyes through the matrix in any direction between northward and eastward: Successive increments in output are achieved with increasing difficulty.

Diminishing return in Table 28-1 is evidenced also by the shape of the curves of equal total variable cost, the **iso-cost** curves. Starting with any number near the northwest corner of the table, trace roughly the locus of this same cost as your eye moves southeastward. Note that this iso-cost line is concave, and so are all the others. Two are sketched as solid lines in Table 28-1. Cost along the upper curve is $300; along the lower one, $50. Other things equal, successive increments of A can be had only at greater and greater sacrifices of B, and vice versa.

TABLE 28-1
IDEALIZED PER-ACRE-PER-YEAR PRODUCTION FUNCTION FOR TWO RIVAL FOREST PRODUCTS, A AND B, AND BEST COMBINATION OF RESOURCES UNDER TWO DIFFERENT ASSUMPTIONS ABOUT PRODUCT VALUES

Output of product B—physical units	Output of product A—physical units										
	0	1	2	3	4	5	6	7	8	9	10
	Total variable input—dollars										
10	275	289	307	329	355	385	419	457	499	545	595
9	225	238	255	276	301	330	363	400	441	486	535
8	180	192	208	228	252	280	312	348	388	432	480
7	140	151	166	185	208	235	266	301	340	383	430
6	105	115	129	147	169	195	225	259	297[a]	339	385
5	75	84	97	114	135	160	189	222	259	300	345
4	50[b]	58	70	86	106	130	158	190	226	266	310
3	30	37	48	63	82	105	132	163	198	237	280
2	15	21	31	45	63	85	111	141	175	213	255
1	5	10	19	32	49	70	95	124	157	194	235
0	0	4	12	24	40	60	84	112	144	180	220

[a]Unit value of A, $40; of B, $40.
[b]Unit value of A, $7; of B, $21.
Solid curves are iso-cost; broken, iso-revenue.

> If Table 28-1 is a map of elevations, the iso-cost curves are contour lines. Suppose a rivulet enters the map at the northeast corner. Approximately what path will it follow to the foot of the hill? Where will it tend to be running most swiftly?

Since diminishing return prevails, you can identify the best combination of A and B (again see page 121). It is that which equates the marginal cost per unit of output of each product to its marginal benefit, or marginal revenue (value), per unit.

Thus, suppose each product is worth $40 per unit. Search Table 28-1 for the combination where the marginal cost of each product has risen toward $40 but not beyond. Your search leads to the combination of 8 units of A with 6 of B, footnoted a. Here total gross revenue is 14 units at $40 each, or $560. Revenue net of variable cost is this minus $297, or $263, the sum available for paying fixed cost and profit, if any. Nowhere else in the whole matrix can a combination be found that yields so high a net revenue at the given prices.

The (8,6) best combination is found on the iso-cost curve at its point of tangency with an **iso-revenue curve,** the broken line. This iso-revenue curve has a negative slope of 1-to-1. Since A and B have the same unit value, all combinations along the curve yield the same total gross revenue.

Can you develop a proof of this?

The combination (8,6) lies on the iso-cost curve at the point of highest revenue. That is, it lies on the iso-revenue curve at the point of lowest cost.

Alternatively, suppose that A's unit value is $7; B's, $21. The iso-revenue curve now has a 1-to-3 slope: Three units of A are worth one unit of B. Tangency with an iso-cost curve is found at the combination (0,4), footnoted b. No other combination in the matrix generates a higher revenue net of variable cost. A's relatively low value excludes it from multiple use with B—that is, the two products are **incompatible.** Such incompatibility always arises, in the last analysis, because of differences in value and cost.

This chapter is interested in incompatible forest uses because it accepts segregated use as multiple use. And, of course, it is interested in uses that are **compatible.** In a way, such uses are its central interest, because they give rise to problems of multiple use in the most widely accepted sense—that is, integrated multiple use. On the other hand, multiple uses that are independent or cooperative hold little interest, for they seldom present social problems. Thus the multiple-use question dwells upon rival compatible products.

To look again at the best resource combination in multiple use, it is the combination at which a small shift of cost in favor of any product or set of products, C, at the expense of any other product or set, D, reduces the benefit from D and increases the benefit from C in equal amounts, so that total benefit is unchanged. (You are asked to *envisage* C and D. They are not graphed or tabulated here.) Bear in mind that the loss of D-benefit in this case is the opportunity cost of the gain in C-benefit. Now if the output of C is raised enough to depart from the best combination, the concavity of the iso-cost curve produces the result that the loss in D-benefit is greater than the gain in C-benefit. That is, the opportunity cost of a unit of C-benefit is greater than before. Recognizing that C is *any* product or set of products, one concludes that the best combination of resources is that which minimizes the opportunity cost per unit of every product in the mix.

You are observing here how a **multimarginal equilibrium** is found. The equilibrium takes analogous forms in production and in consumption.

APPROACHES TO MULTIPLE USE: TRIAL AND ERROR

In these times, when even the best informed are still very much in the dark about production functions and unit values, what means are being found for solving multiple-use problems in forestry? The approaches are wide-ranging. Let's recognize just three sorts of approaches, remembering that we are interested in deliberate, but not inadvertent, multiple use. The three approaches are (1) trial and error, (2) planning for compatible uses, and (3) planning for incompatible uses.

Trial and error is by far the commonest approach to multiple use in every forest region of the world, whether we are thinking of taungya for teak and food in southeast Asia, forest settlement in tropical Africa, shifting cultivation in Latin America, or timber-game recreation in western Europe. Trial and error is not necessarily haphazard; it may make studious use of research, and certainly it emphasizes enlargement of net benefits on the basis of experience. For a case in point, we go to Belgium [3].

Belgium is a $7\frac{1}{2}$-million-acre patch of land with a population of 10 million. Heavily industrialized and urbanized, the country nevertheless is 20-percent forested and has a long-standing tradition of forest improvement and management. Virtually all the forests, both public and private, are multiple-use forests.

Although a few forest areas have been set aside to preserve animals, plants, or historic sites, timber growing has long been, and still remains, the backbone of forestry. Belgium imports half of the wood she uses and is under pressure to be more self-sufficient. The history of forestry has been a twofold response to this pressure. First, land once denuded or given over to coppice has been reforested, and private forests destined for heavy exploitation have been bought by the state. Second, exotic tree species, including several conifers, have been brought in to replace or combine with the native hardwoods. The average rate of output has risen to $56\frac{1}{2}$ cubic feet per acre per year, which, next to the Danish forests, is the highest in Europe. The theme of all this activity is to grow more timber and earn more money.

But it is not only timber that is keenly wanted. In this congested corner of the world, all resources—certainly all forest resources—are at a premium. This is where multiple use comes in. And it's where trial and error comes in. Belgian forestry research is continuously at work to raise the level of output, not just of timber, but of all useful forest biomass. And forest managers and landowners have long been trying systems for combining service activities with commodity production.

Hunting was one of the early joint uses of the forest. Today, the sale of hunting privileges is a major enterprise. Here arises a typical multiple-use rivalry problem: The young conifers that will produce the best timber are subject to damage by the roe deer that are most prized by hunters. Collective trial and error over long years reveals solutions little by little.

In general, Belgians look upon the forest as recreation ground, a place where they are free to walk or to go on bicycle, horse, or skis. They like to bird-watch, enjoy the scenery and the quiet, have picnics, and take photographs. In recent decades, the recreation use of the forest has grown apace. Government agencies have made sincere and largely successful efforts to diminish the adverse effects of rivalry between human use and the safeguarding of the forest ecosystem. Schoolchildren are taught about the forest and are taken there for nature study. As adults, their forest education is continued. Most forest users are cooperative. The theory is that people who understand the forest will better understand themselves and be kinder to one another. Well informed in forestry matters, they can be consulted about forest policy and practice.

One can emphasize two points in the foregoing. First, multiple-use forestry can be deliberate and wholly accepted without being self-conscious: It can be taken for granted. Second, trial and error lends itself well to treating multiple-use choices as political choices, which is what they are whenever they involve equity, allowing these persons a little more of the benefits and those a little less.

> What does "political" mean? How does it relate to "economic"? Is it helpful to see equity questions as political? to see them as economic?

SECOND APPROACH: PLANNING FOR COMPATIBLE USES

Planning for compatible uses—or potentially compatible uses—is the approach to multiple use that is illustrated with an imaginary example in the section on Best Combination in this chapter. Real examples can be found, too, though none so neat as the make-believe case. In the U.S., the most common real examples concern two products: timber and forage for livestock. A few economic analyses have been made, also, of many different outputs from a specific forest watershed. A good example is the Beaver Creek Watershed in central Arizona [2].

Here we will look briefly at a nationwide analysis [4] in which the U.S. is divided into seven regions. The forest and range lands of each region are subdivided into some 5,000 "resource units," consisting of the possible combinations of cover type, owner class, productivity class, and stocking condition. For each resource unit, a set of management levels is recognized, each emphasizing different management objectives and giving preference to different resource outputs. The outputs analyzed are timber growth, timber harvested, forage for domestic livestock, forage for wild ruminants, herbage and browse, dispersed recreation use, water yield, storm runoff, and sediment yield.

Using a complex mathematical model, the analysis seeks combinations of the kind sought in the simple best-combination exercise. In the exercise, the products are potentially compatible rivals, and marginal benefit and marginal cost are equated for each product represented in a combination. Production functions—product interactions and costs—are largely judgment estimates gathered from panels of professional experts: resource managers and research workers. Unit values are taken from a special study made within the Forest Service by methods such as are discussed in Part Five.

The analysis begins with the setting of national and then regional production goals for timber and forage. For example, the 1995 goals for the Southeast region are 3.06 billion cubic feet of softwood timber harvest, 1.35 billion of hardwood timber harvest, and 21.5 million animal-unit-months of range forage. These goals, along with the production functions and unit values, are put into the model, which then provides some forecasts under the assumption that goals are met and marginal benefits and costs are equated. For example, again in the southeast: Dispersed recreation use by 1995 must be cut to 4 percent below the 1977 level. Increased timber harvesting and grazing will necessitate entry into poorer lands more susceptible to erosion, and sediment yields will rise by 1995 to 116 percent above the 1977 level. The marginal cost of producing timber will rise to 24 cents per cubic foot—and of forage to $10.52 per animal-unit-month—by 1995.

The result of this analysis is a set of guides to national and regional planning and a backdrop against which local planning may go forward.

THIRD APPROACH: PLANNING FOR INCOMPATIABLE USE

Now for a look at the third sort of approach to multiple use, planning for incompatible uses.

When a proposal is weighed, within the federal government, for reserving a national forest tract as wilderness, there are alternatives at hand. These are to use the property for timber and other commercial purposes or to use it for hiking, camping, and the like in a wilderness environment. The alternatives are seen as incompatible, and so the question is whether to have a segregated wilderness or not to have it.

The accepted criterion for probing the question is that the tract should be reserved or not reserved according to which choice is expected to result in the greater net benefit to society. The tract should be reserved if, but only if, the benefits or reservation exceed the costs. These costs include the opportunity cost of forgoing the benefits of timber and other nonwilderness uses. This is an efficiency criterion.

Costs may be stated as capitalized values or as annual values. One analyst estimated annual costs for the proposed 35,500-acre Cranberry Wilderness on the Monongahela National Forest in West Virginia (Table 28-2). The costs for coal, oil, natural gas, and private land were calculated by (1) estimating the total costs of acquiring property rights and (2) converting these costs to an annual basis at $6\frac{7}{8}$-percent interest. The estimated total of all costs, put on a per-visitor-day basis, comes to about 54 1990 dollars in 1978; it is about $36 in 2000, when use of the wilderness is expected to be comparatively heavy.

And what about the benefits? How does one arrive at a response to this comparatively elusive question? The most common method is political: Let the legislature be the market in which values are arrived at. Ordinal values will do. The members of the congressional subcommittee ask themselves if the Cranberry Wilderness is worth $54 per visitor-day. They ask if it will be worth $36. They go over the affirmative and negative arguments and take a vote. They may attach importance to estimates, in dol-

TABLE 28-2
ANTICIPATED COSTS, IN 1990 DOLLARS, OF THE CRANBERRY WILDERNESS IN 1978 AND 2000

Cost category	1978	2000
Opportunity costs:		
Coal production	389,527	389,527
Oil and natural gas production	38,953	38,953
Timber production net of mgt. cost	208,387	1,159,519
Wilderness management cost	31,053	65,254
Private land acquisition and other costs	77,033	77,033
Total direct social cost	744,953	1,730,286
Average cost per anticipated visitor-day	53.60	36.20
Average cost per acre	20.99	48.67

Source: Guldin, Richard: "Predicting Costs of Eastern National Forest Wilderness," *Journal of Leisure Research,* second quarter 1981, Table 2, p. 127.

lars, of the social value of wilderness. For the region that includes West Virginia, the Forest Service has *recommended* a value of $12 per visitor-day ($21 in 1990 currency).

A bill authorizing the Cranberry Wilderness was signed into law in January 1983.

Why do you suppose the $21 value was disregarded? To what records would you turn to find out?

GROUNDS FOR SKEPTICISM

Through all our talk about the multiple use of forest resources, there runs an invitation to skepticism, whether or not the topic is manageable with the customary tools of economic analysis. We may be skeptical mainly because the two rocks of refuge that we are used to finding near at hand are, with multiple use, inaccessible: (1) We are mostly unable to specify the production functions; and (2) we are largely ignorant of the unit values of outputs and even of variable inputs.

Thus, in multiple-use analysis, the problem solver is crippled in the effort to be specific about an efficient combination of resources. To be sure, this sort of handicap is familiar: We always run into it because of our uncertainty about the past and our dire uncertainty about the future. But multiple use compounds uncertainty because it compounds the variety of productive outputs, extending them further into the regions where values are hard to judge—values of recreation, scenery, water, wildlife, and the like.

However, skepticism about problem solving in the realm of multiple use is not only, or even chiefly, focused on the efficiency question. It concerns equity, too. Indeed, a large part of our self-doubt on the score of multiple-use analysis arises out of uncertainty about the distribution of benefits and costs among persons and groups. One doubts that equity is any more difficult an issue in multiple-use questions than in others. But in multiple use, equity is a prominent issue, publicly aired, and widely and vehemently argued.

The manifold multiple-use problems that give rise to our skepticism have been addressed repeatedly, and often in an effort at comprehensive treatment [1]. But solutions to the problems are persistently elusive.

Go back to the three sorts of approach to solving multiple-use problems in forestry. The first is trial and error, with Belgium as an example. The point is made that trial and error lends itself to treating multiple-use choices as political choices: Shall these persons or those be granted a little more of the net benefit? Here is the equity question, which must be resolved in recognition of cultural tenets.

A second sort of approach is "planning for compatible uses." A nationwide Forest Service analysis is taken as a case. A "complex mathematical model" is used, and alternative resource combinations are described in numbers for 1995. The closing comment is that the result of this analysis is a set of guides to national and regional planning and a backdrop against which local planning may go forward. This comment means that the Forest Service's complex mathematics, which addresses efficiency, may prove to be of minor interest. The real problem solving will be tackled on the political, equity front, where "planning" will be accomplished by representatives of the varied contending interests in forest use.

The third sort of approach is "planning for incompatiable uses," with wilderness versus multiple use of a national forest tract as the case in point. Analysts in pursuit of efficiency spelled out the numbers, which purported to show that the wilderness use of the tract would entail costs markedly higher than were justified by the benefits foreseen. The tract was then (surprisingly?) declared a wilderness. Equity considerations prevailed, and no arithmetic was called out in support of them.

SUMMARY

This chapter defines multiple use and the rival, cooperative, and independent relationships among forest outputs. It recognizes that product relationships depend upon circumstances that may differ from case to case. It draws a distinction between multiple uses that are integrated and those that are segregated in space or in time. It touches on the economic theory of joint products and takes up three real cases of multiple-use management, involving (1) trial and error, (2) planning for compatible uses, and (3) planning for incompatible uses. The chapter closes with a philosophical essay on the validity of our economic problem solving as applied to multiple use of forest resources.

REFERENCES

1 Driver, B. L., compiler: *Contributions of Social Sciences to Multiple-Use Management, an Update,* U.S. Department of Agriculture, Forest Servive Gen. Tech. Rep. RM-196, 1990, 125 pages.

2 Fox, Kel, and Lawrence D. Garrett: "Beaver Creek and Multiresource Management Forestry"; see also the companion paper by Lawrence D. Garrett, W. Wallace Covington, and Aregai Tecle: "Research Needs in the Southwest Ponderosa Pine Type, and Other Related Papers," *In* Tecle, Aregai, W. Wallace Covington, and R. H. Hamre, coordinators: *Multiresource Management of Ponderosa Pine Forests,* USDA Forest Service, General Tech. Rep. RM-185, 1989, iii + 282 pages.

3 Gathy, Pierre: "How We Use the Principle of Multiple-Use Forestry in Belgium," *In IUFRO-MAB Conference, Research on Multiple Use of Forest Resources,* Flagstaff, Arizona, May 18–23, 1980; U.S. Department of Agriculture, Forest Service Gen. Tech. Rep. WO-25, 1980, pp. 97–101.

4 U.S. Department of Agriculture, Forest Service: *An Assessment of the Forest and Range Land Situation in the United States,* FS-345, Washington, 1980, pp. 508–526.

CHAPTER 29

SUPPLY OF FOREST OUTPUTS

DEFINITIONS

To wrap up our studies of genetic production, let's look at the supply of forest services. The **supply** of a service is a statement of the quantity of the service that producers stand ready to release to users during a given time, depending on circumstances such as price or the guiding rate of return. Supply is a functional relationship: quantity as a function of a determinant such as price or guiding rate (glance at Fig. 29-2). The function is termed a **supply function.** Thus supply as a function is analogous to demand (page 63, item 4). And analogous to consumption (item 5) is production, the output that actually flows to consumers during an interval, the determinants being what they are.

As a function, supply is a statement of simultaneous alternatives. Differences in quantity supplied are attributable to differences in determinants, not to changes over time. All readings on a given supply function apply to the same point in time.

In order to describe a supply function, one must explain what *service* is supplied, in what *quantities,* in what *place or market,* during what *period of time,* and what *determines the quantities* supplied. This last is the chief question here.

The problem of resource supplies and their determination is the fundamental forestry problem. The purpose of forestry is to supply forest services to people. As people become more numerous and ravenous, the forestry profession seeks to respond. Where does it look for the means to respond? What is required to grow more timber—or to make more recreation services available? These are policy problems, and an understanding of supply is requisite to their solution.

FLOW VERSUS STOCK

Supply is a **flow** of service, a movement of a resource at a rate of thus-and-such a *quantity per unit of time.* In contrast is a **stock,** an idea used frequently in Part Three and used now in the same sense as before: a reservoir of a resource (example: timber growing stock) that can give rise to a flow (example: timber harvest) or receive and store a flow (example: timber growth that is not harvested, but is added to stock). Stock is measured as a *quantity at a point of time.*

The supply of forest genetic outputs is a flow out of stock. The part played by stock is never insignificant, as it often is in the supply of manufactured goods. At the paper mill, for example, newsprint may be loaded for shipment directly from the end of the production line. But in the forest, there is always a stock of finished goods from which the product is supplied. This stock is characteristically large, which is what makes forestry such a capital-intensive business. Capital-intensiveness is the rule when timber—or water or range forage or fish or wildlife or aesthetic values or recreation opportunity—is supplied from the forest.

> For each type of forest output, explain what the stock is from which the flow of supply is taken. Why is the stock held? Is there a financial advantage in it? a physical necessity? Is it an aspect of our culture?

SUPPLY CONDITIONS

A supply of forest resource may be generated under any of three **supply conditions.** Remember that supply is a function: a set of simultaneous alternative quantities of output for release to users. The three supply conditions are: (1) **Stock supply:** A set of quantities that exists because the circumstances of stock holding change while all other inputs are fixed. (2) **Short-run supply:** a set of quantities that exists because some inputs other than stock are varied, though some inputs are fixed and technology is fixed. (3) **Long-run supply:** a set of quantities that exists when as many as all the inputs are varied and technology can be changed. To a degree, it is the element of time that distinguishes among these three conditions of supply. If, for illustration (1), you want more timber from your forest today, you can only change the "circumstances of stock holding" and draw down your stock by cutting. But if (2) you can wait a while, you can get more timber by using fertilizer, an input other than stock. In the still longer term (3), you can plant land now bare, thus varying inputs that hitherto were fixed. Compare the analogous *demand conditions,* page 81.

In each of the foregoing illustrations of timber supply, there is a **supply response,** a change in the rate of resource flow. Such a response can come about in either of two ways (compare the *demand* case, page 82). First, it can be a **change in quantity supplied,** a shift along a given supply function. Thus the prospect of high real prices for timber may prompt you to use some variable input such as fertilizer, as prescribed on a given short-run supply function. But second, a supply response can result from a **change in supply:** a shift in the whole function, as when your creation of a new forest in the long term then presents you with new opportunities to vary the quantity supplied in the new short run. Supply responses are **positive** if they result in more being supplied; **negative** if less.

STOCK SUPPLY

Now to examine in more detail the first of the three supply conditions, the stock supply:

Suppose that a forest changes hands, from an owner with a high guiding rate of return to one whose rate is low. For the new owner, the forest is understocked, and the owner begins, therefore, to build up stock by seeding, planting, and holding timber off the market. This is a negative supply response.

Or suppose that the state parks department's budget is increased, and the department behaves like any other firm with bettered finances and a lowered guiding rate of return: It takes on some investment projects that were previously unattractive, opening up some new recreation facilities and thus increasing the quantity supplied. Here is a positive supply response.

You will note in the illustrations a difference between timber and recreation in the way that stock changes are related to stock-supply responses. In the timber case, the relation is inverse: A negative response is effected by enlarging the stock. For recreation resources, on the other hand, the relation between stock change and supply response is direct.

> How is it with water, forage, wildlife, and aesthetic values: Is there an inverse or a direct relation between stock change and supply response? Why the difference?

What are the independent variables in stock supply? This question is virtually the same as asking what determines the best stock. The variables are in hand in Chap. 24. They are, first, the firm's guiding rate of return and, second, the stock's marginal value growth percent (or, in general, since growth in a literal sense may not be involved, the marginal rate of return on the stock). The latter is affected, not by the magnitude of unit value, but by an anticipated change in unit value or a change in size or quality premiums.

So, for example, in the case of timber, firms are apt to give a positive stock-supply response if (1) their guiding rate of return rises, (2) size premiums on timber are reduced, (3) the firms come to envisage a drop in real timber prices greater than they had previously anticipated or a rise less steep, or (4) the past growth of timber has led to accumulating stock in excess of the optimum.

Right from the start of this chapter, supply is referred to as a function: magnitude of resource flow in relation to some independent variable. From this it appears that one should be able to draw a **supply curve** on coordinate paper. For short- and long-run supplies, this is possible. But stock supply is not so readily graphed. The optimal stock can be pictured, as in Fig. 17-1; but what about the flow from or into stock? The flow depends on current as well as optimal stock and the determinants of these: (1) past departures from optimum, (2) current guiding rates and changes, (3) prospective value changes. The relationships can be described with symbolic mathematics, but not so readily with geometry. We'll not try to draw stock-supply curves.

SHORT-RUN SUPPLY

Now for short-run supply. This is the same short run as on pages 81 and 115. Technology is fixed, and some of the inputs are fixed, some variable. Thus diminishing return caused by crowded fixed inputs is characteristic of the short run, and marginal cost per unit of output rises with output in the relevant range [Table 17-1 (bottom half)].

Indeed, the curve of marginal cost per unit of output may be taken as the short-run price-supply curve. Turn to Fig. 17-1(f). Look at it as a general statement of short-run output in relation to price or cost. Picture the price level (marginal revenue) moving up or down through the graph, determining series of intersections with marginal cost and series of total outputs. You can see that marginal cost is the locus of quantities that producers stand ready to produce in response to price. The supply curve, by definition, is exactly this locus.

In genetic production, there *may* be a considerable time lag between incurring cost for inputs and receiving revenue from outputs. To understand the effect of a lag, consider the fertilizer case of Chap. 26, where the lag is 10 years. We will depart from two of the Chap. 26 assumptions in order to make the case more instructive for short-run supply.

First, assume that you know enough about fertilizer technology so that you can draw a production function, a curve of timber yield per acre, (item 6, page 191) in relation to quantity of fertilizer per acre (item 1). For a given unit cost (items 2 and 3), you can then graph the relation between board-foot yield and marginal cost of fertilizing per board foot. *This is timber's short-run price-supply on account of fertilizing,* the lowermost solid curve in Fig. 29-1. Demand, based on items 8 and 9, is sketched as a broken line.

For a second departure from the terms of Chap. 26, let Fig. 29-1 refer to the tenth year, when the lag has run its course, the effect of fertilizer is complete, and the timber is ready for harvest. The cost of delayed returns has been paid for at the guiding rate of interest. Now what if the guiding rate isn't 4 percent? The result is sketched in Fig. 29-1. The higher the guiding rate, the more costly the 10-year time lag, and the lower the quantity supplied at any given price. Furthermore, the higher the guiding rate, the higher the timber price below which supply ceases altogether. The supply curves may be viewed as discontinuous, hopping over to coincide with the vertical axis at prices too low to warrant the use of fertilizer.

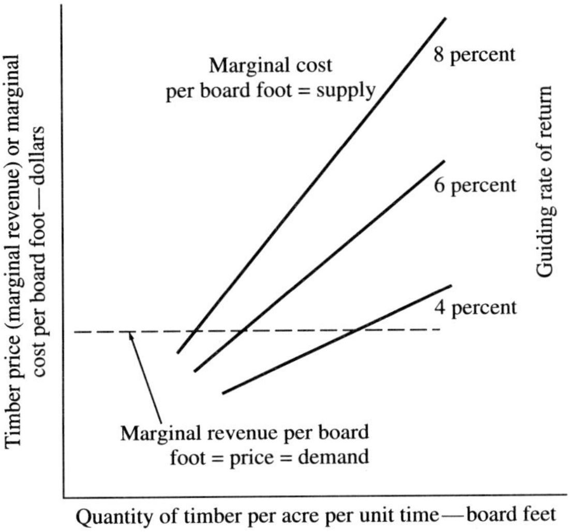

FIGURE 29-1
Short-run price-supply of timber on account of fertilizing: quantity per acre per unit of time in relation to price at three illustrative guiding rates of return (solid lines); and price-demand for the timber (broken line). Hypothetical data.

Just as there may be short-run supply of timber on account of fertilizing—or fire fighting or insect control or seeding or planting or stand improvement—so there may be supply of water on account of artificial impoundment, or a supply of range forage on account of seeding, or of wildlife through habitat improvement, or of recreation resources through road building. In each case, there are variable inputs other than stock, such as labor and materials, but some inputs are fixed, and diminishing return is in force. Each case, too, contrasts with stock supply in this respect: There are variable inputs that appear on the cost side of the ledger but not on the revenue side, and product price therefore can affect quantity supplied. In each case, there may be, between input and output, a time lag long enough to call for use of an interest rate, which then becomes another supply determinant.

A common instance of the short-run supply of timber (on account of logging, in this case) is found in timber's **margin of availability.** If end-product values are low enough, they fail to cover the costs of extracting and converting the stumpage. Thus stumpage value is zero or negative, and the quantity supplied is nil. But let end-product values rise relative to costs, and a supply curve emerges into the positive quadrant. Some of the timber becomes available. At successively higher product values, the marginal log or bolt becomes smaller and of lower quality. Again, the marginal tree changes in size, quality, and species. The marginal stand is a lighter or otherwise less valuable stand. The marginal logging chance moves further back from the market, or into less accessible ground. The marginal supply regions of the nation and the world shift outward toward higher-cost situations. The marginal firm is a different, less efficient firm. And new margins are set for all the agents and instruments of production that the firm uses.

One concludes that forest output may show a positive supply response in the short run to any of the following: First, a higher unit value expected for the output as of the time of its occurrence, resulting in higher marginal revenue per unit of output and an *increase in the amount supplied.* Taking Fig. 29-1 as the model, the driving force is an increase in demand, a higher broken horizontal line. Second, where there is a time lag, a lower guiding rate of return, the substitution of a lower-rate for a higher-rate supply curve in Fig. 29-1, which is an *increase in supply.* Third, a lower unit value of input as of the time of its occurrence, resulting in lower marginal cost per unit of output, an *increase in supply* at any guiding rate. Fourth, any difference in the physical relation of output to input that will make for either higher marginal revenue (other things equal) or lower marginal cost.

Anticipated changes in value, influential in stock supply, have no bearing on short-run supply. In the short run, it is value *difference* (value at a point of time or value applicable to a given period), *not change* in value (over time) that matters.

> Why? What are the alternatives?

Before leaving the short-run supply, we owe it to ourselves to remember that our analysis implies rational responses by genetic producers to the unit values of things. Such a reminder is perhaps not essential with the stock supply, because there the guiding rate of return is dominant, and the guiding rate is an index of human action that requires only the analyst to be rational; it places no such requirement on suppliers. But with the short-run supply, unit values become dominant. Though these values may be subjective, as with many recreation opportunities and aesthetic experiences, often they are objectively set in the market, and if they are, one is attributing rationality to suppliers if one supposes that they respond to values as "economic men" would do.

What, for instance, will induce laborers to work faster or longer? Greater material rewards? Perhaps so, but perhaps not. During World War II, there was a greatly increased demand for ginseng root, gathered in the northern forests and used for medicine. Buyers sought to increase the output of the pickers and decided to raise the price per pound offered to the pickers. But when the buyers raised the price, the result was the opposite of what they expected. The pickers gathered *less* than before! The price having been raised, they could now satisfy their desire for income by four days' work instead of five. Apparently if the buyers wanted more ginseng, they should have lowered the price! Any sort of supply response in which human beings are involved—that is, every supply response—can be a complex phenomenon.

> Sketch the price-supply curve for ginseng and the price-supply curve for the labor of the ginseng pickers. What is the difference between your curves and those in Fig. 29-1? What gives rise to the difference?

LONG-RUN SUPPLY

Long-run supply differs from short-run supply in the proportion of inputs that are variable. The greater the proportion of variable inputs, the longer the run of supply. The last word in long-run supply is the supply that society looks forward to if it imagines or plans that all inputs to genetic production are variable. The long-run supply necessarily lies in the future. When it has materialized and major forest inputs are again fixed, the supply is short-run or stock supply.

The simplest example of long-run supply involves the forest versus nonforest use of land. Where currently open land is planned for afforestation, as in China and in parts of South Temperate nations, major forestry inputs are seen as variable. And the supply responses, by and large, take a long while to materialize. But long-run responses are not invariably long delayed. The negative responses following deforestation show up soon after the initial (positive) stock-supply responses have played themselves out.

The less simple, but more challenging and exciting, example of long-run supply involves technological change, change in knowledge, which is the boundless resource. What is the long-run timber-supply potential of U.S. forests if genetic improvements greatly increase trees' efficiency in converting light, water, air, and nutrients to wood? What is the multiple-use potential if constraints of incompatibility are broken down by cultural and engineering changes? What is the potential if the technology of forest-resource consumption is revolutionized in favor of full utilization? The answers depend on our research and development.

But to return to the simpler case of land use, what are the determinants of long-run supply responses in the form of afforestation or deforestation? The financial **theory of location** or of **land use** is this: The lines of activity that can afford to pay the highest price for a given piece of land, or the highest **rent** for its use, occupy the land. The tendency is for the most intensive land uses to locate nearest population centers and for the less intensive uses to locate farther and farther out. Forests, which for many uses are most tolerant of location, take what's left. Small forest parks take what's left in and around the town. Woodlots take what's left on the farms. State, county, and national forests and parks and industrial and other sizable forest tracts take what's left farther back.

Looking around, one finds that financial land-use theory explains a good bit of the forest-location pattern that one observes. Now comes the long-run forest supply question: What changes the pattern?

Although we hear a great deal about the impact on forest area of urban, residential, right-of-way, and reservoir development and the like, by far the principal determinant of forest acreage in the U.S. is its interchange with agricultural acreage. In New England and the mid-Atlantic regions, forests at first gave way to early primitive farming, then rebounded as the fertile midwest called out to eastern agriculturists. Change has slowed but still continues. In the past quarter century, forests expanded by about 8 percent in the northeast and shrank by some 6 percent in the midwest [4].

In the Southeast, clearing for corn, cotton, and tobacco produced a positive timber-stock-supply response. Then came an immense reversal with farm abandonment. This phase is over for now, and there are signs of agricultural revival.

In the West, one finds, not cycles of land use so much as a long-persisting downward trend in forest area. In the southern part of the Rocky Mountain region, pinyon-juniper woodlands have been converted to nonforest range—perhaps half a million acres in the past decade. In California and Oregon, 3 percent of the forest has given way to pasture, cropland, and reservoirs in the last 25 years. In Washington, urban expansion in the Puget Sound area has pushed the forest back; this is essentially the only major interplay of this sort involving long-run forest supply in the U.S.

Now consider another kind of long-run supply response arising from changes in forest use. This is the response that follows shifts in emphasis among the services of the forest. The clearest illustration is a public forest that has been managed for multiple use but now is set aside for recreation use, timber harvesting excluded. There is an immediate long-run negative timber-supply response. (Note that there is not an initial positive stock-supply response, as there always is when forest use is changed to nonforest.) Plenty of public forest lands have been "preserved" from timber use since the environmental revolution of the 1960s.

But it doesn't take a revolution in forest use to produce a supply response. Shifts in emphasis invariably affect the supplies of all forest outputs, no matter how slightly. Slight changes over wide areas can be significant. In any event, here again are the resource interactions studied in Chap. 28. These are biological and social interactions set in motion by cultural forces.

What, then, have been the independent variables in the U.S. long-run forest supply? They are determinants lying more clearly outside forestry than any dealt with to this point. They are, in brief, culture and agriculture.

Isn't the unit value of forest outputs a determinant of the quantity supplied in the long run? Take forest recreation opportunity as an example. Hasn't the positive long-run response of recent decades been the result of heightened unit values? It seems likely, but in the absence of any very good measure of these values, perhaps it is better to do as we have been doing: speak of a cultural revolution rather than a price revolution.

> It has been said that "the law of supply and demand is the best forester we have." What do you think of this statement?

TIMBER'S LONG-RUN SUPPLY

And what about timber? Are there significant relationships between quantity supplied and its determinants in the long run? In a study [1] made some years ago, supply functions were built for Douglas-fir west of the Cascade crest in Washington and Oregon. The functions were labeled "long-run," under a definition a little different from ours: the future time when the timber forest will have been converted to young growth in full production.

The study findings are displayed in Fig. 29-2. Regional timber output is somewhat feebly related to timber price [Fig. 29-2(a)]. The price-elasticity of supply at the anticipated level of 40 1957–1959 dollars per M bd ft is about 0.2. It is less at higher prices. The principal element in elasticity is the incentive to fertilize if the price is higher. We would identify this as a short-run phenomenon.

Quantity supplied is related to the guiding rate of return in Fig. 29-2(b), where price is held constant at 40 1957–1959 dollars. One's intuitive impression is that quantity is more sensitive to the guiding rate than to price. This is indeed the case. At the assumed average rate of 4 percent, elasticity is (negative) 0.4, about double the price elasticity. The guiding rate works on timber output through all the same influences as does price. Additionally, it affects output through its control over growing stock and rotation length.

The moral of all this? If society wants more timber from the Douglas-fir region, it should concentrate on lowering forest owners' guiding rates of return. One suspects that somewhat the same moral holds in other forest regions.

SUSTAINING THE SUPPLY

Before leaving the subject of genetic supply, I am obliged to raise a question that runs deeply into forestry planning and the carrying out of plans. How stable are our supply

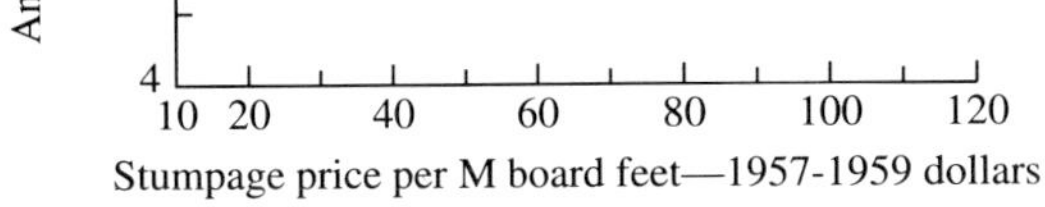

(*a*)

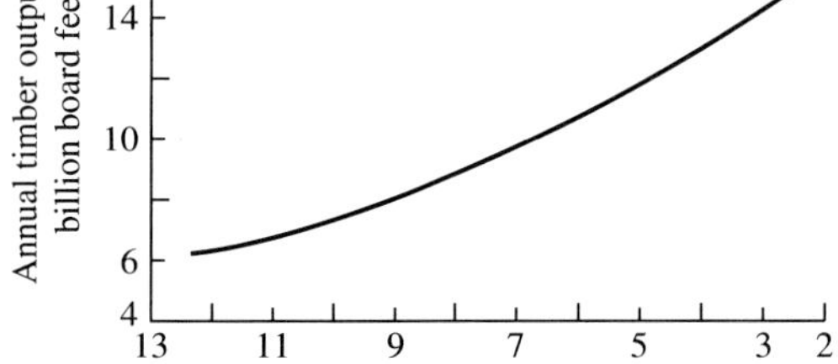

(*b*)

FIGURE 29-2
Long-run timber supply in the Douglas-fir region, in relation to (a) stumpage price and (b) guiding rate of return. (*Source:* Division of Forest Economics Research: *Timber Trends in Western Oregon and Western Washington,* U.S. Department of Agriculture, Forest Service, Res. Pap. PNW-5, 1963, p. 76.)

functions over time? If we plan for a given output of forest resources, is it possible to hold firm to the plan as years go by? Can there be such a thing as sustained yields of forest services? Or is there an inevitable loss of the forest's momentum—its productivity?

Elsewhere in this book, notably in Chap. 39, productivity is discussed as a parameter of the manufacturing economy. It is defined, in brief, as output per unit of input per unit of time. One can apply the same definition in genetic production.

In manufacturing, the sustaining of productivity is a relevant question, though by the nature of manufacturing technology, social concern lies rather with building higher productivity over time. In genetic production, with its typically longer period of performance, society is of course concerned with enhancing the level of productivity, but, in the interim, while productive efforts are awaiting fruition, the nagging question of maintenance thrusts itself forward. How are forestry efforts going to sustain pine output in southeastern U.S.? In the Pacific Northwest, how are the efforts to sustain the yields of Douglas-fir, hemlock, and pine going—and how *will* they go? How can we sustain the yields of water, wildlife, and recreation opportunity? Concern for sustaining resource productivity extends across the resource fields—land, water, air—and across national boundaries. It takes a long time to find answers and to develop technology that will ensure encouraging answers [2, 3, 5]. Meanwhile, the question persists: How can we develop, or even hold constant, the technology of genetic forestry?

SUMMARY

Here is the argument in brief: In forestry, genetic supply is conspicuously a flow of services out of stock. Such a flow can occur under any of three supply conditions: (1) The flow can result solely from a change in stock. This is the stock supply, and its chief determinant is the guiding rate of return. (2) The flow can come from inputs other than stock, when some inputs are fixed and technology is fixed. This, the short-run supply, is controlled by the physical relation of productive outputs to inputs and by the unit values of outputs and inputs, including the guiding rate of return. (3) The flow can result from changed technology or land use or from a changed relative emphasis on the types of forest output. This is the long-run supply, and the chief related variables in the U.S. have been culture and agriculture.

All in all, the most effective influence on forest output is the machine that produces it: the forest stock. Thus the guiding rate of return comes to the front as a supply determinant.

REFERENCES

1 Division of Forest Economics Research: *Timber Trends in Western Oregon and Western Washington,* U.S. Department of Agriculture, Forest Service Res. Pap. PNW-5, 1963, pp. 74–81.

2 Ince, Peter J., John Fedkiw, H. Edward Dickerhoof, and H. Fred Kaiser: *National Measures of Forest Productivity for Timber,* U.S. Department of Agriculture, Forest Service General Tech. Rep. FPL-GTR-61, 1989, 14 pages.

3 Powers, Robert F.: "Are We Maintaining the Productivity of Forest Lands? Establishing Guidelines Through a Cooperative National Program," *In* Schweitzer, D.L., and M.J. MacNaughton, compilers: *Proceedings of the National Workshop on Monitoring Forest Plan Implementation, Minneapolis, Minnesota, May 14–17, 1990,* U.S. Department of Agriculture, Forest Service Land Management Planning, 1990, pp. 98–112.

4 U.S. Department of Agriculture, Forest Service: *An Analysis of the Timber Situation in the United States 1952–2030,* Forest Resource Rep. 23, Washington, 1982.

5 Van Kooten, G. C., and George Kennedy: "An Economic Perspective on Sustainable Agriculture in Western Canada," *Canadian Journal of Agricultural Economics,* December 1990, pp. 741–756.

CHAPTER **30**

HIGH POINTS IN PART THREE

1 Genetic production. The output of wood, water, wildlife, forage, recreation and aesthetic values, and sundry nonwood products that takes place in the forest apart from manufacturing activity constitutes genetic production.

2 Production function. The process of genetic production can be described by production functions. Their main ingredients are fixed inputs, notably land; variable inputs, commonly stock and sometimes labor and materials; outputs of forest services from stock; and technology, the setting and procedure of production.

3 Supply. The flow of output that is conveyed to users is the amount supplied per unit of time. For any type of output, the relation of amount supplied to its determinants is the supply of this type of output.

4 Supply conditions and determinants. The principal determinant of stock supply is the guiding rate of return; of short-run supply, unit value of outputs and variable inputs, including the guiding rate; of long-run supply, culture and agriculture.

5 Short-run supply. Examples of short-run supply responses are decisions about the intensity of management measures such as fertilizing and protecting the forest, where there are inputs other than stock.

6 Stock supply. Examples of stock-supply responses are decisions about timber stocking, financial maturity, and allowable cut. The model that supports these decisions is the equating of marginal value growth percent and the guiding rate of return.

7 Marginal growth. This index of the output from stock is, in the case of indivisible capital, the current growth. In the case of divisible capital, it is the derivative of current growth with respect to stock.

8 Guiding rate of return. This is a measure of what the firm regards as a satisfactory rate of return on its marginal investment. It is a rate of compound interest suitable for the firm's use in transposing values from one time of reference to another, as in compounding, discounting, and capitalization.

9 Compound interest formulas. There is a set of formulas for transposing values in time. Applied with appreciation for their limitations and with thoughtful interpretation of results, these formulas can be a useful tool for solving problems in forestry economics.

10 Koenig-Faustmann model. One of the compound-interest formulas is a simplified statement of the Koenig-Faustmann model: $V_0 = r \div [(1 + i)^t - 1]$. This model is useful for identifying the best combination in the case of indivisible capital, but, strictly speaking, is inaccurate in the case of divisible capital.

11 Guides to the best combination. Two guides are generally useful in identifying the best combination of resources. One is the principle of unity of the firm: that the choice among alternative combinations should be made in light of its effect on the firm's affairs in their entirety. The other is the principle of relevance: that the choice among alternative combinations need consider only the circumstances that vary with the alternatives; those circumstances that are fixed are irrelevant. The rational firm tends to behave in such a way as to capture revenues that are variable and escape variable costs.

12 Economic criteria for choice. Two such criteria are equity and efficiency. Among measures of efficiency that serve as guides to the best combination of resources are net revenue, internal rate of return, current net worth, and benefit-cost ratio. These guides can be used to weigh silvicultural alternatives and to evaluate investments in fertilization, fire management, and multiple use. However, multiple-use decisions, commonly in the public domain, are strongly swayed by equity criteria.

13 Does forestry pay? The more closely the woods enterprise is integrated with other enterprises and interests of the firm, the more clearly affirmative is the response to this question.

PART FOUR

MANUFACTURING PRODUCTION

CHAPTER **31**

UNCERTAINTY AND SIMPLE FORECASTING

HOW CHAPTERS 31 AND 32 FIT IN

My aim in this chapter is to introduce the subject of forecasting for the forestry economy. As a forecaster, I am deeply aware of my uncertainty about the future: of the risks I take in trying to foretell the future with any reliability. Indeed, so pervaded is prediction by risks that I will bring my topic on stage with the help of a little sketch of uncertainty and ways of coping with it. Then I'll conclude the chapter by displaying some primitive forecasting models. In Chap. 32, I'll talk about some less primitive models.

With these two chapters on forecasting, we interrupt the sequence followed in other parts of the book, which open with a preamble: an introduction and a look ahead. In Part Four, the preamble is delayed until third place, so as to leave the forecasting topic on neutral ground between Parts Three and Four.

The purpose of the changed sequence isn't to declare that forecasting is an orphan, but to emphasize its belonging, not to any one part of forestry economics, but to the subject in total. It has been postponed this far only so that we could first acquire useful background.

> Do you have any appreciation now for the subject of forecasting that you didn't have before starting this book? Explain.

Forestry economics concerns choices, and these exist only in the future; the past is a closed book. The problem solver's data must take the form of predictions. The consumer's data are predictions; the producer's data are predictions; marketing and planning demand predictions.

In passing, let's set the record straight about forestry economics as compared with other forestry subjects. Often we hear the remark that the social scientist is a forecaster, whereas the biological and physical scientists escape the burden. Reflection will convince you that this is not the case. Biological and physical scientists often deal with probability distributions that are higher and narrower than those the economist must cope with, but all are in the forecasting game together. Your ultimate purpose in studying silviculture is not to learn what happened in the past: It is to be able to foretell with some confidence what *will* happen to this stand of woods if it is treated in this way or in that way.

To say that the future is everything, or nearly so, is not to deny that information about the past is valuable. After all, the past is our only source of information. Forecasts are rocket journeys launched from the past.

UNCERTAINTY AND ITS DEFENSE

Consider, then, uncertainty and ways of coping with it.

Uncertainty is a state of mind and thus an attribute of persons and not of external objects or events. For example, I am uncertain about the price my stumpage will bring: I have multivalued expectations about the price. The probabilities that I attach to different levels of value may be subjective, based on a large element of judgment. One may be uncertain about some past event, but our interest here is in the future. If I estimate that stumpage price will fall within the most probable range, $31 to $35, I run a **risk** that the price will turn out to be lower or higher. The extent of uncertainty, or risk, is measured in terms of the scatter of values, notably the standard deviation of the probability function. Surrogates for this measure include the payout period (Chap. 23). Among ways for a firm to cope with uncertainty are these:

1 Disregard it, and make a **deterministic** forecast, one that implies single-valued expectations.

2 Make a deterministic forecast, but be explicit about the error attached to it. A related form of this model is sensitivity analysis (page 193).

3 Take a deterministic approach, but allow for uncertainty by altering one or more estimated values so as to be on the "safe side"—for example, upping the guiding rate of return (page 146) or lowering the estimate of anticipated net benefit (page 180).

4 Take a deterministic approach, but allow for uncertainty by making alternative forecasts under alternative assumptions, such as high, medium, or low population, GNP, income or cost rates, or site quality (page 181). This is approach No. 2 without error estimates.

5 Take a **stochastic** approach, working with probabilities.

6 Regardless of the approach, reduce uncertainty by shortening the planning period.

7 For certain types of forecast, take refuge in special provisions that may be available. One such special provision is insurance, appropriate where there is a small probability of a large loss.

In the years before World War II, a lot of research effort was spent on the feasibility and terms of forest fire insurance, notably for southern U.S. After the war, a number of firms began to offer such insurance. But the demand was not great enough to keep the program alive: Premiums were too high to attract many woods owners. Today, only a few insurance companies issue policies covering forest fire or storm loss. They provide coverage only to industrial owners that they already insure. They protect themselves by carrying reinsurance with Lloyd's of London, an international insurance market.

In 1980, Congress amended the Federal Crop-Insurance Act to include timber losses from fire and storm. The amendment provided for pilot test programs in selected states, and the Forest Service, in cooperation with insurance companies, made an intensive study of the acceptability and feasibility of this publicly subsidized coverage. However, Congress appropriated no funds to activate the system [1].

8 Akin to insurance in its effect upon the firm is a device used from time to time in the U.S., in which the government acts as the insurer. The decision to act is taken only after loss has occurred. Thus, during the heady months of high construction activity and lumber values in 1978–1979, the Douglas-fir industry was buying federal stumpage at accustomed soaring prices. Then the Douglas-fir market crashed, and the industry was left with stumpage-purchase contracts that it could not afford to fulfill. The industry's pleas for clemency generated a relief bill, signed into law in the fall of 1984.

9 Still another special provision for coping with uncertainty is **hedging** in the **futures** market. Price forecasts for softwood lumber can be hedged on the Chicago Mercantile Exchange—and, in the past, for softwood plywood, too, though this market was at times inactive. In the futures market, contracts for future delivery of the commodity are bought and sold at currently published prices (see the financial section of a major newspaper).

To illustrate: A yard wholesaling firm, when it acquires an inventory, can protect itself against risk of a price drop in the cash market by selling contracts for delivery in the futures market. This is termed a **short hedge.** If cash prices fall, the firm is protected against loss because, as delivery time approaches, it can buy futures back at the low price and thus cancel its futures obligations at a financial gain that offsets its loss in the cash market.

Conversely, a construction firm that agrees to build and sell a house at a set price can protect itself against an interim rise in the cost of building materials by executing a **long hedge:** buying contracts in the futures market. When delivery date nears, if prices have in fact gone up, the firm's cash loss is countered by the gain it enjoys in the futures market when it sells contracts at the higher price.

What happens to the short hedger when prices rise, or to the long hedger when they fall? The firms then forgo the fortuitous gains they would have made if they had not hedged. The gains are made, instead, by the speculators who play the futures market not as hedgers, but as risk-taking investors. These speculators grasp the opportunity to gain by their shrewdness and luck, but they are prepared, too, to lose part of the time and thus make the futures market a refuge for their conservative brethren.

This is all we'll try to say about uncertainty and ways of coping with it: a brief sketch of a large subject. We'll turn now to our main program: forecasting.

> Do you think that foresters will become more conscious or less conscious of their uncertainty as time goes on? Why? How do you expect the profession to respond?

FORECASTING METHODS: INSTINCTIVE; CUSTOMARY

Three basically different forecasting methods are in common use. Everyone uses all three, and often the three are used jointly. The one a person emphasizes for the question at hand depends on the question and on the person's constitution and upbringing. The methods are

1 Instinctive forecasting
2 Customary forecasting
3 Deliberated forecasting

Instinctive and customary forecasting are touched upon in the present chapter. Deliberated forecasting is the subject of Chap. 32.

Instinctive forecasting need keep us only a moment. It is built into the nervous system and is performed by the reflexes. If my hand gets too close to a flame, something tells me instantly that if I persist, I'll be burned, and I snatch my hand away. Again, I interpret a sudden loud noise as a forerunner (predictor) of trouble, and I bat my eyes or duck or run.

Customary forecasting, too, is built in: not into the nervous system, but into the culture. Every firm makes innumerable predictions by rote or rule or custom—according to how the members of the firm have been *taught* to think or behave. I have been taught to expect a high probability of reward if I behave in certain ways, of punishment if I reverse my behavior. I expect to be fed if the dinner bell sounds. I expect to be frowned at if I throw my food on the floor.

In *Fiddler on the Roof,* the story of a Jewish community in czarist Russia, the principal character, Tevye, is a practiced customary forecaster. He says, as he goes the rounds with his milk cart, "in our village, we have a custom for everything"; and he sings a song about it. He is the minstrel of a traditional society, where estimating the future is easy, on two counts. First, life is changing very little. Second, everyone is taught, from childhood onward, what the consequence of any given line of action will be, and since they *believe* what they are taught, and act in expectation of it, it most surely comes to pass. For a forecast, one need only turn to the appropriate section of one's memorized learning.

And, remembering *Fiddler,* it strikes one as ironical and prophetic that all of a sudden Tevye's assurances for the future disintegrate. The czar's soldiers organize a pogrom, and the village's traditions and customs are turned to dust. Having their works and words turned to dust is an occupational hazard for forecasters, even in a traditional society!

SUMMARY

To put it all in a nutshell, this chapter begins a two-stage overview of forecasting, placed between the book's Parts Three and Four to emphasize the wide connections of forecasting to much of forestry economics. The chapter opens with a sketch of the forecaster's central concern, uncertainty and ways of coping with it. The story goes on to two kinds of prediction models, instinctive and customary. A less primitive kind, deliberated models, is reserved for Chap. 32.

REFERENCES

1 Siegel, William C.: Information on insurance given to the author by letter dated July 2, 1987.

CHAPTER **32**

DELIBERATED FORECASTING IN THE FORESTRY ECONOMY

DELIBERATED FORECASTING CHARACTERIZED

This chapter turns to our central interest in forecasting, the use of deliberated models. **Deliberated forecasting** is marked by the same formal intellectual process that was identified in the Introduction as the procedure for generating a reply to a question: (1) Clarify the question, (2) find a suitable model, (3) put in data and derive a reply, (4) follow up.

Let's look in on four different forms of models for deliberated forecasting:

1 Judgment model
2 Projection model
3 Leading-variables model
4 Coincident-variables model

These four model forms are not mutually exclusive: Elements of two or more may be present in a single prediction model.

Any of the four models can contain stochastic (probabilistic) elements, the forecaster explicitly recognizing uncertainty by showing expectations as multivalued. Or all variables can be dealt with as deterministic (single-valued), the single value being, in effect, some form of average, with an implicit error.

The data that go into a prediction model and the estimate derived from it may be expressed in numbers or in qualitative terms. They may describe a situation at a point of time or a flow of change during an interval. Here are some sample forecasts:

1 You are going to like this person's work.
2 Sixty percent of our recreation visitors will be hikers and campers.

3 The water yield during next calendar year will be 35 billion gallons. As I figure it, the standard error is 3.2 billion.

4 There is a 0.2 probability that wholesale lumber prices will rise less than 5 percent per year, a 0.5 probability of a 5 to 10 percent rise, and a 0.3 probability that the rise will be greater than 10 percent.

JUDGMENT MODEL

In the **judgment model**

$$v_n = j \tag{1}$$

where v_n is the value of the forecast variable n periods of time from the base (the present or some past time of reference), and j is a magnitude arrived at in some way that can't be explained precisely. The vagueness of the matter often arises from its complexity.

John Putnam, a member of the research branch of the federal Forest Service, worked for years with southern bottomland hardwoods. He developed the ability to make all sorts of predictive judgments about the forest stands where he spent his life. He could tell you where you would find the water table and whether the site would be favorable for cherrybark oak. He could predict what sort of veneer a given tree would produce: the configuration of the grain and the extent of defect and decay. He did this simply by looking at the trunk of a standing tree with his "X-ray eyes." He could judge what it would cost to log a given stand of hardwoods and what sort of return would be obtainable at the mill. He was not a person to do a lot of figuring or to deal explicitly with the many variables that must have lurked behind his judgments. Some of his colleagues tried to convert his knowledge into writing and into quantitative relationships, but they were often unsuccessful.

Some persons have an uncanny ability to judge how people, singly or in groups, will respond to given products or ideas or approaches. Consequently they make good politicians or salespersons or business executives or the like. They use a multitude of guides, not all of which they can describe. They are able to refer to analogous situations that they have encountered before, and so experience is important. But experience accounts for only part of their talent, for they often exhibit their skills early in life. They show a precocious sensitivity to human signals.

Judgment on the forecaster's part usually has in it an ingredient of personal conviction, or faith. Often this is a large ingredient. The forecaster may be constitutionally pessimistic or conservative. In the years before reliable statistical methods had been developed for predicting the yield of timber stands, foresters' estimates often turned out to be low, sometimes amazingly low. But foresters are capable of optimism. One, who became corporate vice president of a large, diversified firm in the southern pine industry, had an abiding faith in the soundness of timberland investments. At times when his colleagues in the front office were expressing skepticism and urging caution, his progressive convictions won out. They proved to be accurate for those times.

When a group of persons is called on to judge the future, they may evince more unanimity than did those just referred to. If I want a judgment forecast, I may consult a group, either assembled for discussion or one person at a time. I may find a consensus with which I can be happy—even though, later on, I may discover that all my consultants were mistaken!

Every good forecast in forestry economics has some judgment elements, because of the complexity and the dialectical quality of social prospects. Many good forecasts are derived entirely from judgment. The smaller the number of persons whose behavior is forecast or the larger the value of n or the less reliable the given information, the greater the need for the judgment model.

> Explain why each of the foregoing three circumstances favors using the judgment model.

PROJECTION MODEL

The **projection model** is

$$v_n = f(v_0, n) \tag{2}$$

where v_0 is the value of the forecast variable at the base time or in the base period (the present or some past time of reference; the time when n = zero). In its pure and simple form, the projection model derives a forecast magnitude of v from its starting magnitude, all other influences on v being tossed into a single ragbag, the passage of "time." If one has a past trend to go on, so much the better.

As we talk about the projection model, let's have Fig. 32-1(a) in the corner of our eye. Disregard Fig. 32-1(b) for now.

Figure 32-1(a) shows some projection forecasts of U.S. total consumption of four wood-products categories. These forecasts were made by the author in 1967, for a period of years *(n)* extending to 1980. The base year (time 0) was 1960, the latest year for which a 13-year moving average (page 95) was to be had. The averages for 1900–1960 are real data identical to those in Fig. 14-1(a), except that the arrangement of the product categories is reversed. The "all-products" line is the same in the two figures.

Here we have a good illustration of the projection model. The illustration is good because the analysis takes advantage of the following presumed sources of strength:

1 The analysis rests on a reliable principle: pulverization. Read over pages 23 and 99.

2 The forecasts are strengthened by leaning upon both total and per-capita consumption data. Only totals are shown in Fig. 32-1(a), but the forecaster had per-capita information, too, to help decide whether the results were sensible.

3 The forecaster had an absolute target in cubic feet that guided the relative trends deducible from the pulverization principle. This was the forecast, independently developed, that wood-products consumption would reach 20 billion cubic feet by the

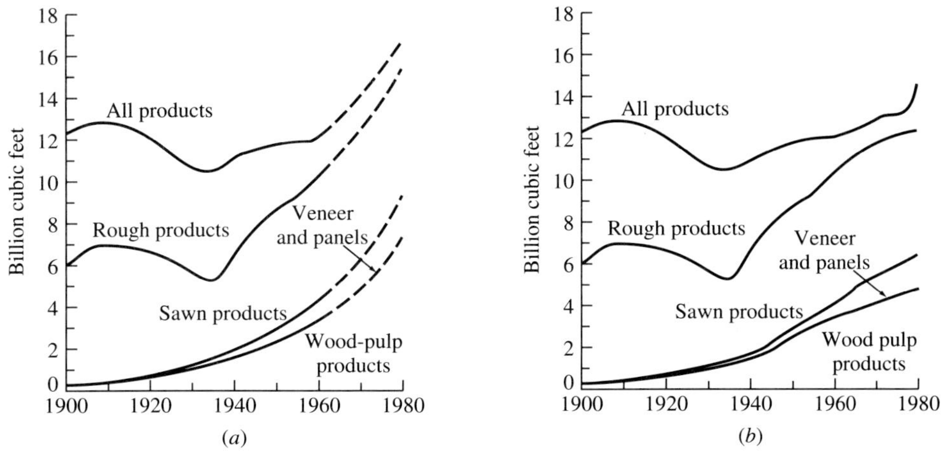

FIGURE 32-1
Apparent total annual consumption of wood products in the U.S., 1900–1980; 13-year moving averages, roundwood equivalent. (a) Actual consumption, 1900–1960; forecast, 1960–1980. (*Source:* Duerr, William A.: "The Long Trend in Wood Use," *Forest Engineering Conference Proceedings, Michigan State University, East Lansing, September 25–27, 1968.* 1969, pp. 99–101.) (b) Actual consumption, 1900–1980. *(Source: Federal Government Historical Series.)*

year 2000. This total was put into per-capita terms with the Census Bureau's Series B forecast of population in 2000.

Figure 32-1(a) is a good illustration of the projection model (or, for that matter, of *any* forecasting model) for an additional reason: It is likely to be in error. Figure 32-1(b) shows how consumption actually developed during 1960–1980. Total consumption did not grow as briskly as had been expected. Primarily, a pulp-products acceleration failed to materialize. Lumber, veneer, and panels performed as anticipated. Rough products continued their long downward way—until the mid-1970s, when the unexpected OPEC surge quickly doubled their use.

All in all, it looks as though the 20-billion-cubic-foot target for 2000 is too high. Wood is getting to be—and, in the eyes of many, *promising* to be—too costly in *real* dollars to stand up to competitors such as plastics to the degree envisaged in Fig. 32-1(a).

> In evaluating Fig. 32-1(a), why emphasize plastics as a competitor to wood? Why not include other substitutes—for example, stone, clay, glass, and metals?

How many, many disruptive events lurk in the wings to spoil the forecaster's neatly plotted drama! Changes in U.S. forest policy or in genetic, manufacturing, or consumption technology may hop out on stage. A war or an energy crisis (neither of which was foreseen in the illustration) can confuse the story. The forecaster may be able to study how such episodes influenced the past. But a study of history can be misleading. After World War I, the trend in American rural life, including the forestry economy, was seriously disturbed by prolonged depression. People expected the same thing to happen after World War II, but it didn't.

You may well imagine that the projection model is popular, especially for long-term predicting. It is certainly popular in forestry. It is simple—indeed, naive. The forecaster can putter around with a sheet of graph paper and a pencil and *see what is being done.* It is possible to shift the line a little to suit judgment. The forecaster can bracket the line between other lines representing outer limits. The value of seeing what one is doing, and of comprehending its significance, is worth stressing.

SHORT-TERM PROJECTIONS

If one wants a short-term projection—a matter of days or months, or even a year or so—one will not want to work from the historical trend: One will want to take into account the cycles and the seasonal and irregular variations. One will seek to discover where a variable stands currently amid these short-term changes, how it is moving currently, and what the prospects are for continuance or reversal.

Perhaps the most commonly used form of the projection model in short-term predicting is the simplified form $v_n = v_0$. That is, the forecaster assumes that there will be no change. In nonvolatile markets, tomorrow's price can be assumed to be the same as today's. Next month's production is assumed to be the same as last month's if there is no seasonal adjustment to be made. People who make such assumptions may misinterpret what they are doing. "I'm not forecasting," they may say, "I'm just using the current value." But if they are working with the future, they are forecasting, by definition. If they see announcements of furniture sales for next week, they won't assume that next week's prices will be the same as last week's.

Nevertheless, the **persistence forecast,** as $v_n = v_0$ is sometimes called, enjoys full acceptance in some circles. Our courts favor this sort of forecast if n is not excessively great, and so do our government bureaus. The Forest Service, when it appraises (predicts) stumpage value by subtracting conversion costs from end-product prices, customarily uses its most recent (past) record of prices. Chances are, the appraiser does not even think of the value estimate as a prediction. In cases such as this, the bureau and the court are playing it safe. They are using data that in our culture are unassailable even when obviously incorrect: *"facts,"* not speculation.

Cite an example of the successful use of a persistence forecast in forestry.

The length to which a public agency may go to avoid the onus of forecasting is illustrated by Forest Service practice. In this agency's planning, long looks ahead are inescapable. But they are not called forecasts. They are called "projections"—"estimates for the future under assumed conditions," or words to that effect. Thus "projection" is not only our name for a forecasting model, but also a name for a forecast that shuns forecasting.

LEADING-VARIABLES MODEL

The **leading-variables** model is

$$v_n = g(a_0, b_0, \ldots, y_0, z_0) \tag{3}$$

where a_0, b_0, and so on are values, in a base time or base period, of variables that are thought to have certain relationships, g, to the forecast variable v, at the time n. In other words, the magnitude or behavior of v at a given time is thought to be related to the magnitude or behavior of the other variables at an earlier time. The other variables are thought to **lead** v (v **lags** them), so that by keeping an eye on them, the forecaster can use them as predictors of v.

It's fair to assume that the leading-variables model is best suited to short-term forecasting, for surely in a dynamic society no manageable set of variables can be found that will predict the state of another variable very far ahead: Too much can transpire in the interim to mar the relationship, break the halter. The best-known uses of the model are in general economic forecasting. Regression models containing hundreds of variables have been built to represent the economy at large. Simple examples of the leading-variables model are to be found, too. An increase in the rate of contracting for new single-family dwellings predicts, some months ahead, an expansion of the lumber and panel markets. Such expansion is, under certain circumstances, predicted even further ahead by a reduction in interest rates, which makes mortgages—that is, new dwellings—less expensive to the consumer. Since the ultimate consumer is, in a sense, the starting and ending place for economic activity, changes in final demand are predictors that set chains of further predictors in motion: An increase in lumber demand leads to more active retailing, wholesaling, transportation effort, sawmilling, logging, stumpage marketing, and land marketing; all along the line, the demand for labor and capital increases in these areas of activity, leading, in turn, to further changes. An appreciation for the working of any system that takes time to perform its function carries with it a power to predict with the leading-variables model.

> Picking up the last thought, describe some "working" of the silvicultural system that takes time to bear fruit, and show how the leading-variables model may be applied to predicting the outcome.

Consumer activity being a predictive key, one wants to know whether there is any reliable predictor of consumer activity. Indeed, there are a number of such predictors.

An early one that demands recognition is consumer plans. George Katona, founder of the Survey Research Center at the University of Michigan, seized upon this relationship in the 1940s. He predicted changes in the economy by analyzing consumer plans. These he learned through sample polls [1]. A current consumer-poll survey is the monthly "Index of Consumer Confidence" compiled by the Consumer Research Center of the Conference Board. The Board also compiles a survey of consumers' feelings about the present situation, including the outlook for jobs, wages, and business conditions, and plans to buy a car, a house, or major appliances—or to take a vacation [2].

A WISP OF PHILOSOPHY

The leading-variables model has long been used with some success (and much failure) to forecast changes in the stock market and in general economic activity in the nation. One of the earliest and most famous such predictors was the Harvard Index of General Business Conditions, developed not long after 1920. At first very effective in forecasting the course of the business cycle, the Index gradually lost its power and, after the crash of 1929, failed utterly. The structural frame of economic life had changed, and formerly good predictors had become worthless for the purpose. Life changes always, a little at a time, and then periodically it comes to a fault line when there's an upheaval. After the Great Depression, there was another upheaval in the form of World War II, and another early in the 1970s, when the energy and basic-materials position of the country was seen to have become transformed. In the late 1980s and early 1990s, widespread rejection of communism created another upheaval.

The story of the Harvard Index is helpful. It illustrates the difficulty, the uncertainty, of the forecaster. It *is* naive to assume that history repeats itself. But one is unable to use the opposing assumption, that history does *not* repeat itself, for this opposing assumption forces one to the ultimate conclusion that the future bears no relation to the past. Yet the past is all we know.

What we are saying, really, is that the more sophisticated forecasting models are revealed as special cases of the projection model: pushing a variable out into thin air, guided only by its own previous circumstances. Even the most reliable predictors are not perfectly reliable. An increase in lumber demand may not affect sawmills. Wholesale and retail stocks may absorb the change. Or substitutes may come in. Or consumers may be thwarted.

Although the leading-variables model finds its direct use in short-term forecasting, it has an indirect use in long-term forecasting and planning. Forest-resource stocks offer an example. Timber stock today and subsequent changes are predictors of the stock a year hence:

Stock this January 1:	5 billion cubic feet
Growth this year:	0.4 billion cubic feet
Removals this year:	0.35 billion cubic feet
Stock next January 1:	5.05 billion cubic feet

The process can be repeated for next year, the year after that, and indefinitely. Where does the information on growth and removals come from? It can be predicted. Or it can be *assumed.* Various levels of each variable can be assumed and thus the development of the forest **simulated** as it would occur under alternative management programs. Such a procedure is widely used for planning timber harvests and silvicultural measures.

COINCIDENT-VARIABLE MODEL

The **coincident-variables model** is this:

$$v_n = h(a_n, b_n, \ldots, y_n, z_n) \tag{4}$$

where a_n, b_n, and so on are magnitudes, at the forecast time *n*, of variables thought to be significantly related *(h)* to *v*. The functional relationship is derived from analysis of historical data. Usually the forecaster does not go into the past any further than necessary and avoids periods thought to be abnormal. Information for all variables is taken for the same points or periods of time, unless variations of the leading-variables model are combined with this one. In that event, the independent variables can be measured at points of time different from the time of reference of *v*. Consumer behavior in July of recent years may have been better correlated with May or June incomes than with July incomes.

When the model is ready for use in forecasting, where does the forecaster get the values of a_n and the other independent variables? He forecasts them or assumes them.

To forecast a_n and company in order to get a prediction of v_n is not quite so ridiculous as it sounds; v_n may be harder to forecast except with coincident variables, and forecasts of these latter variables may be easy to come by. For example, the Forest Service has shown [3] that paper consumption in the U.S. has been very closely related to current total population and gross national product, except for the abnormal Great Depression and World War II years. For the remaining years, a regression equation can be derived, expressing the quantity of paper consumed as a function of current population and GNP. Then, if one wants a forecast and wishes to accept the assumption that history repeats itself (and, as before, what better can one do?), the next steps are easy: Turn to the federal government's predictions of population and GNP for the year of interest, pop the figures into the equation, and crank out a solution.

Or one may make one's own forecasts of a_n and company. For this purpose, one may turn to a big model of the economy at large, such as was mentioned in connection with the leading-variables model. To that model, one may attach, as a "module," a coincident-variables model of a specific industry such as lumber or plywood. Again, all the data that are built into these models come from the past.

The coincident-variables model may be used in either short- or long-term forecasting. Indeed, it is in many respects the most precise, informative, and flexible of all the models.

When we raised the question of where a_n and company come from, the response was that their values may be predicted or *assumed*. The essence of such assuming is this: The forecaster may take any sensible sets of independent variables, work out values for v_n, and say to listeners, who may be a board of directors trying to decide on expenditures, "If you put in this much money, here is what we can accomplish; or if you put in that much, there is the likely result." This use of the coincident-variables model amounts to simulation, in the same sense as earlier.

CONCLUDING THOUGHTS

Four thoughts by way of conclusion.

First, forecasting models may be put to a test of sorts. All the models may be given this test, but the test is easiest for the leading-variables and coincident-variables models. The test is to try out the model on a "forecast" that might have been made years ago. Put the forecaster 20 years back. How well would this person have done, using this model, with a 15-year prediction (5 years back)? You can answer because you know the answer.

Second, there is an interesting difference between *forecasting* the future and *determining* the future. Some forecasters are in a position to determine the accuracy of their forecasts. The consuming firm is a case in point, when it comes to the firm's forecasts of its own behavior. To this extent, the firm is not only forecasting, but also determining, its future. If an analyst to whom many people listen with confidence says that the stock market is going to take a deep plunge, stockholders will rush to sell, and the market will go into a deep plunge.

Third, one speaks of *good* forecasts. Does this mean forecasts that come true? Often, but not always. A good forecast is one that fulfills its purpose. Perhaps its purpose was to arouse people to a danger. They responded; the danger was averted, and the forecast did not come true. But it was a good forecast.

Fourth comes a query: How is one to choose among models for deliberated forecasting or among combinations of models? In general, one may choose on the basis of the following considerations: (1) the forecaster's technical experience, skills, and preferences; (2) the degree of the forecaster's uncertainty about the question—based, for example, on familiarity with the problem or on the term of the forecast; (3) the data available, including related predictions; and (4) the forecasting effort justified by the values at stake.

SUMMARY

The approach to forecasting of primary interest here is the deliberate use of an intellectual process. For such deliberated forecasting, one may recognize four sorts of models. In increasing order of sophistication and emphasis on quantitative relationships (but not necessarily in increasing order of accuracy), the four are the judgment model, the projection model, the leading-variables model, and the coincident-variables model.

REFERENCES

1 Strumpel, Burkhard, James N. Morgan, and Ernest Zahn, eds.: *Human Behavior in Economic Affairs,* Jossey-Bass, San Francisco, 1972, especially pp. 3–9.

2 *The Wall Street Journal,* Eastern ed., September 3, 1991, p. A18.

3 U.S. Department of Agriculture, Forest Service: *Timber Resources for America's Future,* Forest Resource Rep. 14, Washington, 1958, p. 426.

CHAPTER 33

PREAMBLE TO PART FOUR

Let's turn, now, to the central subject of Part Four, manufacturing production. *Manufacturing* means, literally, making by hand. For us, it means making by machine and hand, under the procedures of an increasingly elaborate **technology of production.** In forestry, **manufacturing production** relates almost exclusively to timber.

MANUFACTURING VERSUS GENETIC PRODUCTION

The kinds of questions that arise in timber-products manufacturing enterprise are very much the same as those in genetic enterprise. The questions include where to locate, what products and outlets to aim for, the scale of production, whom to employ, the technology, the intensity of management, and so on. Again, the questions about the timber-manufacturing economy at large have much the same general character as those about the rest of the forestry economy. And the models and economic criteria are, in general, the same.

> Consider the first of the foregoing questions about forestry enterprise: where to locate. See if you can identify the considerations that enter the location decision for genetic production and for manufacturing. In what respects are the two alike? dissimilar? Think about site quality, transportation, raw materials, power, labor, attractiveness of the community, markets for end products, and other considerations that you judge should be included.

But there are some marked differences. One is a difference already implied. Manufacturing is primarily a timber matter. It includes what might be termed multiple use, but the uses, the joint products, are all timber products.

Another difference between manufacturing and genetic production is that manufacturers generally plan ahead for shorter periods than genetic producers do. A favorite *long-term* planning horizon for the logging or milling manager is 10 years. This is because, in manufacturing, the plant wears out, both physically and technologically. A manufacturing enterprise is part of the culture, exposed to a thousand man-made storms. A forest, though culturally influenced, is largely determined by biological and other natural forces that are unmindful of human behavior.

Related to the difference in planning period between manufacturing and genetic production is the difference in tempo of technological change. Think what is involved in establishing a breakthrough in such genetic-production areas as fertilizing or seed-orchard management. A research worker gets an idea, then devises a test. The test is put into effect experimentally. Then comes a waiting period, perhaps a good many years. If the results leave something to be desired, there must be reruns that consume periods of time comparable to those already endured. By contrast, what about the manufacturing breakthrough? If a researcher has a good idea, it can be tried out, either on the production line or on a model. The result may make itself apparent in a jiffy.

Still another difference between the manufactory and the forest is that the factory is governed by simpler decision rules. Financial considerations are more apt to be paramount. There are a number of causes for this difference. (1) A plant's inputs and outputs are more readily described by numbers than those of the forest are. (2) Differences between plant and forest ownership are crucial. Very few sizable plants are held either by nonindustrial private forest owners, with their complex aims and motivations, or by the public. (3) There are great differences between factory and forest in the degree of their exposure to public scrutiny. The forest lies out in the open for many to see and to respond to in political terms.

> Is it possible that the author sees genetic decision rules as more complex than manufacturing decision rules simply because of his bias as primarily a woods-going forester? How would a pulpmill or sawmill manager compare the two decision systems with respect to complexity? Ask such a manager for an evaluation. Or ask a faculty person in forest products or pulp and paper technology. How would you rewrite the foregoing paragraph?

There is a telling retort to the contention that the factory is governed by simpler decision rules than the forest is. This is that the factory is more largely propelled by people—human beings. Surely no one will argue that human propellants bend to simple rules!

Just one more difference between manufacturing and genetic production can be mentioned here. Chapter 16, where genetic production is introduced, speaks of the *principle of genetic identity* and stresses its consequence, that the guiding rate of return is generally a more potent determinant of output than product price. That is, stock supply gets top billing among the supply conditions. In manufacturing production, too, there is a stock supply, and there is a "principle of manufacturing identity." Glance ahead at page 304. But short-run supply holds ascendancy, and product price is a paramount determinant of production.

HISTORY: SOFTWOOD-LUMBERING, PECKERWOOD-MILL, AND PULP-PAPER ERAS

A little history will furnish a useful introduction to the chapters that follow. The wood-using industry of the U.S. has seen a number of clearly defined eras. These eras overlap, and their start and end are better termed fade-in and fade-out; but each era's character is sharply striking.

The earliest was the *softwood-lumbering era.* It began in New England, where old-growth eastern white pine fed the logging operations and sawmills. Then the industry moved into the pineries of the mid-Atlantic states as the more northerly resources began to play out. By the turn of the twentieth century, Lake States white and red pines had been drawn upon heavily. In another decade or so, the southern yellow pines had become the chief source of lumber for the all-devouring nation. The Pacific Northwest and Inland Empire were the last to be reached by the lumbering wave. On the coast, the industry moved southward from the Puget Sound region, and there are parts of the old-growth timber forest of southwest Oregon and northern California that were not reached at all before the character of the industrial march had changed profoundly. The softwood-lumbering era was marked by industrial and labor migration, the impermanence of the logging and milling communities, the creaming of the resource in the absence of incentives to do otherwise, the dominance of lumber among U.S. timber products, and the acquisition by the industry of a reputation for ruthless waste that it has not yet lived down.

The aftermath of big-scale lumbering in each region was a cleanup carried out by hordes of little sawmills. Many of these were family- or farm-operated mills. Their daily capacity was minimal, perhaps 10,000 bd ft, and their rate of output was typically much less than this over the year. An operator would buy the timber on a tract, often at an amazingly low price. If there was as much as 30,000 bd ft of sawtimber, the mill might be moved to the tract so as to avoid transportation of "waste," of which there was plenty.

Across U.S. forests east of the plains and prairies, there dawned the *peckerwood-mill era.* The era reached its height in the 1940s. Some of the southeastern states, at that time, each provided homes for 2,000 or more sawmills. These mills ran intermittently. Some of them did mainly custom sawing. Many others were stored in the barn loft when the market was slow. When the situation brightened and the stable big-mill element in the industry saw hope for a period of good profits, the haywire mills came forth like bees from the hive, and the market was soon satiated, even though the product of the fly-by-night outfits was poorly manufactured and seasoned. The little firms gained the enmity of the larger ones for undermining lumber's reputation among consumers and for flooding the market. They earned the mistrust of landowners whose timber they bought cheap and whose woodland they often disfigured with their careless logging. Few of their fellow citizens were sorry to see the great majority of little operators fade from the scene as an era of new technology began in the years after World War II.

But in the meantime, another era had opened, the *era of pulp and paper expansion.* Like the peckerwood-mill era, this one came in the aftermath of big lumbering and was a time of cleanup. However, it was much more than this, and it was more enduring. In the 1930s, when land and wood were inexpensive, farming depressed, tax delinquency widespread, and new technology available for pulping resinous pines, the pulp-paper industry started moving into the Southeast. Firms bought land, in many cases from the lumber companies. They put the land under management and began to build back the timber growing stocks. This they succeeded in doing by foraging for their raw material in the **open market,** mainly the farm forests and other small holdings. The already widespread reputation of these little woodlands as a number-one forestry problem was thus reinforced. The pulp-paper companies developed an elaborate system of wood procurement and a correspondingly successful system of product merchandizing. The southern momentum gathered by the industry carried it into other regions, even into the northern tier of states where it had first arisen and where it had subsequently languished at the hands of the southern expansion.

ERA OF NEW TECHNOLOGY

Let's mention just one other era in the history of wood-products manufacture, one already named in passing, the *era of new technology.* This era opened not long after the close of World War II. Like the pulp-paper era, it is still in progress, though its time of greatest momentum appears to have passed. Logging became greatly mechanized, so that large firms came to enjoy special economies and competitive advantages. Between 1974 and 1980, the largest pulpwood loggers in the Southeast (annual production, 10,000 cords or more) increased in numbers by perhaps 85 percent, while the numbers of smallest firms (less than 1,000 cords) fell off by about 50 percent [3]. It is generally believed that these trends persisted during the 1980s and into the 1990s.

In sawmilling, improved utilization, sawing, and other processing devices and methods have given the industry great economies of scale that have favored big firms and further suppressed the little fellows. In the state of Washington between 1982 and 1984, sawmill output jumped 22 percent, while the number of mills dropped 11 percent [2]. The South saw shifts in sawmill size and sophistication that were even more dramatic [1]. In the softwood veneer and plywood industry, new technology for using small logs has opened up new resources, notably in the southern pine region, but in other sections of the country, too. And now the panel industry as a whole is picking up new momentum with the help of partially or wholly reconstituted products that can be made from the smallest merchantable trees.

Commencing, perhaps, in the early 1980s, there has been an accelerating movement of timber activity from the Pacific Northwest back toward the southern pine region. One may speculate about the causes of this migration. (1) Southern pine timber has some inherent advantages over Douglas-fir—its rapid growth in youth, its natural pruning. (2) The southern pine region is less vulnerable to environmentalist activity because of its long-standing acquiescence to the wood-using industry, its scarcity of wilderness, and its comparatively level topography, obscuring the cutover landscape. (3) When recession would hit the lumber and allied products industry, the South, with its good seaports and inland water transportation, had pulp-paper to fall back on. (4) All in all, the advantages of southern industry in terms of labor and raw material had been vigorously publicized and had become recognized.

Nationwide, between the mid-1960s and the mid-1980s, the real dollar value of output of the average pulpmill increased nearly three times. That of the average logging operation and of the average sawmill increased between three and four times.

Thus the new-technology era has seen bigness emphasized in wood-products manufacture. The era has been marked by big, high-speed, costly plants run by diversified corporations able to invest in major raw-material sources and processing and marketing complexes. The pulp-paper industry, with its high rank in the pulverization hierarchy, has been the leader. The industry manages land and operates a system for procuring additional wood raw material in the open market and in the government timber market. In 1985, three paper companies accounted for more than half of total U.S. paper sales. The industry operates, not only pulp and paper mills, but also sawmills, panel plants, and transportation facilities. It channels its raw material to advantage. Its lumber and plywood plants send chips to the pulpmills and finer residues to the reconstituted panel mills. The system produces power and heat from wood leavings.

But the new-technology era has carried the timber-products industry well beyond wood. The ultimate firm is no longer simply a wood-products firm, but a conglomerate with interests that may extend into metal containers, hotel service, and food products. It is a corporation that is not only big enough to invest in such far-flung enterprises, but also big enough to become a tempting target in the game of takeover and diversification.

> With the possible exception of northern aspen and birch for particleboard, all the great wood-industry migrations in America have been based on the use of conifers. Why is this? Can you think of any migrations, perhaps regional or even local, based on the use of hardwoods?

THE DRIVE FOR BIGNESS

What lies behind the drive for bigness? What factors are at work to reduce unit costs of production with increases in scale **(scale economies)?** And are there any offsetting cost disadvantages in bigness **(scale diseconomies)?** We are obliged to consider both the costs borne by the firm itself—its own opportunity costs—and the costs borne by the community and largely escaped by the firm—external costs, or **externalities.**

Marketing influences. Up to a point, larger firms have cost advantages over smaller firms in procuring inputs and selling outputs. Such advantages arise on the input side through quantity discounts and opportunities for vertical integration and on the output side through reaching larger and more specialized markets and through managing market activities. The advantages of scale are diminished and even reversed when costs rise with enlargement of timber sheds, with loss of personal contacts in the market, and with the overtaxing of capacity of some market outlets.

Technological advantages. Larger firms may be able to cut unit costs by utilizing specialized processes and workers. They may install efficient machinery and methods that are usable only on a large scale. They may make more profitable use of wood residues than smaller firms can make.

Advantages in financing. The initial financing of a large firm may be difficult. But once the firm is established, its financing costs are comparatively low because of the firm's prestige and access to nationwide securities markets and, in general, because the per-dollar cost of negotiating large loans is less than for small loans.

Management influences. Larger firms may be able to recruit more ambitious and talented managers than smaller firms can, and to support them with a diversified staff. However, beyond a point, further increases in scale may entail diseconomies, such as awkward inflexibility and delay in decision making and loss of the managerial genius of persons who enjoy modest-size undertakings in a rural or small-town setting.

Externalities. These are effects of economic actions for which markets have not been organized. They are primarily social costs and benefits that vary with the scale of private firm or plant. External diseconomies of scale may be found in the pulp-paper industry: water and air pollution, deterioration of scenery, very large plants that are regarded as eyesores and thus entail some social cost. A major social cost that industry has had some success in escaping is the cost created by the pulp-paper industry and the chemical industry when the lethal substance, dioxin, used in bleach manufacture, is introduced into waters that come directly or indirectly (as through fishing) into human consumption. The Environmental Protection Agency has set rigorous standards for admissible dioxin concentration. The industry has funded a tooth-and-nail rebellion against these standards [4]. In regions where large wood-using firms procure raw material from small holdings through "independent" loggers, such firms are able to disassociate themselves from the havoc and resource waste attendant on the harvesting. On the other hand, socially beneficial forest management and timber harvesting can be, and often are, practiced by large-scale firms, which can carry out their own operations and have easy access to skilled planners and well-adapted mechanical equipment.

ARRANGEMENT OF PART FOUR

The scope, variety, and complexity of the timber-products economy are impressive. Part Four selects just a few illustrative questions, uses some appropriate models for responding to them, and banks on the wider usefulness of these models to do some small justice to the subject of manufacturing production in the forestry economy.

In the chapter that follows, a problem in logging is addressed. The problem is to integrate the harvesting system with the silvicultural system, a problem that tides us over from genetic to manufacturing production.

After that come four chapters, each concerning a different timber region of the U.S.: the West Coast, Southeast, Appalachian highlands, and Lake States. The questions discussed move, chapter by chapter, from the macroeconomics of wood industry in a state to the microeconomics of machine operation in a specific plant.

Part Four then turns to a topic of wide national concern: productivity. How fruitful a use is being made of human effort and other resources in wood-products manufacture?

The last two chapters in Part Four parallel the bow-out in Part Three. There is a discussion of supply intended to tie together preceding studies and a statement of high points intended as a review.

TERMINOLOGY

Finally, a guide to terminology. The decision-making unit in manufacturing production, as in any economic activity, is the *firm:* the lumber firm, the lumber-and-plywood firm, the conglomerate that makes these products and metal containers. And so a firm may have one or more *enterprises* (page 111). An enterprise may be housed in one or more *plants,* or *mills,* each with *machines* and workers carrying out recognized manufacturing steps. At the head of all this series is the **industry,** the aggregate of firms engaged in specified lines of production. Commencing, then, with the *economy,* our terminology for the units of manufacturing production is arranged in this hierarchy: economy, industry, firm, enterprise, plant or mill, and machine.

> Consider the analogous hierarchy for genetic production. How does it differ from the foregoing? Why?

While we are building a vocabulary, it will be well to set down the terms that describe the wood raw material that enters the manufacturing process. Three terms are common. **Logging** converts standing timber into logs and bolts, using equipment that is typically portable. **Primary manufacture** works upon the logs or bolts with equipment that is typically stationary. Examples: sawmilling, veneer and plywood production if integrated, pulping, perhaps pulp-paper manufacture, and the shaping of roundwood into poles, piling, posts, and firewood. **Secondary manufacture** processes the output of primary manufacture. Examples: lumber finishing where separate from sawmilling, furniture making, plywood production where separate from veneer manufacture, and paper-products fabrication.

SUMMARY

The foregoing preamble talks mainly about two topics. First is the character of manufacturing production in forestry and especially how manufacturing differs economically from genetic production. Second is the U.S. economic history of wood manufacture. Four historical eras are recognized: (1) softwood lumbering, (2) the peckerwood mill, (3) pulp-paper expansion, and (4) the new technology. The chapter closes with a preview of Part Four and a note on terminology.

REFERENCES

1 Granskog, James E.: "Southern Sawmills Gain in Size," *Southern Lumberman,* March 1, 1979, pp. 15–16.

2 National Forest Products Association: In *Focus.* October 1986, p. 8.

3 Stuart, W. B., and S. Shartle: *Predicting Forestry, Harvesting, and Pulpwood Procurement Conditions for the Years 1980 and 2000,* American Pulpwood Association Surv. Rep., 1977.

4 *The Wall Street Journal,* Eastern ed., February 20, 1992, p. A1.

CHAPTER **34**

EVALUATING A HARVESTING PROCEDURE

UNITY OF THE SILVICULTURAL-HARVESTING-PROCESSING-MARKETING SYSTEM

The first steps in timber manufacturing are the topic of this chapter. These steps, which together constitute **timber harvesting** or **logging,** may include (1) felling, (2) limbing, (3) topping, (4) skidding or yarding, (5) bucking, (6) loading, and (7) hauling. Some steps, such as step 7, may be classified as other than logging. Some, such as step 5, may be taken earlier or later in the series.

The question here concerns the best combination of resources: the best **harvesting system** of labor, equipment, and technology. What is best in harvesting depends partly on the silvicultural system for growing the crop and the processing-marketing system for creating the end products. All three systems may be operated by a single firm. But commonly harvesting is the work of a separate firm under contract to the processor that owns the forest land or controls it in connection with stumpage procurement. In any case, one may view the finding of the best silvicultural-harvesting-processing-marketing system as a problem for the landowner or processor to solve, either as logger or as the one who selects a logger.

This chapter recognizes interactions primarily between logging and the thinning and harvest-cutting phases of the silvicultural regime. We will not approach the problem in every respect as the landowning or processing firm would approach it. Our interest is in analyzing the problem in such a way that *we* can most readily grasp what variables are involved and how they work.

PROBLEMS IN THIS CHAPTER

In Chap. 22, where the thinning regime is first analyzed, the firm approximates the financial ideal by making successive thinnings that equate the stand's marginal value growth percent with the firm's guiding rate of return. The financially ideal rotation is set for the time when the marginal value growth percent of the whole forest equals the guiding rate.

What is *value* in this case? It is the value of wood in the standing tree. Generally speaking, it is the value of the tree's end product minus costs of converting the tree into this product. If the end product is wood at the mill or point of concentration, the costs of conversion are logging costs. The comment that appears on page 174 is that if logging cost per unit of measure diminishes with increasing tree size, the result is a rising tree value per unit of measure, even if end-product value per unit is constant. Thus logging cost may give rise to a timber-size premium and so affect marginal value growth percent, which in turn affects the thinning regime and rotation length.

But cause and effect run in the opposite direction, too. Thinning regime and rotation length determine the size and spacing of trees, which in turn determine logging costs and the choice of a logging system.

Here, then, are two problems. One is to choose a logging system; the other, a silvicultural regime. To solve the first problem, one must be aware of the second and, ideally, solve it simultaneously.

THE ILLUSTRATIVE FOREST AND ITS SILVICULTURAL ALTERNATIVES

For case material, let's turn to a forest that somewhat resembles the one in the loblolly-pine section of Chap. 25. However, the present forest differs from that one in some respects: (1) It is on a higher site than Chap. 25's high, and timber prices in the locality are somewhat lower than Chap. 25's high. (2) The forest is regulated to the extent that there are stands of widely ranging ages, and the firm pictures a regime of thinnings and final harvests that goes forward pretty much continuously year after year. (3) The firm envisages a first thinning in each stand at about 15 years of age, followed by second and third thinnings and by a harvest cut at the time of maturity, after which the land is replanted. (4) The firm does not aim for a perfectly regulated forest—in fact, knows that it cannot achieve one. But it does want a reasonably good combination of its resources, and it wants to understand the interactions of logging and silviculture and accommodate to them in an appropriate degree. (We'll see as we go along what "appropriate" signifies.)

The firm owns the land, manages the timber, and sells logs and bolts loaded on truck. For the logging, it relies upon contractors, telling them merely what procedures it wants followed in order to safeguard its woods-management program.

For example, there is the thinning procedure. The firm has studied three alternatives: (1) Remove every third row of trees in the plantation and let the pairs of intermediate rows go unthinned. (2) Remove every fourth row and enter the row on each side for a partial cutting, leaving one row in every four untouched. (3) Remove every fifth row and enter the two rows on each side, thus leaving no row untreated. Cost differences are large. On some sample plots where thinning was fully mechanized and hence the contrasts were extreme, the third-row system cost an average of about $21 per cord; the fifth-row system, $32 [1]. However, the latter system has advantages. Lending itself to multiple thinnings and leaving no row untreated, it produces values high enough, in the firm's judgment, to justify the extra cost.

THE LOGGING CHOICES: SYSTEM 1

The firm is also experimenting with three alternative logging systems.

System 1 uses a crew of three workers, chain saws, and a small flatbed truck. While the truck is taking a load to the point of concentration, two sawyers are felling, limbing, bucking the stems into 5-ft pulpwood bolts, and piling the bolts. When the truck is returned to the woods, the three workers join to load it by hand from the piles. This system is known as the **short-wood system.** Costs per cord are estimated for each of the three thinnings, on the assumption that the size of thinned trees is as follows, measured in terms of d.b.h. and number of stems per cord:

	1st thinning	2nd thinning	3rd thinning
Average d.b.h., inches	5.9	7.2	8.8
Tree stems per cord	25.7	13.8	8.6

To begin with, the crew's productivity in worker-minutes per cord is estimated for each thinning:

	1st thinning	2nd thinning	3rd thinning
Fell	10.8	8.1	5.4
Limb-buck	18.9	15.3	12.6
Pile	27.9	22.5	18.0

Do you picture how work-time data such as these are gathered? Draw up a simple plan for gathering such data. Where would you go to make the study? How many research persons would be required? What equipment? How would you introduce yourself to the loggers? How do you explain your purpose? How do you obtain their respect and permission? What arrangements would you make with the landowner? How would you then proceed? How would you keep felling, limbing-bucking, and piling separate in the record, considering that all three tasks may at times go forward together? What characteristics of the logging operation would you take note of? What measures would you take to improve the representatives of your results?

A round trip with the truck takes about 1 hour, of which 40 minutes is operating time. During the hour, in the first thinning operation, the two sawyers cut and pile 2.08 cords, which the three workers can load in $\frac{1}{2}$ hour. Thus the crew's output is 2.08 cords per $1\frac{1}{2}$ hours, or, say, 10.4 cords per day. The cost of this output in 1987 dollars is:

Labor, 24 worker-hours @ $5.25	$ 126.00
Fringe benefits @ 25% of the above	31.50
2 chain saws operated 2.6 hours each, @ $3.20 initial cost, interest, maintenance, and supplies	16.64
1 truck, operated $3\frac{1}{3}$ hours @ $10.14 depreciation, interest, maintenance, and supplies	33.80
Profit (10 percent of total)	23.10
Total	$ 231.04

(**Depreciation** is a cost per unit of time that is intended to allow for the total investment in any input to production that has a useful life of more than a year. If you, as a logger, buy a chain saw that will last only a year or less and will have no value thereafter, you add the cost of the saw, the interest you pay, the cost of maintaining the saw in working condition, and the cost of gasoline and oil. But if a piece of equipment will last, say, for 3 years, the problem is not so simple. You must find some *fraction* of the equipment's cost that can be added to the interest, maintenance, and supplies. This fraction is depreciation. There are many ways of calculating how much it is. The simplest, used here, is **straight-line depreciation:** If a truck will last 3 years, take one-third of its cost each year as expense. And if you want an hourly expense, divide by the number of operating hours in the year.)

Returning to the tabulation, the cost per cord in the first thinning is the $231.04 total divided by 10.4, or $22.22 per cord. Costs per cord are estimated by the same procedure for the second and third thinnings. The results for System 1 are:

Thinning	Cost per cord, 1987 dollars
First	$22.22
Second	19.48
Third	17.80

These points are plotted in Fig. 34-1(a). They describe approximately a straight line in relation to stems per cord, and this (a solid line) is drawn. Cost is negatively related to tree size because rising efficiency in felling, limbing, bucking, and piling more than offsets the increasing difficulty of loading larger bolts by hand.

You should note that some of the cost items, such as depreciation of equipment and contractor's profit, are so-called "overhead" items. But they are not fixed-cost items, and therefore they are relevant to the firm's choices. All the costs included in the analysis are variable with the choice of logging system and therefore relevant.

LOGGING SYSTEMS 2 AND 3

Next let's turn to the second of the three logging systems. Since you now have an idea of how detailed the data and figuring must be, there is no need to put you through the arithmetic again. The approach is the same as before.

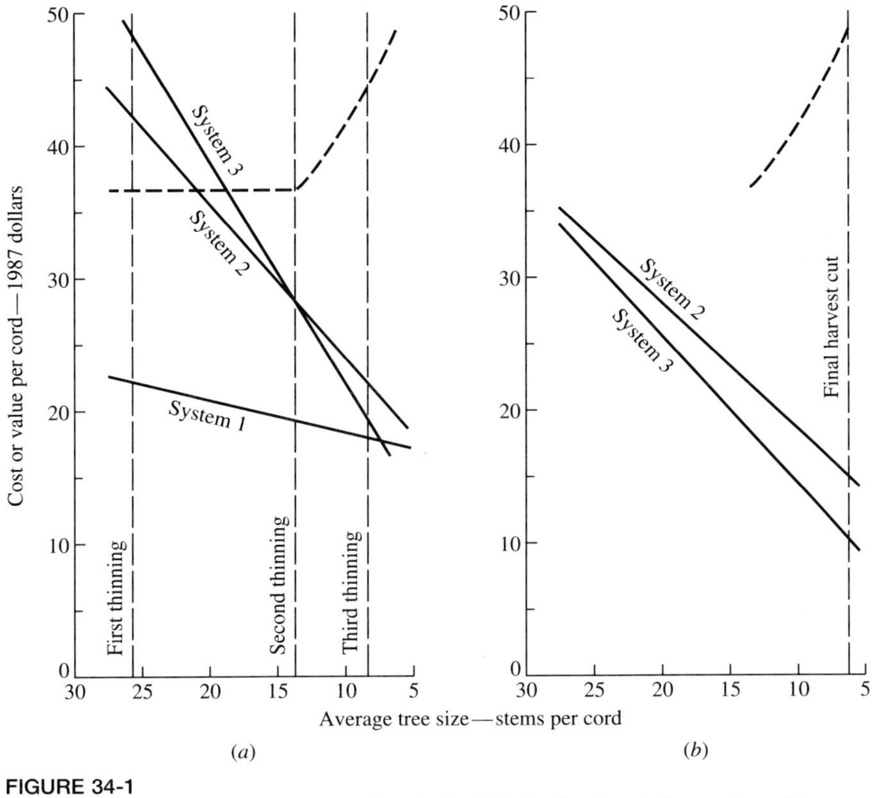

FIGURE 34-1
Logging cost and log value per cord of high-site loblolly pine in relation to size of trees expressed as stems per cord. (a) Thinning operations, (b) final harvest cut. System 1: Chain-saw felling, limbing, and bucking; hand loading on truck in the woods (short-wood system). System 2: Chain-saw felling, limbing, and topping; cable tree-length skidding; chain-saw bucking and machine loading at landing. System 3: Tree-to-tree feller-buncher; grapple skidding; gate delimbing; bucking and loading as in System 2.
——— Per-cord logging cost
– – – – – – Per-cord value of loaded logs
(*Sources:* Adapted from Hypes, Trenor L.: "The Impact of Tree Size on the Performance of Longwood Harvesting Functions and Systems in Clearcut Harvesting of Southern Pine Stands," M.S. thesis, Industrial Forestry Operations Program, Virginia Polytechnic Institute and State University, Blacksburg, 1979; and Lane, Richard A., Jr.: "An Evaluation of Operational Harvesting Techniques and Systems Applicable to the Commercial Thinning of Southern Pine Plantations," M.S. thesis, Industrial Forestry Operations Program, Virginia Polytechnic Institute and State University, Blacksburg, 1981.)

System 2 uses two sawyers with chain saws to fell, limb, and top the trees in the woods. One operator with a cable skidder works behind each sawyer. The operator sets the chokers. (A choker is the end of a skidding cable that encircles and grips the log or bunch. It is equipped with steel fittings that allow it to be "set" in the form of a noose.) The operator skids the tree-length logs to the landing. There, one sawyer on the deck helps unhook the chokers and bucks the stems into products of highest value, whether pulpwood, chip-and-saw logs, sawlogs, or veneer logs. The operator of a mechanical loader, who serves also as foreman, helps the sawyer on the deck as needed and loads the logs and bolts on set-out truck trailers.

System 2 cost per cord for each of the three thinnings is plotted in Fig. 34-1(a) and a curve is drawn. Since this system is able to handle products other than pulpwood, it may be called upon for the final harvest as well as for thinnings. Average tree size in the final cut is expected to be 6.0 stems per cord. In Fig. 34-1(b), a number of cost estimates for final harvesting by System 2 are plotted and curved, including the estimate of $14.35 per cord for 6.0 stems per cord. The final-harvest costs of System 2 are lower than the thinning costs, primarily because of the skidding function. The productivity of machine skidding hand-felled stems through a plantation is much lower than skidding over cleared ground. Under either circumstance, cost declines with increasing tree size. The decline is more marked in System 1 because System 2 puts less demand on labor.

System 3 uses two tree-to-tree accumulating feller-bunchers. A feller-buncher is brought up to each tree individually. The tree is severed from the stump by means of shears and is then held upright while other trees are sheared and held, up to the limit of the accumulating device. The bunch of trees is then oriented and laid down, to be grasped at the butt end by a skidder equipped with a grapple, one skidder for each feller-buncher. On the way to the landing, the skidder backs the trees, tops first, through a limbing gate. This is a mesh of bars held in a vertical plane; the stems pass through it, but not the limbs, which snap off just as they do, in the case of loblolly pine, in an ice storm. At the landing, a sawyer removes any limb stubs that are left, and from then on, the procedure is the same as in System 2. The unit costs are plotted and curved in Fig. 34-1. In thinning young stands, System 3 has a cost disadvantage because of the difficulty of maneuvering the feller-buncher through dense woods and the relative inefficiency of the equipment when used on small trees. The limitation is removed in the final harvest cut, where System 3 shows a clear advantage over System 2 throughout the range of tree sizes studied [Fig. 34-1(b)].

> For all three systems, Fig. 34-1 shows cost per cord descending as average tree size increases. The rate of descent is slowest for System 1, steepest for System 3. Why? What is the principal characteristic of each system that leads to this result?

GROSS AND NET REVENUE

So much for cost estimates. The firm now tackles the question of revenues, gross and net.

The price of pulpwood delivered at the local concentration point by the small truck of System 1 is $37 per cord. This price is marked in Fig. 34-1(a) as a finely broken horizontal line. Somewhere about 15 stems per cord, the opportunity to make lumber (2-by-4s) from some of the logs begins to affect the average price, and the line bends sharply upward.

As your eye moves to the right in Fig. 34-1(a), you are scanning timber stands in which there are increasing proportions of sawlogs and then of veneer logs. The firm estimates these proportions, studies the local markets, and arrives at weighted average prices in relation to stems per cord. The result is the up-sloping section of the finely broken curve in Fig. 34-1(a), repeated in Fig. 34-1(b). The section is convex because the scale of stand age is more and more compressed toward the right side. The ordinate labeled "First thinning" represents an age of 15 years. The next three labeled ordinates fall at 20, 25, and 30 years.

You can now identify *tree value,* equal to *stumpage value,* per cord in 1987 dollars. It is the vertical distance from the logging-cost line to the log-price line:

Cut	Logging system	Cost per cord	Log price per cord	Tree value per cord
First thinning	1	$22.22	$37.00	$14.78
Second thinning	1	19.48	37.00	17.52
Third thinning	3	19.80	45.00	25.20
Final harvest	3	10.40	50.00	39.60

THE BEST COMBINATION

The last question facing the firm's manager is to identify an appropriate resource combination. The manager, let's say, begins by checking whether the 30-year rotation makes sense in his forest. A thinning regime and alternative products and logging systems have introduced large timber-size premiums. Can it be that a 30-year rotation, which has been traditional with the firm, is too short?

You and I can help the manager with this question. After being told that the firm's guiding rate of return is 6 percent, we leaf back to Table 22-1 and observe that the procedure followed there for setting a Douglas-fir rotation can be used equally well for loblolly pine. We calculate stocks and yields in dollars of tree value, so as to take timber-size premiums into account. We find that the forest's marginal value growth percent drifts downward from 20 percent in the years before $t = 30$ and then plunges to below 6 percent. (Compare Fig. 22-3.) A 30-year rotation is clearly appropriate. The headlong descent in the rate results from the fact that as the thinned stand becomes more and more lightly stocked and the stems removed continuously larger, the logging cost of thinning approaches that of the final cut. Meanwhile, the rise in end-product values has slackened. Little gain (size premium) remains to be made.

The manager's attention now turns to the thinning regime and to the fact that the final harvest is by far the most profitable cut per unit of output. The pulpwood thinnings, on the other hand, are least profitable. Why not restructure the thinning regime so as to downplay the early thinnings and augment the final harvest? The manager experiments (on paper) with a number of alternatives and comes up with Combination B, which contrasts with Combination A, the one illustrated in Fig. 34-1. The two combinations are described in Table 34-1. If the forest were regulated so that thinnings and final cuts were made every year:

Annual net return of Combination A would be $2,357.00 per 30 acres.
Annual net return of Combination B would be $2,382.80 per 30 acres.

In Combination B, one thinning is eliminated, and the total net revenue is at least upheld.

TABLE 34-1
LOGGING AND TIMBER-STAND CHARACTERISTICS, COSTS, AND REVENUES UNDER TWO ALTERNATIVE COMBINATIONS OF LOGGING SYSTEM AND SILVICULTURAL REGIME

Item	First thinning	Second thinning	Third thinning	Final harvest	All harvests
		Combination A			
Stand age, years	15	20	25	30	
Avg. d.b.h., inches	5.9	7.2	8.8	11.0	
Stems per cord	25.7	13.8	8.6	6.0	
Logging:					
System no.	1	1	3	3	
Cords per acre	10	10	10	45	
Cost per cord, $	22.22	19.48	19.80	10.40	
Log price per cord, $	37.00	37.00	45.00	50.00	
Tree value:					
Per cord, $	14.78	17.52	25.20	39.60	
Per acre, $	147.80	175.20	252.00	1,782.00	
Per 30 acres, $ per year					2,357.00
		Combination B			
Stand age, years	15	23		30	
Avg. d.b.h., inches	5.9	8.1		11.0	
Stems per cord	25.7	11.0		6.0	
Logging:					
System no.	1	3		3	
Cords per acre	10	15		50	
Cost per cord, $	22.22	24.00		10.40	
Log price per cord, $	37.00	41.00		50.00	
Tree value:					
Per cord, $	14.78	17.00		39.60	
Per acre, $	147.80	255.00		1,980.00	
Per 30 acres, $ per year					2,382.80

Short-wood contracting is not financially appealing. If the contractor serves as a crew member, his total annual take, wages and profit, is not likely to reach 9,000 1987 dollars. Suppose that no one willing to take on a short-wood contract is to be found in the locality, so that resort must be sought in System 2 for the 15-year and the 20-year thinning, if any. Figure 34-1 shows System 2 costing $42.50 per cord for the first thinning and, like System 3, $28.50 for the second. What does the absence of short-wood contractors do to the choice between combinations? Why?

SUMMARY

Summarily speaking, the logging system and silvicultural regime are closely intertwined because each affects the productivity of the other. Ideally, the two are planned simultaneously. This chapter offers a simple example of such planning for loblolly pine plantations. The logging variables recognized are thinning procedure and logging method. Silvicultural variables are rotation length and time and intensity of thinning.

REFERENCES

1 Lane, Richard A., Jr.: "An Evaluation of Operational Harvesting Techniques and Systems Applicable to the Commercial Thinning of Southern Pine Plantations," Master's thesis, Industrial Forestry Operations Program, Virginia Polytechnic Institute and State University, Blacksburg, 1981, p. 61.

CHAPTER 35

WOOD INDUSTRY IN A WEST COAST STATE

OREGON AS THE ILLUSTRATIVE STATE

This chapter talks about the following questions: Suppose that I undertake to describe the primary wood-using industry of an area, the industry that uses roundwood and mill residues as raw material. How do I go about the job? What sorts of information do I collect? How do I interpret what I learn?

We look at these questions from the viewpoint of the **Renewable Resources Evaluation Project,** a nationwide undertaking of the federal Forest Service. Originally authorized as the Forest Survey by the McSweeney-McNary Act of 1928, the Project now looks for its authority to the Renewable Resources Planning Act of 1974 and the Renewable Resources Research Act of 1978. The Project is charged with inventorying and reporting on forest and related resources as a basis for policy. Let's study the periodic report, released in 1984, on Oregon's primary wood industry as it was in 1982 [1]. The primary industry includes the manufacture of lumber, veneer, plywood, pulp, hardboard, particleboard, insulation board, shakes, shingles, posts, poles, and piling. The industry also includes the export of logs as another user of roundwood.

The year 1982 is a fully satisfactory year for our purpose of getting acquainted with wood-industry analysis. Moreover, the facts that we uncover for that year are roughly illustrative of forest-products activity on the West Coast. Yet it is well to note that 1982 was not "typical" for the Oregon industry (if *any* single year can be termed typical). It was a year of some depression. Wood-using plants had turned toward their own and other private timberlands for log supplies, after many years of reliance primarily on the public forests. Changes of other sorts continued: greater diversification in the industry, greater concentration, a wider range of products, fuller utilization of timber. The total volume of activity was markedly reduced, even in the few years since

1976. Timber consumption was down from 9.3 billion bd ft to 6.0 billion. The number of plants had fallen from 498 to 372. But some of this was only temporary. The basic importance of Oregon as a wood-products wellspring held good.

Now I station myself in the early 1980s and use the present tense for what I observe, the future tense for my plans.

FOREST GEOGRAPHY AND HISTORY

Indeed, Oregon is a fine state to choose for present purposes because its wood industry looms so large—as does its timber resource, the biggest among the states. Oregon ranks first in the production of lumber, plywood, and particleboard. Employment in Oregon's primary wood-using industry is about half of all manufacturing employment in the state.

Consequently, as I work up lists of wood-using plants and develop the questionnaire that I will mail out to them all, I have considerable confidence that recipients will give serious attention to the request for information and will go to their records for the answers wherever possible. If necessary, they will make estimates. In developing the lists of plants and their addresses, I begin with previous lists and update them with the help of the Oregon Division of Employment, industry directories, and other sources.

> As we recount the story of Oregon's wood industry, you will find it helpful to list the questions that plant managers were asked. When your list is finished, try to visualize the questionnaire that was used in the survey.

My canvass is made in 1983. The response is good. Where gaps appear, I use the telephone.

One finding that strikes home at once, and that certainly comes as no surprise, is the general similarity between the geographic pattern of the principal wood-using establishments and that of standing timber resources. The correlation is mapped in Fig. 35-1.

Oregon is bounded by the Pacific Ocean on the west, California and Nevada on the south, Idaho on the east, and Washington on the north. The north-south line that runs through the maps in Fig. 35-1 traces approximately the crest of the Cascade Range of mountains. West of these mountains is the cool, moist Douglas-fir region, where this species predominates and is associated with western hemlock, true firs, red alder, cedar, pine, spruce, and other tree species. Here grows more than four-fifths of Oregon's total sawtimber resource. Here, too, is found an equally large share of its wood-using industry.

Large-scale lumbering in the Douglas-fir region commenced in the Puget Sound and Grays Harbor areas of Washington and migrated to Oregon from the north. Some

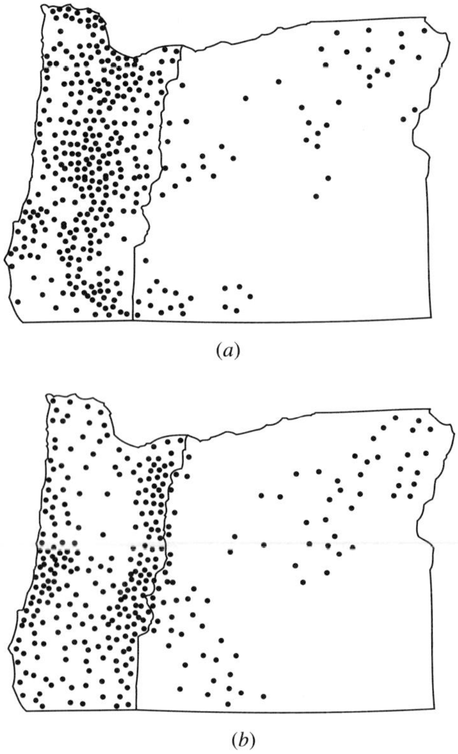

FIGURE 35-1
Distribution of (a) primary wood-using plants and (b) sawtimber stock, Oregon, 1980s. (a) Each dot represents one sawmill, veneer plant, plywood plant, pulpmill, hardboard mill, particleboard mill, insulation-board mill, shingle or shake mill, post or pole or piling plant, or export firm. (*Source:* Howard, James O.: *Oregon's Forest Products Industry, 1982,* U.S. Department of Agriculture, Forest Service, Resource Bull. PNW-118, 1984.) (b) Each dot represents 1 billion bd ft of sawtimber stock, Scribner rule. (*Source:* Gedney, Donald R.: Compilation made for the author from most recent county statistics of the Renewable Resources Evaluation Project, Pacific Northwest Forest and Range Experiment Station, Portland, Oreg., June 1987.)

of the earliest lumbering in Oregon was along the Columbia River, which forms the state's northern boundary, and the Willamette River, which flows northward through the middle of Douglas-fir country and joins the Columbia at Portland, Oregon's principal city and port. There are a few ports directly on the Pacific, afforded by estuaries and bays such as at Astoria on the Columbia; Tillamook, just a bit south; Newport, still farther down; and Coos Bay, some two-thirds of the way to the California line.

However, western Oregon is not well provided with good ports, a fact that has shaped the development of the forest industry. Pulp and paper manufacture, though certainly attracted by abundant supplies of wood, has not thrived as it does where wood, water, and water transportation all come readily to hand. The lumber industry, later joined by veneer and plywood, continued its overland march southward. The wood industries are now heavily concentrated in Lane County, which stretches from the Cascade Ridge to the sea about midway down Oregon's fir belt, and in Douglas and Coos counties, just to the south.

Mills in these three counties in 1982 consumed nearly $2\frac{1}{2}$ billion bd ft of logs, more than four-tenths of the state total. Lane County led all others with 57 mills, or 15 percent of the total for the state. To the north of Lane County is most of Oregon's young-growth fir forest, regenerated on the burns and cutovers. To the south is most of the old growth, aged more than 100 years and commonly 200 or 300. Old-growth timber on good site is an amazing spectacle and an incomparable logging chance; but in many ways more notable, and certainly more auspicious, are the heavy stands of young growth that thrive under management and protection.

Inspection of Fig. 35-1 reveals that much of Oregon's fir-belt wood industry is located in the towns and lowland settlements, while the timber is in the Cascades to the east, the Coast Range to the west, and the generally broken jumble of mountains at the southern end of the belt.

Figure 35-1 also makes clear how greatly the forestry of eastern Oregon differs from that west of the Cascades. These mountains throw a rain shadow over eastern Oregon, creating a different world. One can drive eastward toward the Cascade ridge in cool rain and fog, pass the mountain summit, and descend in the warmth of bright sunshine. This is ponderosa pine country, where the forest is intermixed with juniper woodland areas, and in some regions such as southeast Oregon, replaced by vast stretches of rangeland.

Although eastern Oregon cannot match western Oregon in forest abundance or industrial concentration, the wood industry is a major segment of economic life in many an east-side community. Some of the sawmills rival those of the west side in daily capacity.

More than half of Oregon's forest land is publicly owned—as in national forests and in the public domain and other lands administered by the Interior Department. Most of the privately owned forest is held in large tracts by the wood-using industry. These domains were staked out, for the most part, before World War I. The opening of the Panama Canal in 1914 made possible the first westward moves of the softwood lumber industry from the southern pine region. Lumbering went forward on private lands along the waterways, entering Oregon and expanding rapidly in the 1920s and 1930s, which were decades of recession for wood-using firms in most circumstances. The comparatively inaccessible public forest began to come prominently into industrial life in the 1940s, and their share grew in the 1950s. Thereafter, the rise in industrial use of the public lands was slowed and then halted and, at least temporarily, reversed. Private forests did not fill the gap. In the 1970s, the national wood-using center of gravity had started to shift back to the South. This was the first time in our history that a once widely abandoned timber region was later widely reentered.

SAWMILL SIZE AND INDUSTRY CONCENTRATION

Oregon possesses not only big timber, but also big mills to process it into lumber. Some Oregon sawmills have racked up an output of more than a million board feet per day. In 1982, 44 percent of Oregon's 161 sawmills had a capacity of 120,000 bd ft or more per 8-hour shift. Essentially all these big mills operated log debarkers, chippers for residues, and lumber planers. Two-thirds of them put their lumber through dry kilns. Most were equipped with band headsaws, supplemented in some cases by chipping saws. More than 80 percent had been at the same location under the same ownership for more than 10 years.

Table 35-1 recognizes four mill size classes based on lumber-production capacity and gives the tally of mills for Oregon and its two regions. Not surprisingly, the big-

TABLE 35-1
PERCENTAGE OF ALL SAWMILLS IN OREGON AND OF MILLS WEST OF THE CASCADES AND EAST OF THE CASCADES, BY SIZE CLASS, 1982

8-hour-shift capacity (1,000 bd ft)	Percentage of Sawmills		
	Oregon	West side	East side
120 or more	44	47	36
80–119	26	24	31
40–79	16	15	19
39 or less	14	14	14
All mills	100	100	100

Source: Howard, James O.: *Oregon's Forest Products Industry, 1982,* U.S. Department of Agriculture, Forest Service, Resource Bull. PNW-118, 1984.

gest mills are more common on the west side of the Cascades than on the east. However, the smallest class of mills is equally common in the two regions. Such mills abound at the north end of the Douglas-fir belt, where young-growth timber predominates and the region's small forest tracts, such as farm woods, are concentrated.

Ever since World War II, the size-class distribution of sawmills has been changing steadily. Mills average larger than they used to. Most of those that fell by the wayside were the little sawmills, though some were larger mills with outdated equipment. The newcomers and remodeled mills are large. The furthest comparison is with 1968. Between then and 1982, the number of sawmills in Oregon dropped from 300 to 161. The smallest size class of mills shrank from 102 to 23; but the number of big mills grew from 59 to 71. These big mills increased their share of Oregon's total sawmill capacity from 48 percent to 89 percent. Lumber firms want a high-capacity installation if they are to afford, and take full advantage of, the technological gains of recent years: electronic and other sophisticated, high-cost devices for guiding or performing the mill's key functions of log sorting, channeling, and breakdown; lumber finishing and grading; and, throughout, close raw-material utilization. Such technological developments are one of the lumber industry's sources of *scale economies* (page 249), one of its defenses against the forces of pulverization (pages 97 to 100).

The topic of plant size leads into the subject of the industry's structure. Any thorough discussion of structure, and also of conduct and performance, will be postponed until Part Five, concerning the market. But right here, on the subject of Oregon sawmills, is a good place to introduce one aspect of structure: the concentration ratio.

Students of industrial structure have given much thought to an interesting contrast among industries: Some industries are made up of a multitude of very small firms that are rather highly competitive. Others comprise only a very few large firms that have some monopoly characteristics. Others take on various intermediate forms, such as one or two companies that produce half of the industry's output, and many small companies that produce the other half. Such differences among industries are important because they are associated with differences in the industries' contributions to social well-being.

One expression of an industry's structure is the **concentration ratio,** which measures the proportion of the industry's output, sales, or capacity accounted for by the largest firms. The Department of Commerce [2] gives us nationwide information on the percent of total value of shipments by a manufacturing industry accounted for by the 4 largest firms in the industry (data are also given for the 8 and the 20 largest firms). We may recognize four concentration-ratio classes at the 4-firm level and mention some illustrative industries:

Extreme concentration, 75 percent or more: primary lead and aluminum, flat glass, motor vehicles, and car bodies.

High concentration, 50–74 percent: sanitary paper products, explosives, cane sugar refining, pickles and sauces and salad dressings.

Moderate concentration, 25–49 percent: softwood veneer and plywood, creamery butter, blast furnaces and steel mills, weaving mills.

Low concentration, 24 percent or less: sawmills and planing mills, brick and structural clay tile, steel pipe and tubes.

> Notice that concentration is defined in terms of the market share commanded by a given number of the largest *firms* in an industry. Because of the form in which our data are available, we will often be obliged to think of concentration in terms of the share commanded by a given number of largest *plants.* If a concentration ratio is calculated in each of the two ways, which way is apt to show the higher ratio? Will it ever show a lower ratio?

Where does the Oregon lumber industry fall within the foregoing classification? Undoubtedly *low* in concentration, the four largest firms having perhaps between 10 and 15 percent of the total industry capacity. Why does the industry rank so low? Here are two thoughts on the question:

First, the pace of a lumber production line is set by the headsaws, and there are strict engineering limits (given today's technology) to the rate at which the saws can gobble up logs. There is a limit to saw and carriage speeds, and the size of logs is limited—indeed, shrinking. If a firm wants to step up lumber production beyond these limits in the short term, it can increase operating time; but in the longer term it must add more facilities like those it already has, either at the same location or elsewhere. But, second, circumstances come into play to control this sort of expansion. One is the availability, the cost, of raw material. Another is the opportunity for diversification—for **horizontal integration,** in which the firm takes on enterprises that are at the same stage in production—for example, primary manufacturing enterprises such as lumber, veneer-plywood, and pulp-paper. Horizontal integration of lumber production with other lines offers richer rewards in efficiency than mere enlargement of lumber output, mainly because of its opportunities for fuller and more discriminating use of timber supplies.

VENEER, PLYWOOD, AND OTHER NONLUMBER INDUSTRIES

Speaking of veneer, plywood, and other wood-industry lines, let's have a look at the place of these other industries in Oregon. Table 35-2 summarizes the industries' wood consumption and number of plants.

Veneer and plywood's 101 plants use more than a third of the volume of logs that go into Oregon establishments. Like the sawmills, these plants are heavily concentrated in the center of the Douglas-fir belt. They are more heavily represented than sawmills at the old-growth, south end of the belt and less heavily in the young-growth north. The average plant uses about as much roundwood as the average sawmill, even though a fifth of the plants perform only the plywood lay-up, taking no logs as raw material.

> If you define concentration in terms of the share of roundwood used by the largest plants in the industry, will veneer-plywood or lumber probably show the heavier concentration?

Of the plants with lathes, 28 percent are equipped to handle logs of 80-in or larger diameter, a percentage that has been dropping in recent years along with the proportion of old-growth timber.

TABLE 35-2
LOG CONSUMPTION IN OREGON, BY INDUSTRY, 1982; AND NUMBER OF PRIMARY WOOD-USING PLANTS IN OREGON AND ON THE WEST AND EAST SIDES OF THE CASCADES, BY INDUSTRY, 1982

	Oregon log consumption		No. of Plants		
Industry	Million bd ft	Percent	Oregon	West side	East side
Lumber	3,411	57.2	161	119	42
Veneer and plywood	2,035	34.1	101	93	8
Log export	398	6.7	32	32	—
Pulp and board[a]	77	1.3	36	29	7
Post, pole, piling	29	.5	8	7	1
Shake and shingle	11	.2	34	32	2
All industries	5,961	100.0	372	312	60

[a]Board is hardboard, particleboard, or insulation board.
Source: Howard, James O.: *Oregon's Forest Products Industry, 1982,* U.S. Department of Agriculture, Forest Service, Resource Bull. PNW-118, 1984.

Log-export establishments are included as "plants" in Table 35-2 so as to account for their intake of roundwood.

Of Oregon's 36 pulp and board mills, 16 are pulpmills, all west of the Cascades where there's a good supply of water for the manufacturing process. The amazing feature of these mills is their sparing use of wood in the round: only 77 million bd ft in 1982, and of this, less than 10 percent was in sound logs. The residues of other industries—chips, shavings, and sawdust—constituted more than 97 percent of the total tonnage of raw material.

Besides the wood residues used in pulp and board, some $3\frac{1}{3}$ million tons of wood and bark were used by all industries to generate energy. Less than 1 percent of all wood and bark residues went unused. This remarkable level of raw material utilization has been achieved only since the 1960s. In the 1950s, nearly every sawmill and veneer plant in Oregon had an open-air burner, into which residues were dumped to be destroyed. In 1982, only 11 burners were left in the state.

The shake, shingle, post, pole, and piling plants in Table 35-2 are small, specialized installations. Nearly all are located west of the Cascades. Shakes and shingles are made from western red cedar.

SOURCES AND SPECIES OF RAW MATERIAL

Since this inventory of Oregon wood industry is part of the Renewable Resources Evaluation Project, it is concerned, not only with the industrial establishments, but also with the timber they consumed as raw material: the counties it came from and went to; the ownership of the timber; its quality, size, age, and species; and other details. Let's not go into these details here, except to recognize that covering and reporting on them are part of the Project. Here, let's just mention some highlights by way of clarifying the view of Oregon industry.

First, where did the industry's 6.0 billion bd ft of roundwood raw material come from in 1982? Some 96 percent of it came from Oregon; the other 4 percent from adjoining states. Oregon's wood-using economy is as nearly self-sufficient as that of any state or nation. Where nations are concerned, Oregon somewhat resembles Finland in self-sufficiency; it clearly contrasts, say, with Japan.

Second, who owned the standing timber that nourished the industry in 1982? Private timber was the industry's mainstay. It accounted for 54 percent of the supply; 31 percent came from the wood users' own lands, 15 percent from other industry firms, 8 percent from nonindustrial private owners. National forests furnished 34 percent of the total. The share of all timber supplied by the public, 46 percent, was less than the share of all available forest area owned by the public.

> Even in 1976, when public forests provided 56 percent of the industry's wood, this public fraction was less than the public's share of the timber acreage. Why?

Third, what species of trees did the Oregon industry use in 1982? Douglas-fir, of course, was the big leader: 59 percent. East-side ponderosa pine, supplemented by some timber of this species from the tumbled mountain area at the southern end of the west-side belt, came through with 16 percent. Douglas- and true firs, pine, and hemlock together provided well over nine-tenths of Oregon's wood raw material.

SUMMARY

Let's summarize the story in the spirit of its special purpose, which is to consider what inventory to take of a region's wood-using industry, how the inventory may be interpreted, and what means to use to acquire the information. The inventory is based on a 100-percent census of industrial establishments, with a mail questionnaire and telephone follow-up. The report takes the form of text and tables, in which such matters as the following are given attention: (1) historical background of the wood-using economy; (2) review of standing timber resources; (3) wood-using plants: their type, size, location, ownership, equipment, and trends; (4) major branches of the industry: their characteristics, including structure, and their trends; (5) the industry's raw material: species, quality, form (roundwood or residues), geographic and ownership origin, and degree of utilization; and (6) prospects for the industry implied in the analysis.

REFERENCES

1 Howard, James O.: *Oregon's Forest Products Industry, 1982,* U.S. Department of Agriculture, Forest Service Resource Bull. PNW-118, 1984.

2 U.S. Department of Commerce, Bureau of the Census, Census of Manufactures: *Concentration Ratios in Manufacturing,* 1981.

CHAPTER **36**

GETTING ESTABLISHED IN THE SOUTHERN PINE LUMBER INDUSTRY

THIS CHAPTER'S QUESTION

The question is: What is involved for the firm that wants to enter the lumber industry in the southern pinery? Suppose the firm proposes to set up a new mill of reasonably efficient size. What arrangements are to be made, and what are some of the costs?

> We commonly think of efficiency (that is, output per unit of input) as a function of plant or firm size. The big plant or firm is not expected to show the same efficiency as the little one. But surely size is only part of the efficiency story. Can you think of some characteristics of plants or firms that affect efficiency independently of size?

We'll take the state of Mississippi as the illustrative area, dipping into Alabama for some of the supporting data.

MISSISSIPPI'S LUMBER GEOGRAPHY AND HISTORY

By way of orientation, it will help us to glance at Mississippi's geography and review some history of the lumber industry. Here we can lean heavily on the reports of the nationwide Forest Survey and Renewable Natural Resources Evaluation Project [1, 3–5, 7–9, 11–13].

The state lies between Arkansas–Louisiana on the west and Alabama on the east. Its northern boundary is with Tennessee, and on the south is the Gulf of Mexico. The forests are as widely varied as those of any southern state. Bottomland hardwoods are a rich resource, growing in the floodplains of streams and rivers, and, notably, along the Mississippi River, which forms the state's western boundary. This major bottomland belt—Fig. 36-1 (a)—widens out into an extensive "Delta" region toward the north. In the southern tip of Mississippi, which is lower coastal plain, are longleaf and slash pines. Elsewhere are loblolly and shortleaf pines and upland hardwoods in wide variety. Over the years, pine and hardwood have fought a seesaw battle for most of the upland sites of the central and northern areas. Man has intervened on the side of pine when he has planted it, increased his destruction or use of hardwood, or promoted farm abandonment. He has often intervened on hardwood's side when he has stepped up fire control or when he has sought out the pine for harvest and let nature take its course thereafter.

The lumber industry has long held, and still holds, a prominent place in Mississip-

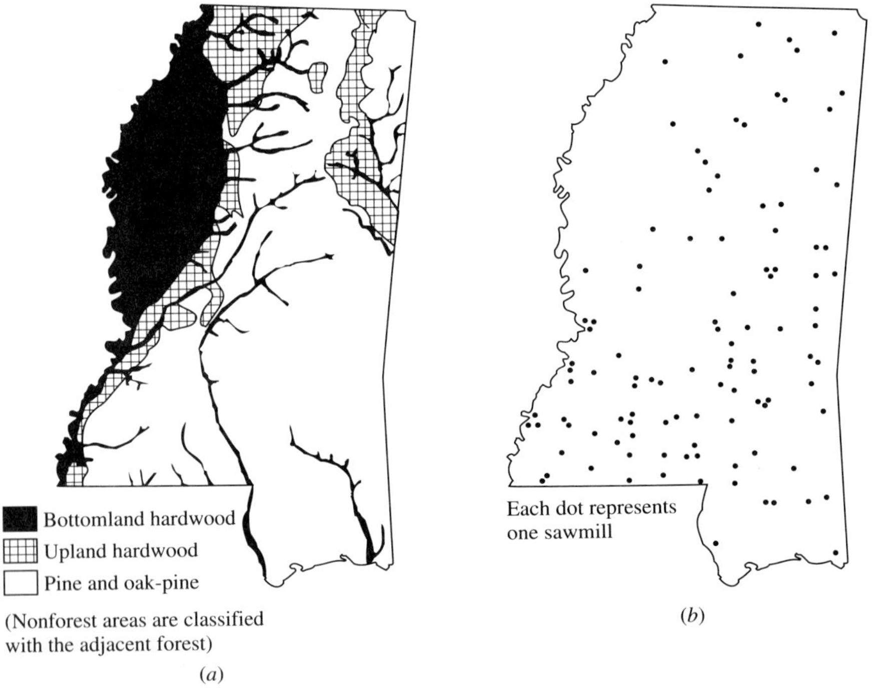

FIGURE 36-1
Mississippi's forests and sawmills. (a) Major types of forest. (*Source:* Anonymous: *Mississippi Forests,* Southern Forest Experiment Station, Forest Survey Release 81, New Orleans, La., 1958, p. 5.) (b) Sawmills that produced 3 million bd ft or more in 1976. (*Source:* Bertelson, Daniel F.: *Mississippi Forest Industries, 1976,* U.S. Department of Agriculture, Forest Service Resource Bull. SO-68, 1978, pp. 17–19.)

pi's timber economy. The industry came into dominance during the softwood lumbering era, toward the end of the nineteenth century, when Lake States pine was running low and the industry turned southward. In Mississippi, access to the old-growth pine was provided by the Illinois Central, Southern, and Ship Island railroads, which were completed in 1902. Softwood lumber output rose rapidly, to 2.2 billion bd ft in 1909 and an all-time peak of 2.6 billion bd ft in 1925.

Thereafter, continuing and often violent fluctuation has characterized Mississippi's lumber industry. Softwood lumber output skidded from 2.2 billion bd ft in the brink-of-depression year 1929 to 400 million in the 1932 nadir. Then it came back in response to war, touching 1.3 billion bd ft in 1942. Meanwhile, hardwood was a relatively minor segment of the industry, though it was a growing segment and by World War II was not far behind pine in some uses.

In the war and immediate postwar years, there were nearly 2,000 sawmills in Mississippi. Most of them were "class-4" mills, the smallest size class, operating just a few weeks in the year with a bare minimum of equipment and producing less than 500,000 bd ft of lumber annually. These little mills abounded both in pine and in upland hardwoods; many were "farm mills" that were brought out when there was useful or profitable employment for them and were put away again when the occasion had passed. In those times, given the operability of little mills, anyone could get into the "lumber industry" for a few hundred dollars and sell his way out again for a comparable sum. The little mill was mobile and could be operated as profitably on a patch of timber as on a big tract; so raw material was no special problem.

Then came what was perhaps the most drastic upheaval that the southern lumber industry had yet experienced. Mississippi was at the center of it. The state's total sawlog output fell from 1.6 billion bd ft in 1946 to 1.0 billion in 1956; and by 1966, output was down to .8 billion bd ft, half of the 1946 level. Then it rebounded to 1.7 billion bd ft in 1976 and 2.2 billion in 1986. Here we are seeing the southern-pine side of the industrial outmigration from the West Coast noted on page 265. Numbers of mills in the first four of these years were

1946:	1,980
1956:	1,000
1966:	350
1976:	218

But during these years, the number of the larger mills held up. It was the ranks of the little fellows that were decimated.

To get a clearer view of what happened, one should stress the fundamental difference between the bottomland hardwood lumber industry and the upland industry, characteristically pine. In bottomland timber, quality is of prime importance; sawmills are obligated by financial circumstances to saw for grade. If they cannot make the most of their logs, they cannot survive against their competitors. A firm must be able to employ the necessary specialists to play the game, which means that it must operate on a substantial scale. Thus the bottomland mills were characteristically large, and for the most part they escaped the trauma suffered by the upland mills.

> The traditionally great difference between bottomland hardwood and upland (primarily conifer) mill practice with respect to "sawing for grade" arose from the difference in end use between the two lumber categories. What was this difference in end use? Why did it have the result that quality was of prime importance for bottomland hardwood timber, but not equally so for upland timber?

THE PINE LUMBER INDUSTRY, 1946–1976

Recognizing the bottomland–upland differences, James Moak [10] analyzed the lumber revolution of the 1950s, confining his attention to pine sawmills. He studied the changes that took place between 1946 and 1959. He found that the number of pine sawmills (including sawmill-planers) in Mississippi fell from 978 to 117 between these two years. The numbers of mills in size classes 1 and 2 (annual output 3 million bd ft or more) hardly changed; but the number of smaller outfits was cut by 91 percent. Figure 36-2 illustrates this dramatic change. The average pine mill, which pro-

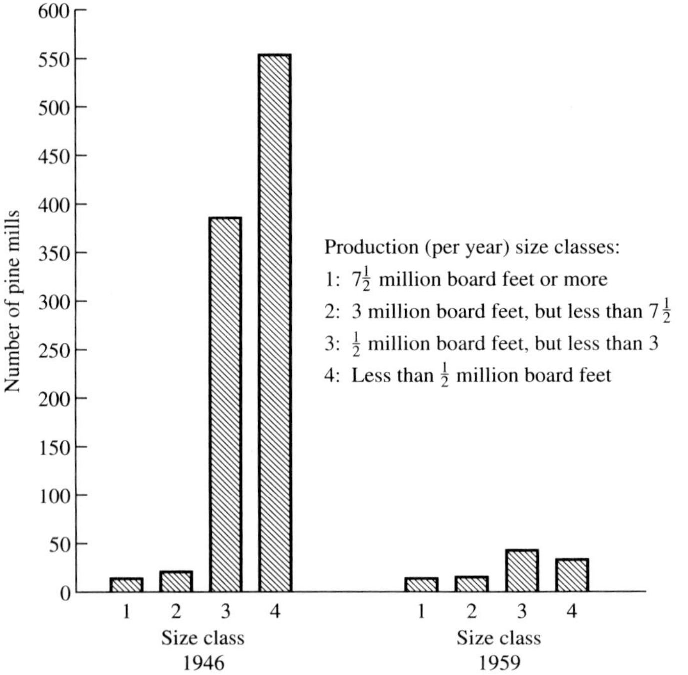

FIGURE 36-2
Number of pine sawmills, by production size class, operating in Mississippi in 1946 and in 1959. (*Source:* Moak, James E.: "The Pine Lumber Industry in Mississippi, Its Changing Aspects," Ph.D. dissertation, State University College of Forestry at Syracuse University, Syracuse, N.Y., 1965, p. 12.)

duced less than 1 million bd ft of lumber in 1946, produced more than 2 million bd ft in 1959. In 1946, the 15 class-1 mills (output 7.5 million bd ft or more) sawed 18 percent of all the lumber. In 1959, class-1 mills, still 15 of them, turned out 49 percent of the lumber. Here was a remarkable increase in concentration! By 1959, the Mississippi pine lumber industry had been reduced to a nucleus of grim survivors in which the largest (and still the fewest) mills were necessarily a dominating force.

But why the big change in the structure of Mississippi's pine lumber industry? What sort of plague was it that wiped out the small mills? In part, the same forces were operating here as were at work in Oregon during the same years: changing technology of lumber-making that put little sawmills at a hopeless disadvantage. But there was a major difference between the western and the southern industries that gave the Mississippi story a special twist: In those years, the total output of Douglas-fir and other western softwoods was on the upswing; whereas the total output of southern pine was hitting the skids. Between 1946 and 1959, while lumber production in the United States was rising by nearly 10 percent, that in the southern pine states was falling by 25 percent, and production in Mississippi was taking a $38\frac{1}{2}$-percent plunge. Southern pine lumber was suffering the ill effects of pinched stumpage supplies and rising costs, a bad reputation for product quality, and keen competition, not just from West Coast *lumber,* but also from Coast *plywood,* for plywood wasn't yet being made commercially from southern pine. Furthermore, freight rates were shifting nationally in favor of the Douglas-fir region. And remember that the region never had the acreage of nonindustrial private forest and the horde of "farm mills" that were encountered in Mississippi and other southern states. The little southern mills went into storage, as they always did in depressed times, but now technological changes in both lumbering and farming were preventing their reappearance.

Why was there a "horde" of little mills in the South, but not on the West Coast?

And so, as Mississippi's lumber industry recovered from depression, the recovery was marked by a continued trend toward concentration. Here are some Forest Service figures for *all* sawmills:

Year	No. of sawmills	Output of average mill
1966	350	Less than 3 million bd ft
1972	241	4.0 million bd ft
1976	218	7.8 million bd ft

In 1976, 107 of the mills sawed at least 3 million bd ft of lumber—see Fig. 36-1(b)—46 sawed more than 10 million bd ft; one mill produced in excess of 93 million bd ft. All but 3 of the 107 large mills made chips for sale to pulpmills, and 41 of the 111 smaller mills did likewise. Southern pine lumber was substantial business. Nationally known names such as Weyerhaeuser, Masonite, Koppers, and Georgia-Pacific were in the roster of big-scale sawmills, all of these operating two or more installations in the state. Interestingly, the status of pine and that of hardwood had become reversed: Two-thirds of the small sawmills were hardwood mills. The market for hardwood pallet lumber had come into its own.

OPTIMAL PINE SAWMILL: DEGREES OF INTEGRATION

And so southern pine is no longer an industry that tolerates firms that hop in and out. Getting established in the industry is a serious matter, requiring much capital. James Buford [6], working with Alabama data, found that an "optimal-size" pine sawmill has a one-shift annual capacity of between 30 and 45 million bd ft. This is a *large mill* by the standards followed in this chapter. It falls in class 1 of Fig. 36-2, with an annual output of at least $7\frac{1}{2}$ million bd ft. Let us borrow Buford's thinking for use in Mississippi and see where it leads.

> How would you go about finding the "optimal-size" sawmill for a given timber area? Primarily, how would you define the optimum? Try using economic efficiency as the criterion.

The cost of the plant itself is only part of the capital investment. The firm is obligated to consider integrating lumber manufacture with other activities. Chapter 35 speaks of horizontal integration between lumber and panels. Surely the advantages that it holds in Oregon are equally operative in Mississippi, and one finds a number of examples here. One also finds examples of horizontal integration between pulp-paper and lumber (and panel) production. It is said that there is a trend toward such integration, with the pulp-paper corporations taking the lead. Surely one route to becoming established in the lumber industry is through the pulp and paper industry. However, let's not pursue horizontal integration any further here.

There are two sorts of **vertical integration** (combining enterprises that are at *successive* stages in production) for a lumber manufacturer to think about. One is **forward integration** toward the consumer: lumber drying and finishing, wholesaling, advertising, and retailing. For large mills, drying and finishing and planing are universal activities. In wholesaling and retailing, a study of 11 major U.S. forest product firms in the early 1980s disclosed that 10 operated their own ("captive") wholesale enterprises or retail outlets, and 4 operated at both wholesale and retail levels, with an average of more than 150 distribution yards each. The proportion of all U.S. lumber, plywood, and millwork sold through manufacturers' sales branches and offices rose from about 15 percent in 1963 to about 25 percent in 1982 [2]. As for the little sawmilling firms, apparently a great many deal directly with consumers, though the total volume is small.

Backward integration extends the manufacturer's business toward the forest. Although lumber producers can be found aplenty that own or control no forest land, they are a rarity among firms of efficient size, and the same is true of veneer and plywood manufacturers. And when one comes to pulp and paper companies, one finds that essentially all are timberland owners. The gains from owning timberland are too great to be neglected. The principal one is assurance of raw material, including price protection. Another is the tax advantage that was effective until 1987. This advantage is mentioned in Chap. 56. The upshot of these advantages is that possibly eight-tenths of the large sawmill firms own forest land.

INVESTMENT-CAPITAL REQUIREMENTS: THE LARGER MILL

Logging, too, is part of backward integration. Quite possibly nine-tenths of Mississippi's large pine sawmill firms have logging operations that they own—or control through contractors. If a firm owns its logging facilities, the **investment capital** required for its plant and machinery may be something on this order:

	Million 1987 dollars
Fully equipped new mill, 30 million bd ft annual capacity	8
Five new mechanized logging enterprises at $400,000 each	2
Total	10

The remaining investment is in raw-material supply. The firm has three approaches, which it can use in combination: (1) buy timberland, (2) control timberland by leasing or obtaining cutting agreements, (3) buy stumpage. It is becoming increasingly common for Mississippi wood-using plants to draw on their own timberland for at least part of their raw material. Between 1952 and 1985, when the state's total timberland acreage declined a bit, forest industry increased its holdings 40 percent. Industry owned 15 percent of all timberland at the start of the period, 22 percent at the end [5].

Let us take a simple case. Suppose that the firm decides on a 50-percent self-sufficiency in timber supply, the balance of its requirements to be bought as stumpage on the open market, largely from nonindustrial private owners. Assume that the forest can be bought at favorable prices: $100 per thousand board feet for pine, $60 for hardwood, and $200 per acre for the land. Prices can vary substantially with time and place. At the foregoing prices, in average pine country, the firm will want to buy close to 70,000 acres for a sustained annual yield of 15 million bd ft, and the cost will be roughly 40 million 1987 dollars.

The firm's three major types of investment, then, amount to $50 million, of which about 80 percent is forest land.

THE SMALLER MILLS

But we must recognize that a firm may enter the pine lumber industry with less than a marching band and all flags flying. Mills with an annual capacity of much less than 30 million bd ft can be efficient and earn a profit. Such smaller mills can lean on contract loggers and the open market for raw materials. Take the case of a 5-million-feet-per-year mill. Like its big counterpart, it produces **commodity,** or **structural, lumber:** 2-in dimension. It has a conventional band headsaw rather than a quad band. It has no "decision assists." It has no kiln, no planer, but sells its output rough-green. Price? $750,000 as of 1987.

If there is no outlay for land or logging equipment, the $750,000 is the sum of the investment.

Besides capital for its initial investments, a firm must have command of working capital to meet outlays for power, supplies, payroll, taxes, open-market stumpage purchases, and the like.

> If you were planning to become established in the Mississippi pine lumber industry and found it hard to raise even $1 million, what further shortcuts could you take, at least in the beginning, to lighten your financial burden?

The foregoing story of the lumber industry, intentionally simple, could have been qualified, here and there, with an *if* or an *on the other hand.* One such qualification may provide a useful postscript: Let's not suppose that the little sawmill is headed for total extinction.

Wherever raw material is low-grade and scattered, small-scale manufacturing may hold advantages. (There may be diseconomies of scale.) So in the South and elsewhere, one finds little mills coming back into use for making local lumber, custom sawing, conserving beetle-killed pine, and producing hardwood pallet stock.

SUMMARY

Here's the long and the short of it. The southern pine lumber industry, once so easy to enter and to leave, is in general no longer so. For an all-out plunge into the industry, investments in land, logging enterprises, and a highly efficient mill can run to nearly $70 million. But a more modest system can be had for as little as $1 million. Working-capital requirements are additional. Besides such optimal mills, there is still a place in the southern sawmill industry for the little peckerwood mill.

REFERENCES

1 Anonymous: *Mississippi's Forests [1957],* Southern Forest Experiment Station, Forest Survey Release 81, New Orleans, La., 1958.

2 Barnes, Celeste Stockwell, and Steven A. Sinclair: "The Distribution and Marketing of Building Products, Past and Present," unpublished manuscript, Department of Forest Products, Virginia Polytechnic Institute and State University, Blacksburg, 1984.

3 Bertelson, Daniel F.: *Mississippi Forest Industries, 1972,* U.S. Department of Agriculture, Forest Service Resource Bull. SO-43, 1973.

4 Bertelson, Daniel F.: *Mississippi Forest Industries, 1976,* U.S. Department of Agriculture, Forest Service Resource Bull. SO-68, 1978.

5 Birdsey, Richard A., and William H. McWilliams: *Midsouth Forest Area Trends,* U.S. Department of Agriculture, Forest Service Resource Bull. SO-107, 1986, p. 14.

6 Buford, J. A., Jr.: "Some Aspects of Competition in the Southern Pine Lumber Industry in Alabama, 1967–1972," Ph.D. dissertation, University of Georgia, Athens, 1974.

7 Christopher, Joe F.: *Mississippi Forest Industry Statistics, 1962,* U.S. Department of Agriculture, Forest Service Resource Bull. SO-4, 1963.

8 James, Lee M., and Albert L. Tofté: *Mississippi's Forest Industry [1946],* Southern Forest Experiment Station, Forest Survey Release 62, New Orleans, La., November 1949.

9 Kelly, John F., and Mike Sims: *Forest Resources of Mississippi [1987],* U.S. Department of Agriculture, Forest Service Resource Bull. SO-147, 1989.

10 Moak, James E.: "The Pine Lumber Industry in Mississippi, Its Changing Aspects," Ph.D. dissertation, State University College of Forestry at Syracuse University, Syracuse, N.Y., 1965.

11 Thomas, Clarence E., and William H. McWilliams: *Midcycle Survey of Mississippi's Forest Resource [1983],* U.S. Department of Agriculture, Forest Service Resource Bull. SO-101, 1985.

12 VanHooser, Duane D.: *Mississippi Forest Industry [1966],* U.S. Department of Agriculture, Forest Service Resource Bull. SO-12, 1968.

13 VanHooser, Duane D.: *Midcycle Evaluation of Mississippi Timber Resources [1972],* U.S. Department of Agriculture, Forest Service Resource Bull. SO-44, 1973.

CHAPTER 37

ECONOMIC FEASIBILITY OF A PANEL PLANT IN APPALACHIA

THE SUBJECT OF FEASIBILITY

A common sight in a growing economy is a community searching for a manufacturing plant. And another, equally common, is a plant searching for a community. Out of such searching grows the sort of question that we'll study here: Is it economically feasible to locate a given type of plant in a certain area?

Economic feasibility is customary jargon for a question about a contemplated investment. The question, almost invariably, is whether the investment will pay off satisfactorily. It is a question involving predictions about technology and about financial matters: all sorts of revenues and costs and how their aggregates compare. Of course, for the community that seeks development and employment and for managers who look for a location, there is much more than just money at stake. But in the last analysis, planners have to face the fact that investors will be asked to take a venture, and if the investors are ordinary human beings, they cannot be expected to comply without good promise of a reward.

Hence the economic **feasibility study.**

For an illustrative study, consider a plant that makes composite plywood (com-ply) panels: 4-by-8-ft sheets, $\frac{1}{2}$-in thick, comprising a flakeboard core sandwiched between veneer faces. And consider an Appalachian Highland community, a 25-county area in north Georgia. Picture either that the plant is looking for a community or that the community is seeking a plant or both. In any case, imagine that before this particular manufacturing operation and location are brought together, the analysts think through a lot of alternatives. Whether they arrive at an optimal combination of resources, one cannot know. The analysts in this case [1] cooperated in a project of the Georgia Institute of Technology at Atlanta. Their inquiry is timed at about 1980.

> Suppose you are the analyst. You have no commitment to a particular location or branch of wood-using industry. What alternatives could you reasonably expect to examine before deciding to study the feasibility of putting a panel plant in north Georgia? What would you try to learn about these alternatives?

COM-PLY VERSUS ALTERNATIVES

In the 1980s, the wood-panel industry and its manufacturing technologies are in the throes of change. Softwood plywood, the traditional structural panel for wall and roof sheathing and floor underlayment, suffers from the increasing scarcity and price of peeler logs, both Douglas-fir and southern pine. In the South, wood cost has been 60 percent of total plywood-production cost and is moving toward 70 percent [2].

Thus the panel industry is pressed to find technologies responsive to the pulverization principle, technologies that can use raw material less expensive than peeler logs. From this pressure have come such products as waferboard and oriented strand board (OSB), reconstituted panels comprising only flakes and adhesive, that can be made from peeled pulpwood-size bolts with negligible residue. And com-ply panels, though they require some veneer-size logs, can fully utilize their wood input. Wood cost can be 30 to 25 percent or less of the total cost of manufacturing reconstituted panels [2].

Com-ply has the disadvantage that its production calls upon two different technologies: veneer-plywood and reconstituted board. Com-ply's capital and operating costs therefore tend to exceed those of a single, uniform product. However, com-ply cost can be competitive if the product can use veneer from an existing, depreciated plant, perhaps one of the plants that are being shut down, both West and South, as reconstituted-panel manufacture advances. And com-ply has the advantage that it resembles tried-and-true plywood in both performance and appearance.

In any event, the north Georgia analysts chose com-ply for their feasibility study. In truth, whether their choice was optimal need not disturb us, for our interest is not in their findings, but in their method: their approach to problem solving.

CHOICE OF THE STUDY AREA

The analysts explain their choice of composite panels and of the north Georgia location. The panels, they say, are a logical choice in north Georgia for five reasons:

1 They are a new product, though the technology of manufacture has been well worked out. At the time of the analysis, 1978, there were three plants in the U.S., all on the West Coast.

2 The panels are used mostly in sheathing and underlayment in construction, where they are satisfactory substitutes for softwood plywood. The long-term market prospect is bright.

3 A panel plant demands a large supply of labor, such as is available in north Georgia.

4 Composite panels are a high-yield product involving virtually no waste.

5 The panels' raw material is a mixture of softwoods and hardwoods, which is what the north Georgia forest provides.

The analysts then turn to the timber resources in the 25-county area of Appalachian Georgia. The area centers on the town of Ellijay, Gilmer County, which they thus regard as a logical plant site. They visualize that the plant will get all its wood from the open market within the area. Of the area's roughly 5 million acres of all land, nearly three-fourths is commercial forest, judging from the most recent Forest Service report on Georgia [3]. About 60 percent of the commercial timber holdings are a mixture of softwood (mostly shortleaf, loblolly, and Virginia pines) and hardwood (mostly oak, hickory, and yellow-poplar). The remaining 40 percent is about equally divided between pine forest and hardwood forest. Total sawtimber stock approaches 10 billion bd ft, growing at the net rate of some 610 million bd ft per year. Annual removals are 340 million bd ft, leaving an excess net growth of 270 million bd ft. Considering that the panel plant envisaged would require less than 36 million bd ft of logs per year, the resource is clearly ample, even though logs of the lowest grades cannot be used as raw material for panels.

What attitude do owners take toward selling timber and letting it be cut? About 60 percent of the owners have sold timber before; more than half of responding owners intend to sell timber in the next 5 years.

MANUFACTURING TECHNOLOGY

The plant under consideration would receive its raw material in the form of tree-length logs. Softwoods and hardwoods could be represented in any proportions, except that, to avoid excessive panel weight, oak and hickory would be held down to 15 percent of total wood volume. As a basis for cost estimates, it is assumed that annual timber receipts would total about 22 million bd ft of softwood and 14 million of hardwood.

Incoming material would be debarked and then bucked into 8- to 12-ft logs. These would be separated according to their suitability for flakeboard or veneer. Flakeboard logs would be chipped, flaked, and mixed with resin adhesives for mat forming and board pressing. Veneer logs would be steamed and rotary peeled. The sandwiches would be put through a conventional plywood press to cure the resin bond between the veneer faces and the flakeboard core.

Residues unusable for flakeboard, such as bark, sander dust, and panel trimmings, would be used as boiler fuel. These residues would supply about 40 percent of the fuel requirements of the plant's two boilers.

Operating 22 hours per day for 300 days per year, with a total labor force of 188 plus 9 administrative persons, the plant would turn out some 127 million square feet of product annually. Four-fifths of this output would be sheathing-grade panels; one-fifth, industrial grade, usable for such purposes as table tops.

TABLE 37-1
ESTIMATED CAPITAL REQUIREMENTS FOR A PANEL PLANT IN APPALACHIA

Type of capital	Amount of capital, in 1978 dollars
1 Investment capital:	
(a) Land, 30 acres	338,000
(b) Buildings, 180,000 sq. ft.	3,059,000
(c) Facilities	2,642,000
(d) Machinery	15,160,000
(e) All investment capital	21,199,000
2 Working capital:	
(a) Material supplies (2 months)	1,364,000
(b) Accounts receivable (1 month)	2,372,000
(c) Cash for payroll (2 months)	474,000
(d) Contingency fund	2,120,000
(e) All working capital	6,330,000
3 All capital	27,529,000

Source: Chiang, Tze I., and Gerald A. Koenigshof: "Economic Feasibility Study, Composite Panel Complex in Georgia," *Forest Products Journal*, November 1980, p. 26.

If you are the analyst, what will you want to learn about the labor supply in the area?

FINANCIAL DATA

Turn now to the financial side of the venture. The analysts estimate *(forecast)* costs and revenues, in 1978 dollars, and derive some measures of the panel plant's financial promise.

1 *Capital requirements* (Table 37-1). Investment capital totals $21 million, a figure that would be a lot higher if the analysts had foreseen a necessity for timberland acquisition. Working capital is estimated at more than $6 million. Material supplies, line 2(a), are raw materials, mainly timber and adhesive, held in stock to ensure a steady flow into the production line. **Accounts receivable,** line 2(b), are the value of panels billed to customers who have not yet paid for them. All capital requirements are $27\frac{1}{2}$ million if the analysts are on the right track.

It is assumed that about a fourth of the capital, $6,332,000, will be raised from the sale of common stock (**equity** in the business). The remainder will be borrowed on a 10-year mortgage at 12-percent interest, thus constituting the firm's **debt.**

2 *Production costs* (Table 37-2). These costs are estimated (forecast) in 1978 dollars, for a year when the plant has achieved a stable operation at capacity. The analysts classify as **variable costs,** 1Ai in Table 37-2, those that vary with rate of output and not necessarily with the alternatives at hand. They classify as **fixed costs,** 1Aii, those that do not vary with output. Depreciation, line 2, is estimated by the straight-line method (page 256): on machinery and equipment, 10 percent of initial cost each year for 10 years, its estimated life; on building and facilities, 5 percent for 20 years.

3 *Product prices.* North Georgia is favorably located in reference to a group of cities in Alabama, Florida, Georgia, North Carolina, and Tennessee that represent about $22\frac{1}{2}$ percent of the domestic plywood market and about the same fraction of shipments from southern mills. The consumption of plywood and other panels in this region has been increasing rapidly in the long trend and is expected to continue. Thus

TABLE 37-2
ESTIMATED ANNUAL PRODUCTION COST AT A PANEL PLANT IN APPALACHIA

	1978 dollars	
Source of cost	Total	Per M sq. ft.
1 Out-of-pocket cost:		
A Manufacturing cost:		
i Variable cost:		
(a) Logs	3,979,635	31.41
(b) Chemicals	3,108,004	24.53
(c) Labor	2,842,500	22.43
(d) Maintenance and supplies	1,516,000	11.96
(e) Production supplies	506,800	4.00
(f) Electricity	442,109	3.49
(g) Heat energy	103,145	.81
(h) All variable cost	12,498,273	98.63
ii Fixed cost:		
(a) Salaries	186,880	1.48
(b) Facilities maintenance	114,000	.90
(c) Grading service	294,000	2.32
(d) Office expense	176,000	1.39
(e) Contingencies	588,000	4.64
(f) Property taxes	202,672	1.60
(g) Insurance	75,192	.59
(h) All fixed cost	1,636,744	12.92
iii All manufacturing cost	14,135,017	111.55
B Interest and debt retirement	3,649,384	28.80
C All out-of-pocket cost	17,784,401	140.35
2 Depreciation	2,086,100	16.46
3 All production cost	19,870,501	156.81

Source: Chiang, Tze I., and Gerald A. Koenigshof: "Economic Feasibility Study, Composite Panel Complex in Georgia," *Forest Products Journal,* November 1980, p. 27.

TABLE 37-3
ESTIMATED ANNUAL NET REVENUES FROM A PANEL PLANT IN APPALACHIA

	1978 dollars	
Item	Total	Per M sq. ft.
1 Gross sales	29,399,040	232.00
2 Sales expense and discounts	1,469,952	11.60
3 Net sales	27,929,088	220.40
4 Manufacturing cost	14,135,017	111.55
5 Operating net revenue	13,794,071	108.85
6 Debt service and depreciation	5,735,484	45.26
7 Net revenue before taxes	8,058,587	63.59
8 Income taxes:		
a Federal	3,868,122	30.52
b State	483,515	3.82
c All income taxes	4,351,637	34.34
9 Net revenue after taxes	3,706,950	29.25

Source: Chiang, Tze I., and Gerald A. Koenigshof: "Economic Feasibility Study, Composite Panel Complex in Georgia," *Forest Products Journal,* November 1980, p. 27.

the prospects for marketing composite panels from north Georgia are regarded as bright, for, as mentioned earlier, sheathing-grade panels are being accepted in the construction industry as fully substitutable for softwood plywood. The analysts of the Ellijay plant take plywood prices as representative of composite-panel prices. Using conservative figures so as to introduce a safety factor, they arrive at a weighted average price of $232 per thousand square feet of $\frac{1}{2}$-in panels. This price, expressed in 1978 dollars, allows for freight to destination, but is gross of sales expenses and discounts to buyers, figured at 5 percent of the gross.

4 *Net revenue* (Table 37-3). Line 6 in the table is the sum of lines 1B and 2 in Table 37-2. The final estimated net revenue after taxes is $3,706,950 in total for the year, or $29.25 per thousand square feet.

> Can you identify the components of the net revenue as calculated that allow for uncertainty? If you were asked to make further reasonable allowances, how would you do it?

EVALUATION

5 *An evaluation of the estimates. Is* the composite panel plant "economically feasible"? Here, let's be reminded of the point stressed so often in the Introduction to this book: The analysts' results are not answers, but *replies* to the question at hand. To the question of whether a north Georgia panel plant is feasible, only the prospective entrepreneurs can give answer. Or at least they can study the evidence and decide whether the venture is attractive. If they judge the venture so risky that no promise short of a 75-percent annual return on their equity will draw them into the business, then they will turn thumbs down, for 75 percent of $6,332,000 is about $4,750,000, and if the analysts are anywhere near right, there isn't going to be enough after-tax net revenue to pay any such return to stockholders.

Then again, the entrepreneurs might reject enough of the analysts' estimates so that the whole thing would begin to look decidedly iffy. For example, the entrepreneurs might view the raw-material cost estimates as too low: They might think that it would be necessary to use less hardwood than contemplated, not only to hold weight down, but also for other technical reasons; and they might foresee that the added timber-buying operations would raise prices in the open market. Or they might judge that the estimated product price is too high, that if panels can be produced so readily in hardwood-softwood areas such as abound in the northeast, a rise in output would force prices down. In any case, they might look on panel manufacture as akin to pine plywood production: risky and mercurial, a marginal venture for any but the big corporation with diversified enterprises. It doesn't matter if time proves these entrepreneurs wrong. The decision is theirs.

But, of course, there may be some potential entrepreneurs with positive leanings, and to such persons, the analysts have a few investment criteria to offer. They figure an internal rate of return: Taking a 10-year period, they express cash proceeds as an annual percentage of return on cash outlays. In effect, they estimate what rate of interest the outlays would earn. The rate turns out to be about 33 percent after taxes.

Furthermore, the analysts point out that the estimated after-tax net revenue of $3,706,950—Table 37-3, line 9—represents a 13-percent profit margin (on net sales), a 13-percent return on total assets (total capital requirements, Table 37-1), and a 58-percent return on equity. They figure a payout period of 4.75 years: the time required for the after-tax revenue net of out-of-pocket costs ($3,706,950 plus depreciation from line 2 of Table 37-2) to pay off the total capital requirement. This is another way of suggesting that the risk is low (page 170).

How sensitive are the estimated revenues to the assumption that the panel plant would operate at 100 percent of capacity? The analysts estimate that net sales would cover all production costs if operations were at 60 percent of capacity. Net sales would cover out-of-pocket costs at 45 percent of capacity; manufacturing costs, at 15 percent of capacity.

And the analysts conclude that "these indicators strongly suggest that the proposed [business] is an excellent opportunity for investment purposes."

> If an analyst is to produce replies and not answers, how far is it appropriate for the analyst to go in drawing conclusions?

SUMMARY

Our review of an economic feasibility study for a composite panel plant in Appalachia impresses us with the mastery that analysts are obliged to attain over a broad range of detail on resources, production technology, and financial matters. The function of the analysts is to supply as reliable a set of predictions as they can. The function of potential investors is then to put their best judgment into interpreting the analysis.

REFERENCES

1 Chiang, Tze I., and Gerald A. Koenigshof: "Economic Feasibility Study, Composite Panel Complex in Georgia," *Forest Products Journal,* November 1980, pp. 24–29.

2 Pennington, William H.: "Market Expansion Forecast for Reconstituted Boards," *Forest Industries,* April 1984, p. 42.

3 Sheffield, Raymond M., and Herbert A. Knight: *Georgia's Forests,* U.S. Department of Agriculture, Forest Service Resource Bull. SE-73, 1984.

CHAPTER **38**

WEIGHING A LAKE STATES PULP-PAPER ENTERPRISE

DISAGREEMENT BETWEEN ACCOUNTANTS AND ECONOMISTS

A question has arisen at a pulp and paper installation in the Lake States. The manager has three principal enterprises to oversee: (1) a pulpmill with a capacity of a little under 1,000 tons per day, (2) a paper machine of somewhat over 1,000 tons daily capacity, and (3) a 100-ton-per-day paperboard machine. In the course of stock-taking by central management, the small board machine is given close scrutiny.

The accountants call management's attention to some negative aspects of the paperboard enterprise. First, the machine is getting old; its remaining (undepreciated) book value is scarcely $200,000. Second, since the large paper machine fully consumes the pulpmill output, the small machine must be fed pulp from wastepaper. The wastepaper market is extremely volatile, and in very recent years the cost of this raw material has been too high for comfort. Third, and most important, for quite a few years the board machine has been failing to earn enough net revenue to pay its share of total overhead (managerial and related salaries, costs of holding land and capital equipment, and other fixed costs). In other words, the board machine is losing money for the corporation and should be shut down, the accountants say.

The company's economists take issue with the accountants. They concede that wastepaper prices are the key factor in board-machine profits, but they contend that the recent high price levels are temporary. The fact of the matter, they believe, is that wastepaper prices have been inflating, and will continue to inflate, at a lower rate than paperboard prices or for that matter, than costs of any other plant inputs. So the prospects are not bad. As for overhead costs, whether they are relevant depends on the question being asked. Evidently the question is whether to (1) operate the board machine or (2) shut it down. Then one wants to know what the plant's total profits would be, given (1) and given (2). These are the two alternatives. The only revenues and costs of interest are those that are different as between (1) and (2)—that is, those that are *variable with the output of the small machine.* All else is irrelevant.

The plant manager decides to resolve the disagreement between the two parties with the help of consultants.

REVENUES AND COSTS OVER A DECADE

The year is 1980. It is a time when general inflation (decline in the real value of the dollar) has been running strong and is expected to continue. Thus the present problem will give us some insight into ways of allowing for the expectation.

The consultant's first steps are to (1) ask management to clarify the question and then (2) find a suitable model that management will accept. The consultants learn that the question is indeed whether to continue operating the paperboard machine or to shut it down. The criterion for the decision is financial: the effect on the firm's net revenue or net worth. And the period to be considered is the decade ahead. Ten years is the customary planning period in the industry. (Compare the period used in computing the internal rate of return to the composite-panel venture on page 286). The consultants also learn that the corporation's guiding rate of return is *22 percent* before tax. This is the firm's chosen annual rate in inflating dollars. Its *real* guiding rate is less, as you'll see.

Why is the real guiding rate less than the current rate?

Now the consultants set to work on a list of revenue and cost categories that will be affected by the decision to operate or to shut down the board machine. These are categories of outputs and variable inputs. For each category, the consultants estimate *(forecast)* the average quantity of the output or input per year over the 10 years ahead, the average unit value, and the total value per year. Their results are laid out in the first five columns of Table 38-1. The dollar amounts are the consultants' estimates of revenues and costs from the machine if it is run. If it is shut down, of course, the corresponding amounts are all zero.

We'll pause over the five columns long enough to observe that the net before-tax annual revenue is estimated at about $2.6 million. This is the revenue that is variable with the decision on whether to keep the paperboard machine running. After federal and state income taxes, the net revenue comes to some $1.28 million. This revenue can be figured as a return on the 1980 investment in the paperboard enterprise, $1.56 million.

TABLE 38-1
ESTIMATED VALUES INVOLVED IN OPERATING PAPERBOARD MACHINE AT CAPACITY, FOUR SHIFTS PER DAY FOR 355 DAYS PER YEAR

			Revenue or cost in 1980				
			1980 dollars		%		
Item (1)	Quantity, in units (2)	Unit of measure (3)	Per unit (4)	Total (5)	Annual escalation (6)	Capitalization ratio (7)	Current worth 1980s decade, in 1980 dollars (8)
1 Output:							
(a) Paperboard	35,500	Ton	306.00	10,863,000	7.5	6.04	65,612,520
(b) Electricity[a]	8,520	M KWH	.18	1,534	10.0	6.56	10,063
(c) All outputs				10,864,534			65,622,583
2 Input:							
(a) Hourly employees	60	Man-year	24,000.00	1,440,000	9.0	6.35	9,144,000
(b) Salaried employees	5	Man-year	34,000.00	170,000	9.0	6.35	1,079,500
(c) Wastepaper	40,000	Ton	82.00	3,345,000	1.5	5.01	16,761,456
(d) Chemicals		Output ton	6.90	244,950	7.5	6.04	1,479,498
(e) Water	53,500	M gallons	.24	12,840	6.0	5.75	73,830
(f) Steam		Operating day	3,800.00	1,349,000	9.0	6.35	8,566,150
(g) Supplies		Output ton	3.20	113,600	7.5	6.04	686,144
(h) Maintenance		Output ton	9.00	319,500	7.5	6.04	1,929,780
(i) Treating waste		Output ton	1.35	47,925	6.0	5.75	275,569
(j) Shipping		Output ton	8.90	315,950	7.5	6.04	1,908,338
(k) Freight, discount		Output ton	25.00	887,500	7.5	6.04	5,360,500
(m) All inputs				8,246,865			47,264,765
3 Net before tax				2,617,669			18,357,818
4 Net after tax				1,282,658			8,995,331

[a]The machine generates about 1,000 KW more electricity (from discharged steam) than it uses. The excess is used elsewhere in the mill and reduces the power bill.

Machine's book value	$205,000	Chemicals	$ 18,000
Supplies on hand	115,000	Product in shipping room	81,000
Reserve wastepaper stocks	85,000	Accounts receivable	950,000
Fuel stocks	106,000	Total investment	$ 1,560,000

And one finds that the after-tax return on the paperboard enterprise is 1.28 ÷ 1.56, or 82 percent. This certainly gives the enterprise plenty of elbow room beyond the corporate 22-percent before-tax guiding rate.

Also notice that no allowance is made in Table 38-1 for the sale of the board machine in the event it is shut down. The machine has essentially no net salvage value.

> If the machine had salvage value, how would this enter the calculation?

CURRENT NET WORTH OF THE BOARD ENTERPRISE

But the consultants' assignment is to make a 10-year analysis, and so they continue in this direction. For each output and each input they bring together a prediction of the annual rate at which the price or unit cost will escalate in current dollars. Such predictions are found in trade journals, government analyses, and the corporation's files. Taking such data and their own judgment, the consultants arrive at the percentages shown in column (6) of Table 38-1 [disregard columns (7) and (8) for the moment].

What do the consultants have in mind when they say, for example, that the price of paperboard will rise 7.5 percent per year? Suppose that the 1980 price is $100 per unit. Then the consultants are saying that the annual averages during the 1980s will be the same as the first few entries in the 7.5-percent column of a compound-interest table such as Table 19-1, with the decimal point moved two places to the right:

Year	Price, current dollars
1980	100.00000
1981	107.50000
1982	115.56250
1983	124.22968
1984	133.54690
1985	143.56291
1986	154.33012
1987	165.90487
1988	178.34773
1989	191.72380

The consultants are aiming for the current net worth of this series: its value *capitalized* at thc 22-percent guiding rate. Each item must be discounted to 1980 [formula No. (3) in Chap. 19] and the series added:

$ 100.00000
88.11475
77.64210
68.41595
53.11833
46.80500
41.24217
36.34039
32.02121
$ 603.98379

The items in this second series all equal $100 discounted for the number of years in question at about 13.5 percent, which is 1.22 ÷ 1.075 − 1 (*not* 1.22 − 1.075, which is 14.5 percent). And the sum, $603.98, is the present worth of a series of 10 annual $100 payments *beginning now,* capitalized at 13.5 percent. This percentage, 13.5, is the corporation's *real guiding rate of return* for items that promise to escalate in current dollar value at 7.5 percent per year.

Note that the capitalized value derived in the second series is roughly 6.04 times the first (uninflated, undiscounted) payment. This ratio will hold for any 10-payment annual series capitalized at 13.5 percent as of the time of the first payment. It follows from formula (10) in Chap. 19, with an r added to it:

$$v_0' = \frac{r}{i} \cdot \frac{(1 + i)^n - 1}{(1 + i)^n} + r \qquad (1)$$

or, for $r = 1$, $i = .135$, and $n = 9$,

$$v_0' = \textit{capitalization ratio} = \frac{1}{.135} \cdot \frac{2.1258}{3.1258} + 1 = 6.04 \qquad (2)$$

All the capitalization ratios in column (7) of Table 38-1 are derived from the escalation percentages in column (6) by the procedure just described. They make it easy for the consultants to find the current net worth of the paperboard enterprise, column (8). The final upshot of the calculations is that for the 1980s planning decade, the paperboard machine is expected to have an after-tax net worth of almost $9 million. That is, in an economy where dollar values are inflating as assumed, the firm can, at 22-percent current interest, invest $8,995,331 and receive a 10-payment annuity of 1,282,658 dollars per year, starting now. The real rate of interest in this annuity (the firm's real guiding rate) is about 9 percent (1.22 ÷ 1.12 ≅ 1.09).

The ratio of the two totals in line 1(c) of Table 38-1 is just over 6.04, the weighted average of the first two figures in column (7). Likewise, the ratio of the two totals in line 2(m) is 5.73, a weighted average of column (7), lines 2(a)–2(k). But the ratio of the two differences in line 3 and in line 4 is 7.01. This ratio implies an escalation of net revenue of nearly 12 percent per year, a rate higher than for any single revenue or cost item. Explain the arithmetic of this result. Explain whether it is the result that the consultants intended to predict.

SENSITIVITY TESTS AND THE FINAL CONCLUSIONS

The consultants supplement their basic calculations with some sensitivity tests. How sensitive are the income and net-worth estimates to the assumption that the board machine would be run at four-shift capacity? They try a recalculation on the assumption of three-shift operation. The three-shift level introduces some inefficiencies, as in shutting down and restarting the machine. And the saving in work force is not quite proportional to the reduction in operating time. Nevertheless, annual revenue and 10-year value with three shifts are about 70 percent of the four-shift estimates.

How sensitive is the four-shift before-tax net-annual-revenue forecast to price and cost assumptions? The consultants calculate that

A \$5-per-ton difference in initial price of paperboard would change revenue \$175,000. In other words, the unit value of paperboard in Table 38-1 would have to stand at about \$231 per ton instead of \$306 in order to snuff out net revenue before (and after) tax.

A \$5-per-ton difference in initial cost of wastepaper would change revenue roughly \$200,000.

A \$1,000-per-year difference in initial wages and salaries would change revenue \$75,000.

How sensitive is the four-shift before-tax current-net-worth estimate to assumptions about inflation? Here the findings are that a 1-percent difference in the annual escalation rate for

Paperboard price (with no difference in cost escalation rates) would change net worth \$2.25 million.

Wastepaper price would change net worth \$575,000.

Steam price would change net worth \$250,000.

Wages and salaries would change net worth \$450,000.

And a 1-percent difference in the guiding rate of return would change net worth by about \$500,000.

Now let's review what the consultants have done. *First,* they have considered the accountants' principal point, that the board machine is not paying its share of general overhead. They have rejected the point on the ground that overhead will continue unchanged regardless of what is done with the machine. Overhead is fixed with respect to the machine decision. To allocate fixed costs to units of output is an error. *Second,* the consultants, taking note of the accountants' other points, have looked at the profitability of the board machine and estimated that its current net worth is a whopping $9 million after tax. *Third,* the consultants have recalculated current net worth, using alternative assumptions about revenue and cost rates. They find that no error that is likely to show up in their assumptions will make a very substantial difference in their estimate of current net worth.

The consultants conclude that the paperboard machine is worth retaining, and management agrees.

> With the benefit of hindsight from the end of the 1980s decade, we know that the consultants overestimated inflation rates. Take an extreme contrasting case. Suppose that the consultants had assumed a zero inflation rate for every revenue and cost item, but still a 22-percent guiding rate before tax. What, then, would be the capitalization ratio for total revenue, recognizing that $(1.22)^9 = 5.9874$? What would be the ratio for total cost? the ratio for total net revenue, before or after tax? Why are all these ratios the same? What would be the current net worth before and after tax? What changes would there be in sensitivity of net revenue to price and cost rates? What changes do you think would be made in the consultants' conclusion?

SUMMARY

Briefly speaking, if one is evaluating an enterprise to decide whether to retain it or close it out, one should take account of all the benefits and costs that are *variable* with respect to the decision, but not those that are fixed. The *principle of relevance* applies to any choice among alternatives.

The wood-using industry commonly evaluates an investment or enterprise by forecasting its performance over a 10-year period. In the present case, the evaluation is in terms of current net worth after tax.

An evaluation can usefully be supplemented by sensitivity tests that tell how different the conclusions would be if different assumptions were adopted.

CHAPTER 39

PRODUCTIVITY IN WOOD MANUFACTURE

PRODUCTIVITY AND ECONOMIC WELL-BEING

The topic of productivity has always been much on the American mind. In recent decades, we have contemplated the topic with a twinge of anxiety. We have seen some of our fellow nations rebuild their industry with new automation and have feared that we were not prepared to keep pace. We have seen them dominate some of the world markets, while we ourselves, once the leaders, have lagged. We have seen ourselves driven more and more into tertiary activities for economic growth. Between 1973 and 1990, our output per person employed increased at the rate of only about one-half of 1 percent per year.

However, as an antidote for our anxiety, we have this item from the newspapers of Japan, presumably one of our keenest rivals: In 1988, GDP per employee in all industry was lower in Japan than in the U.S., France, West Germany, or Britain. The U.S. surpassed Japan in productivity by 31 percent. In forestry and fisheries, we were more than three times as productive as Japan [7].

Productivity, in whatever lines of activity, is a measure of economic well-being.

It is appropriate, then, to look in on the wood-using industry and ask about its productivity.

PRODUCTIVITY DEFINED

Productivity means output, but in the normal use of the term, the idea of a *rate* of output per unit of input per unit of time is invariably implied. Your plant, identical to mine, is more productive if it turns out more product in a year, measured in physical units or dollar value. Or if not identical, we can still compare the plants—say, in terms of annual product per employee. We can count all employees or just **production workers,** those who are actually stationed on the factory floor. And we can use either worker numbers or worker-hours as the denominator in calculating productivity, though, of course, there will be a difference if there is a difference in hours worked. In any of these cases, we are figuring **labor productivity.**

When I estimate labor productivity as described, I must remind myself that the result is *not* a statement of the output attributable to a unit of labor. It is a statement of total output, however attributable, expressed per unit of labor. I could take exactly the same total output and divide it by total capital investment and thus arrive at **capital productivity.**

Ideally, to estimate industrial progress in an *economy,* one should have neither labor productivity alone nor productivity in relation to some other input alone, but productivity in relation to *all* inputs. Students of the subject emphasize this goal. The statistical problems are difficult. Where they are solved, the result is a measure akin to *efficiency,* as it is introduced in Chap. 9 and again in Chap. 23. This chapter clings to labor productivity and works with it only for individual industries or even individual firms. Labor productivity is of special interest to us because it is closely related to GNP per capita, the index of economic growth. Capital productivity, for what it may contribute to understanding, is brought in at chapter's close.

In any case, if I divide total output by total input of labor or total input of capital, the result is an estimate of **gross productivity.** In order to estimate **net productivity,** I'd need to figure the *marginal* product per unit of labor or capital, not the average product. To do this, I would have to know what the production functions are. It would not be simple.

At least one knows that gross productivity is greater than net. Output is always attributable to more than one kind of input. Gross productivity is a comparatively straightforward calculation. It is the one used commonly and in this chapter.

> Suppose you wished to calculate the *net* daily productivity per worker in a given sawmill. What data would you want, and what would you do with these data? Now suppose you set about calculating *net* daily productivity per dollar of capital, defining capital in a special way to include the cost of all inputs other than labor. If you multiplied your labor productivity estimate by the number of workers in the mill and your "capital" productivity estimate by the total daily outlay except payroll, and if you then added the two products, how would the result compare with total daily output? What allowance would you have made for profit or loss? Does the sum of marginal outputs equal total output?

INDUSTRY TRENDS, 1954–1985

Some of the most consistent and reliable labor productivity estimates available for U.S. wood-products industries can be derived from the output and employment data worked up by Dwight Hair [3] and by Robert Phelps [6]. These are the authors whose estimates are drawn upon in Chap. 4, on the timber economy of the world. Their work covers the years from 1954 through 1972, interesting years from the productivity standpoint, since they lie at the center of the "era of new technology" (Chap. 33) that followed World War II. The authors give basic statistics for four branches of primary wood-processing industry: lumber, veneer-plywood, pulp-paper, and logging. In Fig. 39-1, productivity data for the four industries are plotted and then extended from 1972 to 1985 with the help of other federal government statistics. The data are plotted as **indexes**—that is, relative values, obtained from the output-per-worker values with the use of a constant multiplier. The multipler is different for each industry and consequently the position of each on the graph sheet has no significance. But, the graph being semilogarithmic, the *slopes* of the lines may be compared, both from time period to time period and from one industry to another.

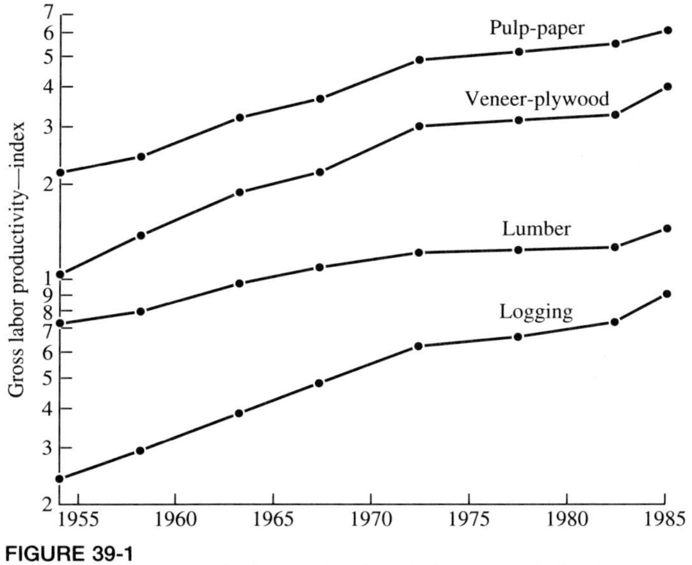

FIGURE 39-1
Trends in gross annual labor productivity in four wood industries, 1954–1985. *(Source of output and employment data for 1954–1958:* Hair, Dwight: *The Economic Importance of Timber in the United States,* U.S. Department of Agriculture Misc. Publ. 941, 1963. *Source of data for 1963–1972:* Phelps, Robert B.: *Timber in the United States Economy, 1963, 1967, and 1972;* U.S. Department of Agriculture, Forest Service Gen. Tech. Rep. WO-21, 1980. *Source of output data for later years:* Ulrich, Alice H.: *U.S. Timber Production, Trade, Consumption, and Price Statistics, 1950–85;* U.S. Department of Agriculture Misc. Publ. 1453, 1987. *Source of employment data for later years:* U.S. Department of Commerce, Bureau of the Census: *Census of Manufacturers.)* Productivity series 1972–1985 is spliced to series 1954–1972.

All the trends from 1954 through 1985 are consistently upward. Between these two dates, the average annual percentage of increase in productivity [Chap. 19, formula (2)] for each industry was

Industry	Annual increase in productivity, %
Veneer-plywood	4.5
Logging	4.4
Pulp-paper	3.4
Lumber	2.3

Lumber, however, showed negligible improvement after 1970.

In each of the four cases, the productivity upswing is at least partly due to technological advancement. In the case of logging and sawmilling, a big part of the productivity increase stemmed from the reconstitution of these industries: the demise or rebuilding of the smallest, least productive units and the multiplication of the most productive. Reconstitution played a part in the veneer-plywood and pulp-paper changes, too, but not a major part.

EARLY YEARS, BEFORE 1930s: HOW TO STUDY THEM?

Students of the subject of productivity have observed that a war can be a turning point in a nation's productivity trends. After a war, productivity growth, on the average, speeds up. For example, in the 30 years before World War I, the average annual increase in physical output per worker-hour in the U.S. private economy was 2.0 percent. In the following years, the corresponding figure was 2.6 percent [2]. What happens is that during a war, at least such wars as we have known in the past, the national genius and energies are fiercely focused, as at no other time, on invention in the interest of military success. After the war, the vast store of new technology is released to the economy and put to peaceful, productive purposes. The economy is given a shot of adrenaline.

Some of the adrenaline built up during World War II found its way into the U.S. forest products industry and was responsible for what we see in Fig. 39-1. Productivity trends in wood manufacture had not looked like that previously.

Let us, therefore, go back before World War II, into the interwar years, for a glimpse of labor productivity in timber processing. This is not altogether easy to do, for there are data problems—the farther back one tries to reach, the more difficult the problems.

The principal source of both output and employment statistics is, in the final analysis, the U.S. Bureau of the Census. In an industry where the majority of firms is very small and mobile, such as was the case with logging and sawmilling before World War II, customary census methods are not well adapted. For the extreme case, you can

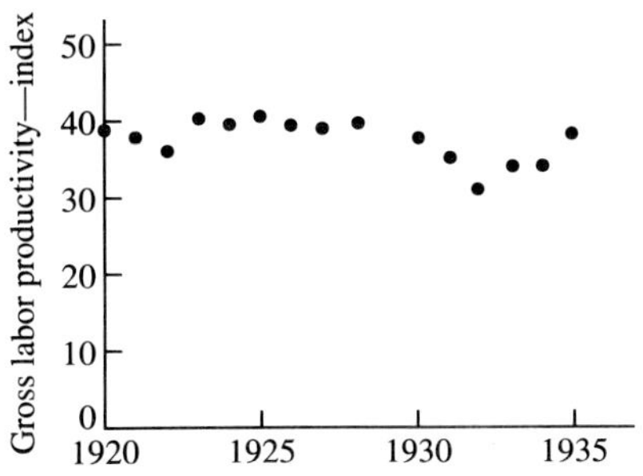

FIGURE 39-2
Trend in gross annual labor productivity in the lumber industry of the U.S. during the interwar period. (*Source of input data:* U.S. Department of Agriculture, Forest Service: *The Outlook for Timber in the United States,* Forest Resource Rep. 20, 1973, p. 347. *Source of employment data:* U.S. Department of Commerce, Bureau of Foreign and Domestic Commerce: *Survey of Current Business, 1936 Supplement,* p. 29.)

picture the problem of sending a questionnaire by mail to loggers and getting replies. You have no list of loggers. Where do you begin? It may be necessary to organize an army of census takers to gather information in person. Even so, how do the census takers identify the loggers? And then, where is the logger to be tracked down? At work in the woods? No. At home in the evening? The logger is relaxing over a beer, and his favorite TV program is coming on. The project is fraught with problems of partial coverage, inaccuracy, and inconsistency. Probably it is best carried out by indirect methods of analysis. But the Census Bureau is not authorized to use such methods.

And so the research worker goes into the Census records and painstakingly amends them. Remember that we are talking about records for the decades before the 1930s, before the Forest Survey project had come into being. Reliable data such as went into Fig. 39-1 were not to be had.

Let's start with the lumber industry, which is the best documented of the wood industries, and look at the interwar period 1920–1935, for which output estimates are available from the Forest Service [8] and employment data from the Department of Commerce, Bureau of Foreign and Domestic Commerce [9]. The labor-productivity trend is sketched in Fig. 39-2. This is not a rising trend, as in Fig. 39-1. The trend is essentially horizontal.

TRENDS, 1899–1939

How does one explain Fig. 39-2 and its remarkable contrast with the later lumber trend shown in Fig. 39-1? Before trying to reply, let's see if other studies have been made of early productivity trends in sawmills and what the results were. The person to whom we turn is Solomon Fabricant, of the National Bureau of Economic Research, who dedicated much of his professional life to studies of productivity. In the early 1940s, he was at work on a study of productivity trends in American industry during the years 1899–1939 [1]. Fabricant found that during those 40 years, while output per wage earner in all U.S. manufacturing industry more than doubled, so that the annual rise averaged 1.9 percent, productivity in "lumber-mill products" *fell* by 8 percent. Thus the annual rate of productivity change for lumber was a negative 0.2 percent.

Here, then, are two stories about labor productivity in the lumber industry before the 1941–1945 war. The stories are consistent. Furthermore, the 1916–1918 war appears not to have given the industry any adrenaline at all. What was happening?

One can speculate that the productivity behavior of the lumber industry during the interwar and earlier years was the outcome of dramatic structural changes in the industry. This was the softwood lumbering era, when the big, more productive mills were migrating and holding somewhat steady in numbers and output. In their wake, the little fellows were entering the industry in increasing numbers. Total labor productivity was sure to sag, just as it was sure to rebound after 1945!

> What about logging in the early years? Do you think its labor productivity trend was probably like that in lumbering or quite different? Explain.

And veneer-plywood? The industry was negligible in most of the early period.

Pulp-paper? Here, Fabricant gives us something to go on. He estimates that in the *60* years, 1879–1939, "paper and pulp" output per wage earner rose an average of 2.7 percent per year. For the period 1909–1938, which is better for our purpose since pulp-paper was more largely a wood industry then, the average annual rise was 2.2 percent, still a better record than that of manufacturing industry as a whole. The hierarchical order was in place at that time!

LABOR VERSUS CAPITAL PRODUCTIVITY

Near the beginning of this chapter, *capital* productivity is referred to. What do you suppose was happening to gross capital productivity in the lumber industry during the eras we've discussed? For a reply, let's take a somewhat different route from the one followed for labor productivity. Good data on capital investment in the industry over time are few and far between. Let's estimate output per thousand 1980 dollars of capital investment for each of three types of sawmills that span almost the whole history of the lumber industry: (1) the pit sawmill, virtually the earliest one that can be called a mill, (2) the portable peckerwood mill, and (3) the state-of-the-art 1980s mill incorporating all the technological developments to date. The analysis will have some errors because all the forms of capital are not equally durable. And it will be deficient in the same sense that labor-productivity data are deficient. Nevertheless, the results will be instructive. I will express production in terms of the net output of rough green lumber per day (8-hour shift).

1 *Pit sawing.* This was the most primitive method of sawing lumber ever widely used in America. At that, it was mostly confined to the seventeenth century. A pit is dug, a log is supported above the pit, and a two-man saw is operated by two workers, one standing on the log and the other in the pit, wearing a broad-brimmed hat for protection against sawdust. Investment in the pit, saw, and hat comes to 315 1980 dollars. Daily production is 200 bd ft [4].

Daily productivity of pit sawing:
Per worker, 100 bd ft
Per $1,000 capital, 635 bd ft

2 *Peckerwood sawmill.* Here is a small circular sawmill such as typified the era of the same name in the U.S.: heavy-duty frame and carriage, rope feedworks, manual setworks, 100-hp motor, 54-in diameter saw, crank-operated jack chain and sawdust chain, all mounted on rubber-tired axle with tow bar. Investment: $18,115. Daily production with three workers: 6,500 bd ft [5].

Daily productivity of peckerwood mill:
Per worker, 2,167 bd ft
Per $1,000 capital, 359 bd ft

3 *Modern high-speed mill.* For illustration, let's glance at Simpson Timber Company's Mill No. 5 in Shelton, Washington [10]. This mill accepts Douglas-fir and western hemlock logs up to 55 ft long and 14-in butt diameter, down to 4-in top diameter. Operators at the infeed deck cull over the logs and debark, trim, and load them. From that point on, the procedure is fully automated by means of electronic, computer-guided devices: bucking for maximum yield, sorting, sequencing into the breakdown process, deciding on the products to be derived from each log, and grading, sorting, and storing the products. The products include, not only 8-ft 2-by-4 studs, but also squares and boards of various lengths. The mill can handle up to twenty 8-ft logs per minute and on many days saws 6,500 logs per 8-hour shift. Investment: $8.7 million. Daily production with nine workers: 82,000 bd ft.

Daily productivity of Mill No. 5:
Per worker, 9,111 bd ft
Per $1,000 capital, 9.4 bd ft

Thus, over the long history of the lumber industry, labor productivity has risen at the expense of capital productivity. One moral: In a poor economy, where capital is scarce and labor abundant, there's much to be said for the peckerwood mill and even for pit sawing. But in a wealthy economy, where the balance of resources is just the opposite, the high-speed, automated mill is the ticket.

> If we had data on *net* labor and capital productivity for the three types of sawmills, how would these data compare with the six foregoing gross figures?

SUMMARY

Two sorts of productivity are touched upon in this chapter: output per unit of labor and output per unit of invested capital.

Labor productivity is akin to GNP per capita and is a measure of economic growth. During the twentieth century, it has risen most steadily in the pulp-paper industry, least steadily in logging and lumber manufacture. However, labor productivity has enjoyed an upward spurt in all branches of wood industry since World War II.

High capital productivity, which in sawmilling is associated with low labor productivity, is advantageous for underdeveloped economies in which capital is scarce, labor plentiful.

REFERENCES

1 Fabricant, Solomon: *Labor Savings in American Industry, 1899–1939,* National Bureau of Economic Research Occasional Pap. 23, 1945.

2 Fabricant, Solomon: *Basic Facts on Productivity Change,* National Bureau of Economic Research Occasional Pap. 63, 1959, p. 11.

3 Hair, Dwight: *The Economic Importance of Timber in the United States,* U.S. Department of Agriculture Misc. Publ. 941, 1963.

4 Hallock, Hiram: "Sawmilling Roots," *In Electronics in the Sawmill, Proceedings of the Electronics Workshop, Sawmill and Plywood Clinic,* Portland, Oreg., March 1979, pp. 7–13.

5 Jackson Lumber Harvester Company: *World's Most Portable Sawmill,* Mondovi, Wisc., September 1976, 5 pages.

6 Phelps, Robert B.: *Timber in the United States Economy, 1963, 1967, and 1972,* U.S. Department of Agriculture, Forest Service General Tech. Rep. WO-21, 1980.

7 *The Wall Street Journal,* Eastern ed., September 12, 1991, p. A12.

8 U.S. Department of Agriculture, Forest Service: *The Outlook for Timber in the United States,* Forest Resource Rep. 20, 1973, p. 347.

9 U.S. Department of Commerce, Bureau of Foreign and Domestic Commerce: *Survey of Current Business, 1936 Supplement,* p. 29.

10 Williston, E. M.: *Lumber Manufacturing, The Design and Operation of Sawmills and Planing Mills,* Miller Freeman Publications, 1976, pp. 316–319.

CHAPTER **40**

SUPPLY OF MANUFACTURED OUTPUTS

SUBJECT OF THIS CHAPTER

In Chap. 29, on the supply of genetic outputs, the principal ideas and terms relating to supply are defined and illustrated. That general discussion applies equally to the supply of manufactured outputs. You may wish to turn again to Chap. 29, especially the first three sections and the opening paragraphs of the sections on stock, short-run, and long-run supply. That discussion is basic. It will not be repeated here.

Manufactured outputs include roundwood and other proceeds of logging, plus more highly processed timber products such as lumber, panels, pulp, paper, poles, bags, and naval stores.

Supply means the functional relationships between (1) the quantity of an output that producers stand ready to release to a market or other outlet in a period of time and (2) influential variables such as the output's price and the supplier's guiding rate of return. Supply does not mean *production,* which is quantity actually released.

Consider the three supply conditions for manufactured products.

STOCK SUPPLY

First, the stock supply. It is encountered wherever stocks are held. Consider a firm that is in the *business* of holding stocks: a **yard wholesaler** of lumber and allied products, which buys the output of sawmills, stores it, and sells it to retailers, building contractors, and the like. The firm performs the functions of concentrating an often wide variety of lumber from scattered sources, guided by its foresight into customers' varied wants and filling these wants on short notice, thus relieving customers of all but minimal stockholding obligations. To assemble, hold, and market stocks is expensive. The firm tries to meet the expense by means of a markup of lumber prices.

When a yard-wholesaling firm fills a customer's order, the firm produces the order. In what factory does the production take place? The factory is the stock. The firm produces lumber from stock. The process is analogous to what takes place in the forest when timber is produced from timber stock. Product and factory are identical. It's true that one doesn't emphasize a "principle of manufacturing identity," but there *is* one, nevertheless.

Name examples other than lumber yard wholesaling where the principle of manufacturing identity operates.

What significance does the principle of manufacturing identity have? Is it the same as in genetic production, that price level is immaterial to stock supply and that other variables govern? For a reply, let's consider what constitutes the yard wholesaler's best stock (optimal stocking).

A yard wholesaler's net revenue is a return on the investment in stock—its procurement, handling, holding, and marketing. The return is an internal return (page 122) from markup. If the firm holds only a meager stock, its absolute return is low, but the rate may be high, assuming the firm can stay in business. But customers are dissatisfied by the restricted choices and the delivery dates. The firm must build up its stock and its total net revenue, even though its marginal percentage return on stock falls. At the stock level and assortment where the marginal revenue in percent equals the firm's marginal cost of capital (guiding rate of return), the firm finds its best combination (Fig. 40-1).

Marginal revenue in this case is not a growth percent, as it is for standing timber. It is a percent return that the stock earns by performing a service: making lumber available for immediate production (delivery to users). However, the model is otherwise the same as for the best stocking of timber, and the same conclusions flow from it (page

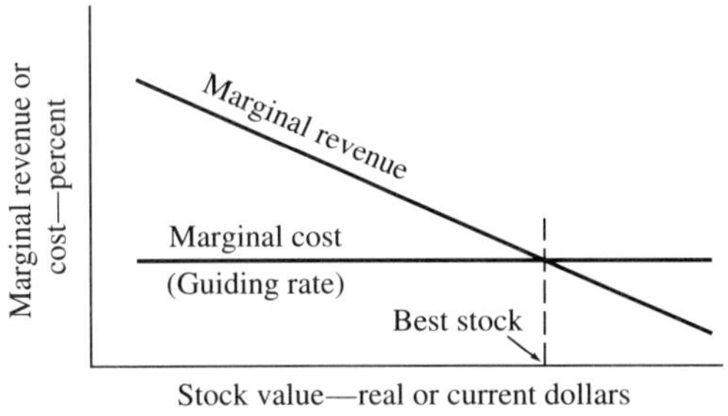

FIGURE 40-1
Concept of a yard wholesaler's optimal stock.

217): The best stock is a function of the firm's guiding rate and of the rate of marginal revenue from price markup. For example, the yard wholesaler tends to draw down stock and thus exhibit a positive stock-supply response (while holding off any increase in the rate of stock replenishment) in any of the following events: (1) The firm's guiding rate (stockholding cost) rises, possibly because its cost of borrowing has gone up, a penalty that can be lessened by carrying less stock. (2) Price markup, which controls marginal revenue, promises to fall, either because the firm faces slackened demand and will respond by cutting its price or because of other market developments, and here the firm may foresee the opportunity to build stocks back later at a lower price. The *level* of price is not influential unless it bears on price markup expectations (marginal revenue) or the guiding rate of return.

SHORT-RUN SUPPLY: GENERAL CASE

In the short run, however, **quantity supplied** *is* sensitive to price level. Short-run production functions are introduced in Part Three, commencing with Chap. 17. It will be helpful to reintroduce them here, with logging as the example. Consider a simple, largely unmechanized operation such as System 1, the short-wood system, in Chap. 34. Labor is the principal input that varies (with output, where the alternatives lie in this case).

Total daily output can be increased by adding more sawyers to the crew. But the operation runs into diminishing (marginal) return from the outset—Fig. 40-2(a)—because of crowding of the fixed inputs. Here the most influential fixed input is the truck. Loading it by hand becomes more and more time-consuming per unit as the load grows higher. And in the end, there is a capacity limit.

In Fig. 40-2(b), the curve in Fig. 40-2(a) is flipped upside down, the axes accordingly reversed, and the variable-input axis rescaled as variable cost (wages and saws), in dollars per day. The slopes along this total-cost curve are then measured and plotted in Fig. 40-2(c) as marginal cost per unit of output. The rise in marginal cost with output is the upshot of diminishing return caused by the inputs that are fixed and become crowded in the short run.

The curve in Fig. 40-2(c) is the short-run price-supply curve. This you can see in Fig. 40-2(d), where the curve is copied over. Since it is the curve of marginal cost per unit of output, the firm's best output is found where the curve is intersected by marginal revenue per unit of output. Assume that marginal revenue is given in the line so labeled. It is a horizontal line because the firm's scale of operations, being relatively small, has no bearing on marginal revenue. Thus the horizontal line is a price-demand curve for logs, equal to average revenue, or price; and the best output is OA. If the price is higher, as in the upper broken horizontal line, output is OB. If it is lower, as in the lower broken line, output is OC. Thus the marginal-cost curve is the locus of *quantities that the firm stands ready to produce at various prices*. This is the definition of **price-supply.**

Find and explain the analogies and differences between Fig. 40-2 and Fig. 17-1.

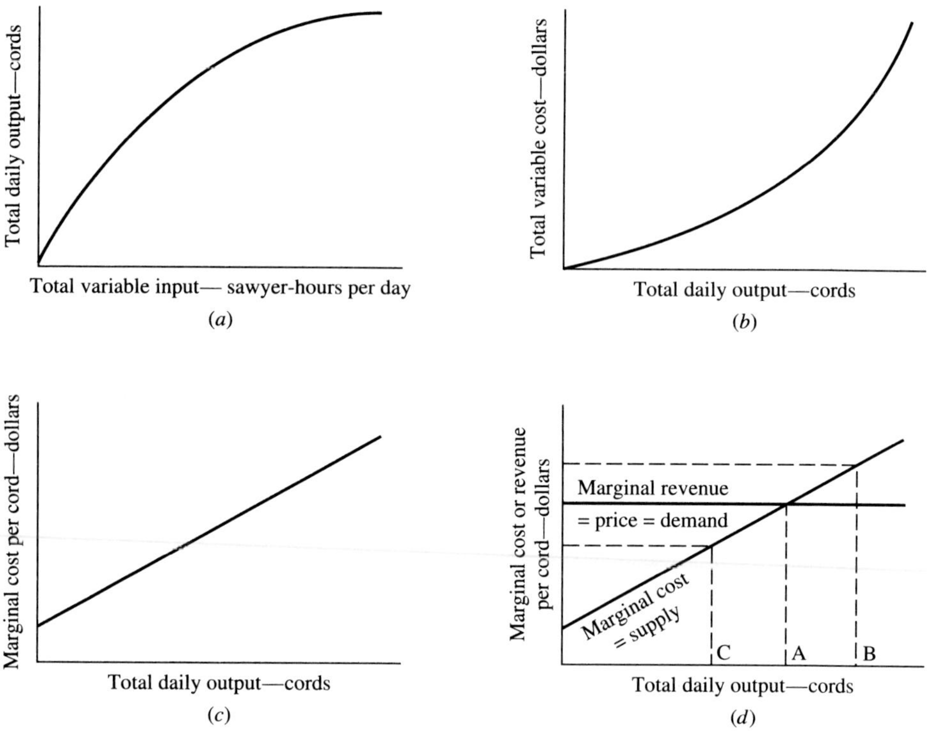

FIGURE 40-2
Derivation of short-run price-supply.

The price to which production is sensitive in the short run is the anticipated price at the time when the firm's output will enter the market. If production is on contract, the price is the contract price. The firm's anticipations of price change have no further bearing on output, because this is the short-run condition: The firm doesn't stock its output or carry stocks for release.

The guiding rate of return bears upon short-run supply through its influence on the cost of inputs such as machines or materials, the outputs of which are delayed. Compare Fig. 29-1. In the logging example, the influence is slight, for almost all the variable cost is labor.

SHORT-RUN SUPPLY: FIREWOOD CASE

An interesting analysis of the short-run supply of wood fuel was made by Michael Liebenow in 1982 [1]. The analysis takes the viewpoint of a pulp-paper firm seeking a source of energy for power and heat. What quantities of firewood are available to the firm at various prices? Liebenow measures quantity in BTUs of steam and price in dollars of marginal cost per BTU.

Liebenow reasons that in order to consider the cost of wood, he must recognize

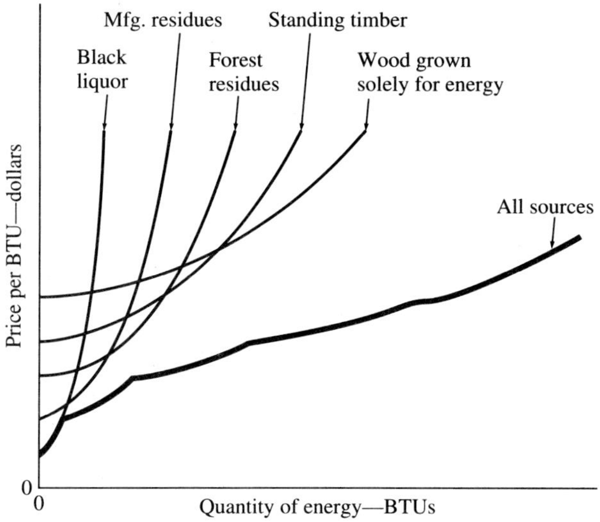

FIGURE 40-3
Supply of energy from wood fuel. (*Source:* Liebenow, W. M.: "Wood as an Energy Source in the Kraft Pulp-Paper Industry of the Southeastern United States," Ph.D., dissertation, Virginia Polytechnic Institute and State University, Blacksburg, 1982, p. 69.)

some different categories of wood. These constitute a hierarchy: (1) black liquor, (2) manufacturing residues, (3) forest residues, (4) standing timber, and (5) wood grown solely for energy. Black liquor is a combustible residue of the pulping process, which may be had for merely the variable cost of removing some water and piping the remainder to the boilers. Manufacturing residues—for example, sawdust—must pay their way from the sawmill to the boilers. Forest residues, such as tops and limbs, must be chipped and transported. Thus their variable cost includes one process beyond what sawdust requires: an alteration of form. Standing timber—for instance, "cull" trees—must bear the additional cost of logging. And, finally, wood grown solely for energy must command a sufficient price to cover the still further variable cost of genetic production.

And so it is that as the pulp-paper firm works its way down through the hierarchy of wood categories, it encounters sources of energy that are increasingly costly, at least in small quantities. However, the supply of each type of source is more and more elastic from top to bottom in the hierarchy. The firm has very little opportunity to increase its output of black liquor beyond a point, even at greatly increased cost. At the other extreme, there is scarcely any limit to the amount of wood that can be grown solely for energy if the user is prepared to bear the cost. The upshot of these cost relationships is a succession of overlapping supply functions, the unbroken curves in Fig. 40-3.

If the firm is prepared to pay for energy a price higher than the one at which black liquor is the cheapest source, it can most economically use some manufacturing residues besides the black liquor. At still higher prices, the other energy sources can be drawn upon in turn. The aggregate supply of energy, represented by the broken curve in Fig. 40-3, is the sum of quantities read from the unbroken curves at each price.

What price *does* the firm pay for wood energy? It can afford to pay a price equal to its cost of energy from coal or oil, minus the cost of modifying its boilers and firing systems so as to accommodate the wood. At this price, the quantity of wood energy that the firm will supply to itself is found on the aggregate supply function.

LONG-RUN SUPPLY

And what about the long-run supply of manufactured products? This is the supply that the firm (or, more significantly, the community) looks forward to when it contemplates opportunities for reconstructing industry organization, technology, and plant. In the last analysis, the long run is the condition under which all types of input—all costs—are variable with output. Together, all form the marginal cost in relation to output: All together form the supply.

The long-run supply of manufactured products is that which might have been envisaged in the decades forerunning each of the industrial upheavals and lesser revisions that have been described in Part Four. When each actually materialized, the industrial organization, technology, and plant that had come into being were largely fixed, and the supply condition was short-run or stock.

What is today's long-run supply of manufactured timber products in America? It is a wide range of quantities and assortments, set in relation to variables whose identity is the eagerly pursued object of our forestry policy. Do Americans expect to have more timber products? Of what kinds? What does "more" mean—a larger physical quantity? more efficient technology of consumption? Do we expect the forces of pulverization to continue in effect, or do we expect them to be softened? Of what products, unknown today, do we catch glimpses? What functions may they fill in our lives? Questions such as these constitute the first part of the long-run supply puzzle. They have been pursued, at least after a fashion, in the periodic Forest Service reviews of timber trends and prospects in the U.S. See the chapters on social planning in Part Six.

An even more challenging set of questions concerns, not the kinds and quantities of wood products supplied, but how these kinds and quantities are—will be—determined. This is the second part of the long-run supply puzzle. What is it that spurs technological development? Is the determinant monetary, or is it mainly inborn curiosity? Or is it chance channeled by selection—not, in this case, natural, but economic? To what extent is well-being in the timber economy ensured simply through providing more standing timber for harvest? Maybe the "solution" of the nonindustrial private forest "problem" is enough for meeting national goals! Or maybe timber in the long run is not a strong point with Americans, and we will look to the supply from other nations—or of other resources!

> Sketch a long-run supply curve for some common manufactured product such as paper or lumber. Fully label the axes. Describe the influences that bear upon the elasticity of the curve.

SUMMARY

The question in this chapter is how to conceive of the supply of manufactured wood products. For models, the chapter turns to genetic production and recognizes three supply categories, or *conditions.* Of these, the short-run supply, in which quantity per unit of time depends primarily on price, is uppermost in our daily affairs. Stock supply is less prominent in manufacturing than in genetic production. As a central concern of forestry policy, the long-run supply involves the most dramatic and perplexing questions. What kinds and quantities of timber products, created by what means, can we foresee and pursue?

REFERENCES

1 Liebenow, W. M.: "Wood as an Energy Source in the Kraft Pulp-Paper Industry of the Southeastern United States," Ph.D. dissertation, Virginia Polytechnic Institute and State University, Blacksburg, 1982, pp. 51–69.

CHAPTER 41

HIGH POINTS IN PART FOUR

1 Manufacturing production. Logging, sawmilling, panel production, and pulp-paper manufacture are the industries emphasized in Part Four.

2 Economic differences between manufacturing and genetic production. Chief distinguishing characteristics of manufacturing are (1) its concern primarily with timber, (2) its relatively short planning period, (3) its faster tempo of technological change, (4) its arguably simpler decision rules, (5) its lesser concern with stock supply and the guiding rate of return and its greater emphasis on short-run supply and product price.

3 Close links between manufacturing and genetic production. Ideally, the integrated firm plans the two as parts of a single system. Logging and silvicultural practice are a case in point.

4 Wood-manufacturing economy of a state. Oregon and Mississippi are chosen from the two prime timber regions of the U.S. In both, we are struck by the sweep of change, commencing with the softwood lumbering era and arriving, after 1945, in today's era of new technology. This is a time of industrial restructuring, particularly in logging and sawmilling, and of steep improvements in raw-material utilization and increases in labor productivity in all branches of wood-products manufacture.

5 Feasibility studies. A study of possibilities for a panel plant in Georgia's Appalachian counties impresses us with the wealth of detail over a broad range of subjects that the feasibility analyst must have at command.

6 Weighing a firm's enterprises. The question is whether a Lake States pulp-paper firm should continue to operate a small board mill. The consultants who are called upon for help account for prospective value changes stemming from the interest rate and the rate of inflation. They test the sensitivity of their results to their underlying assumptions. Their analysis carries a reminder of the relevance principle, that only the costs and benefits that vary with a choice are relevant to the choice.

7 Forecasting. Choices necessitate forecasting, for they exist only in the future. This future is laden with uncertainty, though some means are open to lighten the forecaster's uncertainty. Three broad bases for forecasting may be recognized: instinct, custom, and deliberation. For the last of the three, a number of alternative models are available.

PART FIVE

MARKETING AND VALUATION

CHAPTER **42**

PREAMBLE TO PART FIVE

MARKETING, COMPETITION

It is said of the subject of geography that its existence depends on the imperfection of the earth's surface that all places are not alike. It can just as well be said of social science, including economics and the economics of forestry, that it exists because of the analogous imperfections of humanity. And certainly marketing, the subject of Part Five of this book, would be only a thin shadow if it were not for the imperfections of the economy—specifically, the lack of perfect competition among firms.

Adam Smith, in *The Wealth of Nations* (1776), wrote of trade and of market value, but not of marketing. It was he who said that

> every individual necessarily labors to render the annual revenue of . . . society as great as he can. He . . . neither intends to promote the public interest nor knows how much he is promoting it. . . . He intends only his own gain, and he is in this . . . led by an *invisible hand* to promote an end which was no part of his intention [3].

Smith was referring to what later economists termed "perfect competition." He was making the point that in a perfect economy, if every firm seeks its own best combination, the social good is automatically served.

PERFECT AND IMPERFECT COMPETITION

Perfect competition is a state of affairs useful to the theorist, though believed not to exist in the world. An essential feature of it is the perfectly **competitive seller** who faces an infinitely elastic demand, a horizontal function. A firm can sell as much as it likes at the given price, which is its average revenue per unit of output. This is because the firm's output is a negligibly small fraction of the total for the industry and thus has a negligible bearing on unit value. And since average revenue is unchanging with output, it is the same thing as the marginal revenue per unit of output (page 169).

By the same token, the perfectly **competitive buyer** sees a supply that is infinitely elastic, a horizontal function. The firm can buy as much as it wants at the given price, which is its average cost (equal to marginal cost) per unit of input.

When does average revenue equal marginal revenue or average cost equal marginal cost? Why?

The short-run best combinations for the competitive seller and buyer are illustrated in Figs. 42-1(a) and 42-1(b). You should compare Fig. 42-1(a) with Fig. 40-2(d) and note its derivation in the other parts of Fig. 40-2. Figures 42-1(c) and 42-1(d) illustrate how the "given price" is arrived at. The horizontal scale in Fig. 42-1(c) is appropriate to the entire market; the elasticities of demand and supply are those for the whole market. The horizontal scale in Fig. 42-1(d) is the same as in 42-1(a) and 42-1(b). It is greatly stretched out so as to show the individual firm's choices. Figure 42-1(d) is, in effect, an enlargement of the tiny area in the immediate vicinity of the intersection in Fig. 42-1(c).

Thus, in a perfectly competitive economy or industry, not only (1) are firms small and numerous enough so that none can influence input or output prices; but also (2) the industry produces a homogeneous output, every unit of which is identical to every other; (3) firms acquire inputs and release outputs in markets where all parties have perfect knowledge of what goes on; (4) firms are free to do the acquiring and releasing, unconstrained except by competitive prices; and (5) they are free to enter and leave the industry. An economy comprising perfectly competitive firms would be perfectly efficient.

What is an **uncompetitive economy,** or **uncompetitive market** —that is, an **imperfectly competitive** economy or market? It is one that contrasts with what is depicted in Fig. 42-1. Either because of the wide swath they cut in the total market or because of the tactics they employ against their rivals in the industry, individual sellers are able to affect unit values by the scale of their operations. Consequently, they see the demand for their product (their average revenue), not as indefinitely elastic, but as down-sloping. Thus marginal revenue per unit of output becomes detached from average revenue and lies below it. Marginal revenue and cost are equated at a smaller scale of output than otherwise, and price, determined on the demand curve, is higher than otherwise. The simplest, most extreme case, that of perfect **monopoly,** is illustrated in Fig. 42-2(a).

Uncompetitive *buyers* are analogous. Because of their size or market tactics, they see an upsloping supply (average cost) function and a marginal cost that lies above the average. Their scale of output is smaller and price lower than otherwise. The simplest, most extreme case, monopoly on the buyer's part, termed **monopsony,** is represented in Fig. 42-2(b).

It is easy to misinterpret the behavior of the U.S. economy at large as competitive. One sees, for example, price wars and heavy advertising. In a popular sense, this is competitive behavior on sellers' part. But the competitive seller in an economics sense would get nowhere with such shenanigans. If the firm cut its price, no rival would need to follow suit; the firm would go bankrupt. If it raised its price, it would have no buyers. To advertise would be a needless expense.

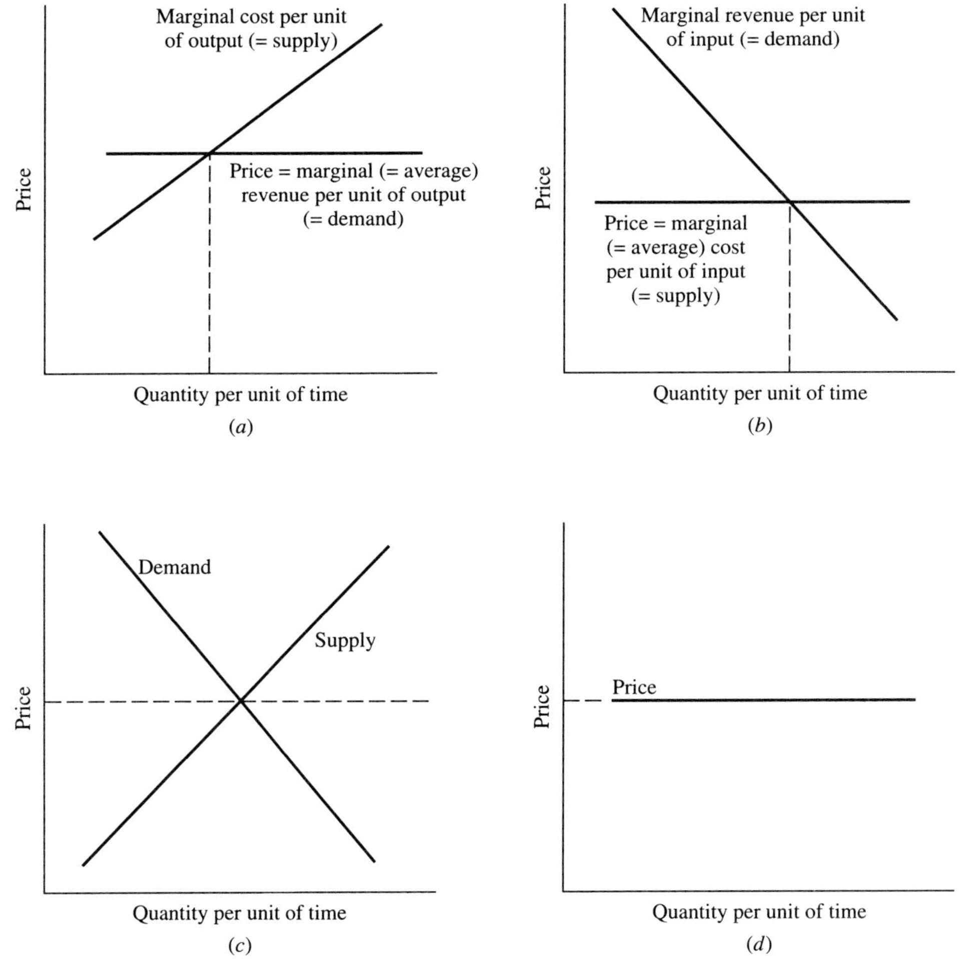

FIGURE 42-1
Concepts of perfect competition: (a) Best output of the competitive seller. (b) Best input of the competitive buyer. (c) Demand-supply intersection as seen by the entire market. (d) Demand-supply intersection as seen by the individual competitive seller or buyer.

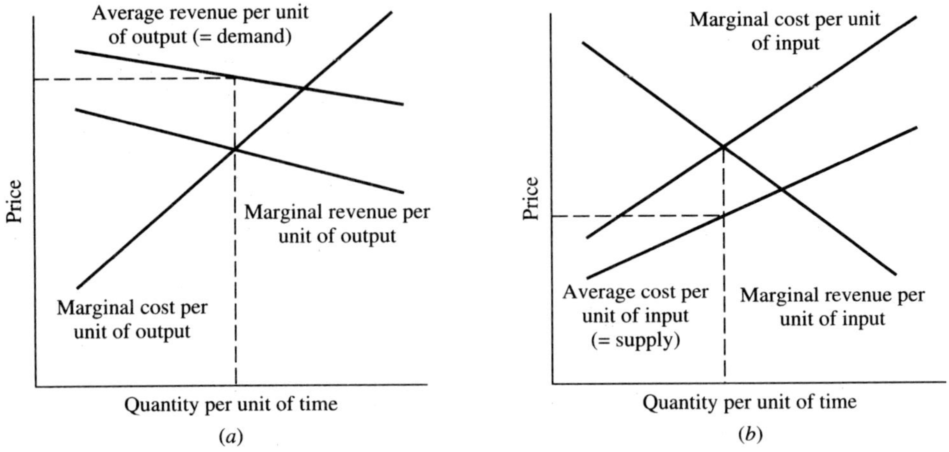

FIGURE 42-2
Concepts of an uncompetitive market: (a) Best price and output of the perfect monopolist. (b) Best price and output of the perfect monopsonist.

> Much advertising aims to convince people that commodity (service bundle) A is of superior quality or price. Surely we can imagine that such advertising would be useless in a perfectly competitive economy. But some advertising aims only to inform people of the existence of a given type of service bundle. Wouldn't we find such advertising in that economy?

MARKETING CHARACTERIZED

Now we're ready to observe how **marketing** [2] is defined. The term has two meanings.

First is the marketing that would take place in a perfectly competitive economy. It would consist of these activities: (1) Exchange of information about the parameters of the market: the resource dealt with prices, quantities, location, time, and participants. (2) Change in title to the marketed resource from sellers to buyers, accompanied by payments flowing in the contrary direction. (3) Performance of facilitating services such as shipment, concentration, dispersion, storage, risk taking, and financing.

Second is the marketing that is performed in our imperfectly competitive economy. It includes, in addition to the foregoing three types of service, the following activities, commonly performed by the producer-seller: (4) Planning: of product lines; channels of distribution; and sales targets in terms of quantity, price, market locations, identification of customers, market tactics, and so on. All such planning may require preliminary research. (5) **Product differentiation:** This is what a firm may do to give its product a distinctive character, an identity, in customers' eyes. It implants in the buyer a loyalty to the seller. This lowers the elasticity of buyer price-demand and allows the selling firm to raise its price while even enlarging its share of the market. Devices for product differentiation are advertising, branding, and offering customers convenience, special services, and a friendly reception in an attractive setting (the "tactics against rivals," page 316). **Monopolistic competition** is a form of imperfect competition characterized by differentiation. (6) Staffing and administering the marketing functions undertaken by the firm, and providing liaison with agencies such as wholesalers and advertising specialists to whom some of the functions may be delegated.

It is the second of the two marketing definitions that is adopted in this book.

We can move on from the term *marketing* to the term **market.** A market is the context of location, time, persons, and activities in which marketing takes place. We will view markets—and market values, too—at times narrowly and at times broadly. In the last analysis, markets are circumstances in which choices are made: the stock market, the grocery, the legislature, the courtroom, the manager's desk. . . . In each circumstance, pros and cons are weighed and values determined. And in the last analysis, values are merely the upshot of weighing, no matter in what units, if any, they are measured.

> What analogies can you find between (1) the market for forest recreation services in a community and (2) the community's courtroom?

We have described marketing activities as being carried out by sellers. But remember that buyers, too, perform the marketing functions. A good example in forestry is wood raw-material procurement by wood-processing firms, which has elements of monopsony in many cases. All the activities mentioned may be undertaken by buyers, including differentiation, which in this case is not of the product, but of the firm: its efforts to win the interest, loyalty, and diligence of buying agents and sellers. We may term this **buyer differentiation.**

Remember, too, that marketing is something that happens, not just to timber products, but to all products of the forest, even those that seldom pass through markets in the accepted sense [1]. Most chapters in Part Five concern the marketing or valuation of timber and its products. But the principles that hold for timber apply to other forest resources, too. And there is one chapter, Chap. 47, that treats pointedly of nontimber and "nonmarket" resources.

Marketing is a productive activity. In some of the traditional thinking, agriculture, forestry, manufacturing, and even some of the tertiary occupations are viewed as production; and then there is activity clearly recognized as consumption. But marketing is left over as unclassified intermediate ground, heavily populated by hangers-on who serve only to provide a connection between the two great groups of first-class citizens.

The fact of the matter is thought-provoking. Marketing has proliferated in our economy in response to our drive for greater productivity in manufacturing and genetic production. In a primitive economy characterized by small-scale primary and secondary producers, many of them self-sufficient, the marketing function is minimal, just as it would be minimal in a perfectly atomized economy. Only as the opportunity for economic growth is envisioned and grasped—and as productivity and material well-being are raised by dint of industrial concentration—is the marketing function called upon for support. If by any chance the marketing sector were not to respond, economic growth and development would be frustrated.

> In rural America before the days of mail-order houses, traveling salesmen were the sellers and individual families the buyers in one of the principal markets. Both parties were numerous and functioned on a small scale. Yet this was not a competitive market. What characteristics stood in the way of competition?

A final point about the definition of marketing. You recall that in Chap. 12, consumers' subjective values are discussed. A distinction is drawn there between subjective values and objective, or market, values, which are played upon by the supply side of the economy as well as the demand side. A promise is made to talk about objective value in Part Five. The promise is kept: The subject of marketing is defined to include market value.

MARKETING AND CLASSICAL FORESTRY

There is an interesting and little-recognized analogy between marketing and forest resource management.

On page 48, under the heading of world forestry issues, one question raised is this:

> What place will the forestry profession find in resource management? How can the profession adapt itself to changing circumstances? The old, absolute doctrines, the world's forestry heritage from western Europe, are breaking down and being rejected, even within western Europe. . . . It is no longer possible for the forester to enforce traditional authority and traditional rules—the primacy of timber, the necessity for sustained yield, the superiority of professional over popular judgment, and so on. What authority and rules shall replace these that are outmoded?

The reference is to those tenets of the classical forestry culture that say, in effect: The profession's charge is to use land wisely, and we are the best judges of what *constitutes wisdom* in this case. Our obligation is to *create resources.* Then the consumer's obligation is to make good use of them.

Only in comparatively recent decades has the philosophy emerged that management is a service to customers, not merely a schooling of customers to adapt themselves to primary production. Economic activity, to the degree that it is not simply steered by an invisible hand, is a mutual effort.

Now what's interesting is that in wood-products manufacture, one finds the same sort of *classical tenets* and the same moves in recent decades to modernize them.

During the long years of the softwood lumbering era, the major (sole?) concern of the forest-products industry was primary production. The challenge was to bring out the logs and saw them into boards and dimension. There were, in normal times, customers aplenty, and their numbers were climbing all the while through biogeny and immigration. The customers would adapt themselves to the resource.

Now times have changed, or largely so. The primary producer is no longer primarily an engineer who turns the output over to salespersons with a confident wave of the hand. The firm, more and more, is customer oriented. It is a big, widely integrated firm. The invisible hand puts little pressure on the steering wheel. The firm is heavily involved in marketing in the "second sense," as defined on pages 318 and 319.

SUMMARY

This preamble is concerned mainly with a definition of marketing. In a perfectly competitive economy, there are marketing functions to be performed: exchange of information, exchange of title, and facilitating services. However, our prime interest is in the additional marketing functions that the firm calls into play in our imperfectly competitive economy: market research and planning, product differentiation, and the staffing and administration of marketing products.

Marketing is a productive activity, not just for sellers, but for buyers, too. As the forestry economy becomes more concentrated, marketing assumes a larger place in the management of firms.

REFERENCES

1 Bentley, William R., and Darius M. Adams. "Dealing with Buyers and Sellers," *In* Duerr, William A., Dennis E. Teeguarden, Neils B. Christiansen, and Sam Guttenberg, eds., *Forest Resource Management, Decision-Making Principles and Cases,* O.S.U. Book Stores, Inc., Corvallis, Oreg. 1982, pp. 430-446.

2 McCarthy, E. Jerome, with Andrew A. Brogowicz. *Basic Marketing, A Managerial Approach,* Irwin, Homewood, Ill. 7th Ed. 1981, pp. 7-10.

3 Smith, Adam: *The Wealth of Nations.* Random House, New York, 1937, p. 423.

CHAPTER **43**

MARKET ORGANIZATION IN FORESTRY

THE BAIN AND CAVES CONCEPT

Chapters 35 and 36, which look into the wood-using industry of Oregon and of Mississippi, give us a glimpse of industrial concentration, the extent to which an industry's total activity is in the hands of a few large firms. The concentration ratio is a useful measure. It gives some insight into an industry's location along the continuum between perfect monopoly and perfect competition. In the Oregon and Mississippi cases, it must be evident, although we scarcely pause to recognize the fact, that concentration is a characteristic of the *market.* In forestry, concentration is part of a revealing story about how well or how poorly the market for forest resources supports national economic goals. This revealing story is the story of **market organization.** Here it is told briefly, following for forestry a line of thinking developed for the economy at large by Joe Bain [1], Richard Caves [2], and others.

MARKET STRUCTURE, CONDUCT, AND PERFORMANCE

Broadly speaking, our national economic goals fall into five classes. Listed in no particular order, the five are as follows: (1) *Full-employment* —or, at any rate, freedom from excessive unemployment, not just on the average, but in each geographic, ethnic, and other major category. Ordinarily, "employment" means the engagement of *persons.* As foresters, we are concerned with labor, but we are also oriented to land and capital resources and their employment and returns. (2) *Price stability:* freedom from cyclical ups and downs and also from inflationary trends. (3) *Equity* in income distribution. Equity, as implied in Chap. 23, is not necessarily equality per capita, but is equality on some basis, perhaps a combination of personal contributions to social well-being and personal want. (4) *Progress* in economic affairs—for example, the growth of real GNP per capita, innovation, new technology of production and of consumption. (5) Economic *efficiency,* by which is meant creating resources in abundance and using them so that their value, measured by value of product, is relatively high. Here again, Chap. 23 elaborates.

One wants a term to describe the degree to which an industry's market behavior contributes to national economic goals. The term is **market performance.** "Good" performance promotes attainment of the goals. "Bad" performance hampers goal attainment.

Behind the market performance and helping to explain it is **market conduct.** This is the behavior of firms with respect to product or input prices and quantities and with respect to their rivals within the industry. Under what circumstances do firms change their prices? their output? the characteristics of their product? their advertising and other selling or buying expenses? their research? Let's recognize three aspects of market conduct: (1) firms' *price policies,* (2) firms' *product or procurement policies,* and (3) firms' *policies toward rivals.*

To explain market conduct in an industry, one can go back to the environment in which the industry performs. This environment is the **market structure** of the industry. Following are the principal elements of structure: (1) *seller concentration* in the industry, (2) *buyer concentration* in the markets where the industry sells its product or procures its inputs, (3) *elasticity of price-demand* for the industry's product or inputs, (4) *changes in consumption* of the industry's product by all users or of its inputs from all suppliers, (5) *barriers to entry* of new firms into the industry, and (6) *product or buyer differentiation* by individual firms.

Barriers to entry are conditions in an industry that make it difficult for new firms to come in. Chapter 36 notes that Mississippi's lumber industry, once easy to enter, became less accessible in the era following 1945. In this case, the barrier is the high cost of land and efficient equipment. Economies of scale are at the bottom of it. Other kinds of barriers are patents, trade secrets, monopolistic control of resource inputs (**natural monopolies**), and hostility of existing firms.

And so one can characterize markets and marketing in an industry in terms of their contributions to economic goals. That is, one can *evaluate* the industry's market behavior. The process of evaluation is outlined in Fig. 43.1 by way of summarizing this introduction to the topic of market organization in forestry.

> Do you think it would be useful to study market organization from the standpoint of ultimate forest-product consumers as well as producers? Recall that analogies between consumption and production are emphasized in Part Two. How would you apply to consumption the parameters of organization named in Fig. 43-1?

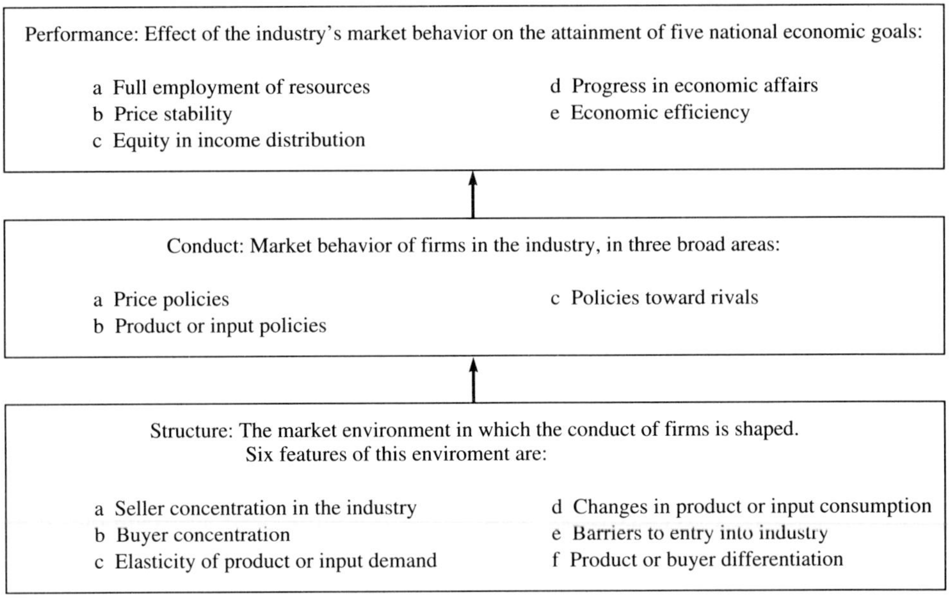

FIGURE 43-1
Elements of market organization in forestry.

COMPETITIVE FIRMS IN AN UNCOMPETITIVE ECONOMY

Let's take some illustrations of market organization in U.S. forestry, commencing with small firms: landholders and timber processors.

As one studies the criteria for good market performance and thinks about the associated features of structure and conduct, one gets the impression that competition is regarded as the economic characteristic most closely associated with the *good,* with the achievement of social goals. Certainly Adam Smith took this view of the matter.

Among economists, competition has been held up as a model for its contribution to the social interest. And economists have always wanted to find illustrations of a perfectly competitive industry. Agriculture in the days of the family farm was said to have many of the features. And, interestingly, the lumber industry has been spoken of in the same terms [5, 8]. Such a view of the industry was partly justifiable at one time—for instance, in Mississippi in 1946—though the thin geographic scattering of the firms put an impassable obstacle in the path of good communication (full knowledge of market events). It is interesting to observe the relation between competitiveness in the industry and the attainment of economic goals. Did competitiveness prompt economic efficiency? progress? equity? price stability? full employment? Generally speaking, it did not, and one is confronted, at first, with a contradiction.

But then one perceives that the theorists' ideal is a competitive *economy,* whereas the lumber industry was a somewhat competitive exception in a largely *uncompetitive* economy. Easy to enter, the market attracted a myriad of small sawmills when lumber prices went up. In the absence of product differentiation, the lumber they produced was as acceptable (more or less) as anybody else's; it entered the market at the going price, checked the price rise, and then reversed it. Their incentive gone, the little lumber makers flocked back to the barn, leaving the market to their stable, disgruntled rivals. Then followed a period of idleness for mills and operators alike, and of price uncertainty. One can look back on that atomized industry as plagued by fluctuation, inefficiency, and technological stagnation. The industry, by virtue of its unique competitiveness, acquired too much of its input resource (land, labor, and some of its capital), kept the resource in a state of underemployment, subsisted in poverty on the verge of overproduction, and subsidized consumers at the cost of forest exploitation.

> If competition otherwise leads to good performance in an industry, why does it not do so where the competitive industry is surrounded by others that are not competitive?

However, one doesn't have to reach into the past for a forestry illustration of a competitive industry in an uncompetitive economy. One can find an illustration today in the genetic production of timber. The clearest case is the nonindustrial private commercial timberland in eastern U.S., some 250 million acres comprising probably more than 4 million holdings, an average of about 60 acres per holding. The only flaw in this industry's competitiveness—and, admittedly, it's a serious flaw—is the same as one observes in the industry of little sawmills: the wide scattering of the firms and thus the geographic barriers that impair their knowledge of the market. Indeed, there is not *a* market: There are thousands.

Consider the six features of market structure (Fig. 43-1). Seller concentration is truly negligible. Product differentiation is nonexistent. True, timber species and quality vary among holdings, but in any timber category, no firm can single out its product as superior, and if it asks much more than the going price, it finds itself out of the market—that is, it faces a highly elastic demand for its product. Holdings change hands continuously in large numbers; owners' average tenure is said to be as little as six years; here is clear evidence that entry and exit barriers are low. Product consumption for the industry as a whole is changeable, but the impact on the firm is nil. Buyer concentration, though vastly greater than that of sellers, is still slight in comparison to most markets for producer goods.

As a consequence of its market structure, the timber-growing industry as represented by eastern U.S. private landowners not integrated with manufacture exhibits a passive, nonaggressive market conduct. Firms are price-takers and not price-setters. Their product policies include no advertising or other promotion, and little attempt to raise output. Their efforts at product improvement, differentiation, and research are understandably negligible. Their policy toward "rivals" is indifference, for firms' actions have essentially no effect on other firms.

Public policy toward forestry in the U.S., as toward agriculture, has long been framed around the proposition that producers are so minuscule in comparison to the industry as a whole that no one firm could possibly afford to do its own research: It could never be able to capture more than a minute fraction of the reward from its efforts. To promote technological and economic progress in the face of this handicap, the federal and state governments have undertaken research programs in the social interest. In forestry, the notable research effort is that of the federal Forest Service, one of whose main branches administers research work in every forest region of the country, through "forest [and range] experiment stations" and their substations. The experiment stations concern themselves not only (and perhaps not primarily) with the problems of the little timber growers, but also with the research questions facing industrial growers, manufacturers, and marketers. The justification has long been that atomization pervades the forestry economy, not simply in timber growing, but in wood manufacture as well.

> How are we to evaluate this viewpoint today? Though the vast majority of forestry and forest-products firms undertake little or no research, quite a few large-scale firms do research or sponsor it and are able to benefit from the results. What is the appropriate research—and extension—policy of government in these circumstances?

PERFORMANCE OF NONINDUSTRIAL PRIVATE FIRMS

Shall we now try to characterize the market performance of the small-scale timber-growing industry [6, 7]?

1 How about full employment of resources? In 1987, growing stock on timberlands of the nonindustrial private owners averaged 1,270 cubic feet per acre as compared with possibly 1,525 for forest industry and 1,560 for all owners, private and public, combined. Not a superlative showing, yet the statistics are brightened by the fact that the private nonindustrials displayed a $14\frac{1}{2}$ percent increase in timber stock since 1977, a dramatically superior performance to that of the other major ownership categories. The conclusion, perhaps, is that the much-berated and widely exploited little holdings, though they represent an underemployment of land, may nevertheless be showing some promise.

Hand in hand with resource underemployment goes the failure to achieve economic efficiency (performance goal 5).

2 Price stability? Nonindustrial private stumpage is open-market stumpage from the viewpoint of wood-manufacturing buyers. Some of these buyers have succeeded in holding *current* prices steady for years at a time. The pulp and paper industry is the outstanding case in point. Other buyers, notably the sawmills, because of their relative competitiveness, have been less successful. In no way has it been possible for them to suppress the persistent inflationary trend in stumpage. The *real* value of, say, southern pine sawtimber stumpage, has been climbing steadily during the past few decades at an annual rate of more than 3 percent. This is the consequence of the rising scarcity of timber, the upshot of all owners' performance as suppliers.

3 Equity in income distribution? Traditionally, and widely still today, buyer and seller in the small-holding stumpage market have greatly unequal knowledge of what they are doing. Often the buyer, whether representing a pulpmill or a sawmill, is engaged in a full-time job and has an experienced eye for timber quantity, quality, and value. The seller, by contrast, may be taking a once-in-a-lifetime step. Equity isn't apt to result. A different sort of imbalance shows up in the phenomenon of the environmental era, the financially comfortable woodland owner who has decided to stay out of the timber market so as to escape logging havoc. But this owner cannot thus govern the future: Sooner or later, land transfer, death, or dire emergency puts the stumpage on the market, with consequences perhaps even less equitable than in the traditional case.

4 Progress in economic affairs? As one surveys American forestry and studies its evolution over the years, one observes some categories of ownership where innovation and constructive change are evident. But nonindustrial private ownership has not, all in all, been one of them. Change, of course, there has been, and some improvement in recent years. The owners are no longer predominantly farmers, but this is true of Americans generally. They are no longer typically poor, but this is in small way attributable to the woods. The little woodlands, despite depletion, have long contributed a large share of their timber stock to the market. In 1976, their contribution was well over 3 percent of their stock. In 1987, the percentage was 2.3. Such rates of timber removal, higher than for other classes of holding, bid fair to continue. A widely held opinion is that the little holdings are a "problem," and that the nation will have to get more wood raw material from them in the future. The question is whether, considering their market structure and conduct, their performance will permit such a contribution.

PERFORMANCE IN INDUSTRY

So much for the small tracts, the least concentrated industry in the forestry economy. We turn now to more concentrated cases, the manufacturing industries. Although these industries have been moving year by year into the hands of fewer, larger, and more elaborately integrated firms, they still retain competitive features. The lumber industry is the clearest case in point.

Lumber, through the agency of the U.S. Department of Commerce and of the trade associations, is standardized into well-recognized species, dimensions, workings, and grades. The industry's principal customers are builders, who are guided by specifications and price. Thus the product is not easy to differentiate. Branding has been tried. Weyerhaeuser Company in 1928 launched its "4-Square" brand, which was stamped on products and at one time widely advertised. Though this brand is still used, it has not been effective enough to warrant continued emphasis. On the other hand, the new panel products (page 281) are being given brand names. Advertising in the lumber and allied industry these days falls mainly into four categories: First, *industry advertising,* aimed at differentiating forest products from alternative building materials such as metal and masonry on grounds of appearance and insulating qualities (Chap. 9), fire resistance, and so forth. Second, *public-relations advertising* by the big, integrated firms, to bring their names and activities to people's favorable attention. Third, *promotional advertising,* which differentiates the individual lumber supplier on the basis of speedy service, wide variety, product quality, and price. Fourth, *informational advertising* in the trade, giving data on quantities available to buyers.

Consider other structural features of the lumber industry—and keep an eye on Fig. 43-1 as we go. Barriers to entry are rising, but still are largely financial and not of huge order. Such barriers as patents, secret processes, and natural monopolies, vital in some industries, are negligible in this one.

> Do you think that the foregoing statement squares with the story in Chap. 36? Explain.

An unexpected connection between natural monopoly and entry was found some decades ago in Weyerhaeuser Company's ownership of so much of the western redcedar old growth, suitable for panels, gutters, and other items requiring deep-clear logs as raw material. Its monopoly of this precious resource did not permit the company to conduct itself as a monopolist in the market. Rather, the result was that society found substitutes for redcedar specialties, which thus became obsolete along with their forest source.

A lumber industry that is growing fast may prove easy to enter because it has plenty of room for newcomers. But, as we see in Chap. 14, the U.S. industry is not growing. However, when it comes to performance, lumber has not done badly in recent times. With the help of the Forest Service's Forest Products Laboratory and allied research groups around the forest regions, the industry has developed some technological aggressiveness (Chap. 39). This has strengthened its productivity and helped it to work against the real-price inflation of its products, in the face of inflationary sawtimber stumpage-price behavior (Fig. 45-2). The industry's forest-management policies—and, in some cases, its labor policies—are becoming more progressive.

MOST CONCENTRATED INDUSTRY: PULP-PAPER

Finally, let's have a look at the most concentrated of wood industries: pulp and paper, along with the conglomerates and big, horizontally integrated firms that so often include pulp-paper manufacture.

The structure of the industry encourages product differentiation by branding and advertising, for this is to a significant degree a consumer-goods industry. Entry is the most difficult of any forest industry, not only because of the large investment required for an efficient plant and forest property, but also because of barriers such as patents and entrenched brands. Product pricing is more nearly an exclusive function of manufacturers than in any other forest-industrial group.

The selling-price policy that characterizes the pulp-paper industry and is found also in veneer-plywood and other panel manufacture is **price leadership.** In its structure as a seller, pulp-paper is an **oligopoly,** an industry with relatively few firms and high concentration ratios. Pulp-paper is clearly not competitive, but surely no one firm is a monopolist. In this intermediate situation, price formation is accomplished by game playing, in which the principal player (price leader) is one of the biggest firms. For some types of output, a single firm retains leadership for years; in other cases, leadership rotates. For each commodity at any time, there is an equilibrium price to which the industry, by and large, is holding. The typical firm refrains from raising its price, because it will suffer a loss of business as long as competitors hold the line. Further, the firm refrains from lowering price, since, unless the firm is of negligible size, rivals may follow suit and all suffer a loss. But upward or downward pressures on prices are always at work in the market, and it falls to the leader or leaders to keep close watch and make the moves to which the industry can willingly, if not always happily, respond.

Ideally, the price leader seeks a combination that will allow the industry an acceptable internal rate of return on new investment (**target-return policy**), or a marginal revenue that will maximize profit or minimize loss (**marginal-cost policy**).

In contrast to the little genetic producers, who tie up more than the socially optimal quantity of some resources and underemploy them, the concentrated industry could be expected to employ less than the social optimum because, with a degree of monopoly power, it could gain by holding prices up and output down, a mark of economic inefficiency. The fact of the matter is less simple than this for pulp-paper. The optimal mill and the long-term prospective paper market have both continued to get much larger. The cost of shutting down a paper mill has also become much larger. As a result, pulp-paper suffers from persistent **overcapacity.** This is another way of saying that the output of the industry is higher and its prices and profits lower than the firms would be thoroughly happy to see. The industry does not perform quite in the manner that one should expect of a well-regulated oligopoly. What this maladjustment means for economic efficiency and attainment of social goals is hard to judge.

The concentrated industry does perform in some respects as an oligopoly should. One respect is that firms are keenly aware of one another's actions and plans. This is an essential part of the game playing. It holds not only for the firms as product sellers, but also for these same firms as buyers of raw material. Since we have spoken little of buyer behavior, let's speak of it now.

BUYERS' BEHAVIOR

As mentioned earlier, buyers are sometimes able to hold local stumpage and roundwood prices steady for quite a while. This is true of the concentrated industry. Of course, there is nothing regrettable about stable prices. The issue concerns the level at which the prices are held. Just as the giant in the market tends, as a monopolist, to boost selling price, so, as a monopsonist in the buying market, the giant tends to cut price. Now the buying market is commonly distinctive in that it is spread out over a wide territory of backcountry and over such a large number of sellers. As compared with the selling market, communication is poor, and the means for firms to observe what the leader is doing are unreliable unless they are fortified by firms' putting their heads together. Of only academic interest today is a case, once famous in Lake States forest-industry circles, of a small group of pulp-paper firms indicted for collusive behavior in the pulpwood market. The case was settled without a trial. There have more recently been analogous cases on the West Coast, with public stumpage prices as the target for alleged industrial collusion in defiance of the Sherman Antitrust Act [4].

Walter Mead [3] of the University of California made an illuminating study of buyers' behavior in the markets for Forest Service and Bureau of Land Management stumpage in the Douglas-fir region. For each tract, the sellers appraise and publish the value of timber, and then hold an auction, at which the bidding starts at the appraised value. The bidders are wood processors. Mead's research centered upon the auctions that were held during a 4-year period.

One of Mead's findings was that the buyers within each district exhibited a type of behavior that zoologists term "territoriality." They defended their district against invasion: If an outsider entered "their" auction, they would see to it that the newcomer was outbid—or, if persistent, was obliged to pay a disastrously high price.

> What is "territoriality" among animals? Does the human animal exhibit territoriality?

Mead studied the number of bids submitted at the auctions, whether only one or more than one. He classified each sale according to whether the successful bidder was a "large firm" or "small." His finding is summarized in the following tabulation. Large firms had the edge over small ones as timber buyers. They enjoyed a higher proportion of single-bid sales, and in other sales they had fewer competitors and paid a lower premium above appraised value. Mead attributed the superior market power of large firms to (1) their relative ease of entry into the market: their ability to finance large purchases and high road-building costs, and (2) their relatively extensive timberlands, which allowed them to be more selective in their buying policy and to avoid high-price sales.

Item	Large firms	Small firms
Percent of all timber sold:		
Single-bid sales	34	25
Multibid sales	66	75
All sales	100	100
Thousand board feet per sale	7,014	4,483
Percent premium, top bid over appraised value:		
Single-bid sales	0	0
Multibid sales	43	68

(Continued)

Item	Large firms	Small firms
Average number of bidders per multibid sale	4.8	5.4
Percent of sales involving road construction	82	68
Dollars average road cost per sale involved	77,330	54,531

Mead also spoke of (3) small firms' "bidder futility," which led them to avoid a confrontation with their larger rivals, who commonly had advantages of plant location, ownership of roads near public lands, superior financial power, and the opportunity to effect large savings in federal income tax.

SUMMARY

This chapter's subject is market organization in the forestry economy. The question asked is how well the markets in this economy serve national economic goals. The model used for generating a reply is the structure-conduct-performance model of general economic theory. The data put into the model are taken from both the genetic and the manufacturing branches of the economy.

Among highlights are these conclusions: (1) A competitive industry, such as small-scale nonindustrial private forestry and small-scale sawmilling, fails to contribute substantially to good market performance when it is set in a largely noncompetitive economy. (2) In any case, one may entertain fears for the competitive industry's ability to achieve productivity gains through research and development. (3) Today's lumber industry, though more highly concentrated than in the past, has retained constructive competitive features. (4) Pulp-paper, an oligopolistic industry, tends to raise output prices, depress raw-material prices, and lower the scale of operations; the tendency is partly defeated by the industry's economies of scale and inherently bright market prospects. (5) The structure of the market for wood raw material is, on balance, unfavorable to good market performance.

REFERENCES

1 Bain, Joe S.: *Industrial Organization,* John Wiley & Sons, New York, 1968.

2 Caves, Richard: *American Industry, Structure, Conduct, Performance,* Prentice-Hall, Englewood Cliffs, N.J., 1972.

3 Mead, Walter J.: *Competition and Oligopsony in the Douglas-fir Lumber Industry,* University of California Press, Berkeley, 1966.

4 Mead, Walter J.: "The Firm and Its Competitors," *In* Duerr, William A., et al., eds., *Forest Resource Management, Decision-making Principles and Cases,* O.S.U. Book Stores, Inc., Corvallis, Oreg., 1975 ed., vol. II, p.34-12.

5 Muench, John: *Softwood Lumber and Plywood Prices,* National Forest Products Association, Washington, 1971.

6 U.S. Department of Agriculture, Forest Service: "An Analysis of the Timber Situation in the United States 1952–2030," Appendix 3, *Forest Statistics of the U.S., 1977,* Washington, 1980.

7 U.S. Department of Agriculture, Forest Service: *An Analysis of the Timber Situation in the United States: 1989–2040.* Gen. Tech. Rep. RM-199, 1990, pp. 43–50.

8 Zaremba, Joseph: *Economics of the American Lumber Industry,* Robert Speller, New York, 1962.

CHAPTER **44**

MARKET GEOGRAPHY IN FORESTRY

A CHARACTERIZATION OF MARKET GEOGRAPHY

In order for an exchange to occur in a market, something moves from one place to another. The buyer moves, or the seller, or the resource that they exchange, or some combination of these three. When stumpage is marketed, the buyer travels to the resource and its seller, and the same thing happens when hunting privileges or other recreation opportunities are opened to consumers. When roundwood is exchanged, the wood is shipped, let's say, to the miller-buyer, and the seller may come with it. Or if the resource can be described accurately, as with hardwood lumber or softwood plywood or water, so that the terms of the exchange may be worked out at a distance, only the resource need be moved.

> Is it possible for services to be marketed without any movement of anything? If so, give an example. If not, why not?

If things are moved, there must be carriers and distances and costs, and the results can be mapped. They can be described in terms familiar to the geographer. And if mapping is employed, the location of market activity becomes essential information, including how market prices and quantities are distributed over an area.

This chapter looks at four questions in forestry market geography and glances at models for responding to them. The questions concern how to explain (1) transportation costs, (2) the location of activity, (3) the geographic distribution of market prices, and (4) the geographic distribution of market quantities.

TERMINAL, LINE-HAUL, AND TOTAL COSTS

The transportation of commodities or persons entails two kinds of cost (Fig. 44-1). One kind is **terminal cost,** that which is fixed with respect to transportation distance. Consider a firm that operates 18-wheel flatbed trailer trucks. The firm's terminal cost is that of meeting its fixed financial obligations, maintaining and staffing its offices and terminals, loading and unloading the trucks, and other activities thought to be more appropriately chargeable per ton of freight hauled than per ton-mile. **Line-haul cost,** on the other hand, varies directly with hauling distance: fueling, lubricating, servicing, and depreciating the trucks; paying the operators; and the like. In Fig. 44-1, line-haul cost is added to terminal cost to represent total cost per ton. The slope of the total curve is the line-haul cost per ton-mile.

Different types of carriers have different terminal and line-haul costs. If carrier B has a higher per-ton terminal cost than carrier A, it must have a lower ton-mile line-haul cost; otherwise, as we'll see clearly in a moment, carrier B could not survive A's competition. The types of surface carriers can be arranged in a hierarchy:

Terminal cost	Carrier	Line-haul cost
Low	**A** Light truck	High
	B Medium truck	
	C Heavy truck	
	D Railroad	
High	**E** Boat, barge, log raft	Low

If you want to move your household belongings one block down the street and must rent a truck, you may well be better off to make several trips with a pickup than one trip with a van. But if you are moving a considerable distance, the van is probably to be preferred. The story of transportation cost by carrier type is idealized in Fig. 44-2. Carrier type A, with the lowest terminal cost and the highest line-haul cost, excels at

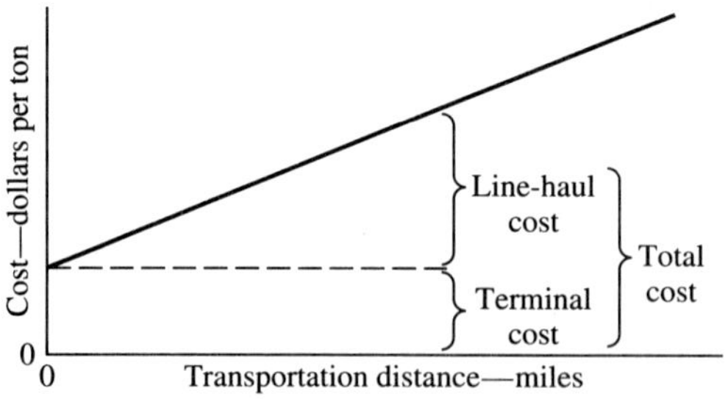

FIGURE 44-1
Terminal, line-haul, and total costs of transportation per ton for a given commodity by a given type of carrier.

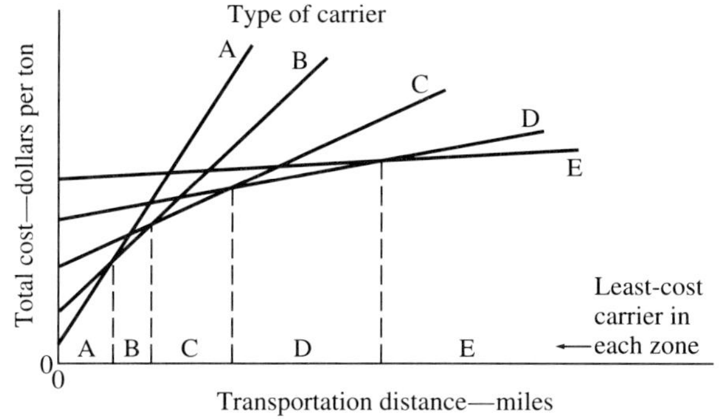

FIGURE 44-2
The hierarchy of least-cost carriers, arising from differences in carriers' terminal and line-haul costs.

the shortest distances. Then, at the distance marked with the first vertical broken line, comes a **break-even point** between A and B, where the choice between them is a matter of indifference. B's zone of excellence then extends to the next vertical line. And so on.

Of course, the ideal pattern is much distorted in real life. Boats cannot be used unless there's water. Railroads require track. Heavy trucks do best on well-built multilane highways. Nevertheless, the concave cost function traced out by the least-cost carriers in the successive zones is a reality, and when the carriers' costs are translated into transportation charges, or **tariffs** (costs to shippers) by adjusting for carriers' profits, if any, the result is still a concave function such as that traced by the least-cost segments in Fig. 44-2. The cost of shipping a ton of *any given resource* increases with distance, but at a diminishing rate.

GEOGRAPHIC ORIENTATION OF ACTIVITY

Now, given the tariff function, what can we say about the location of economic activity? We want to emphasize the location of focal points in marketing. Let's think about timber and timber products and consider the location of the first major market along the path from forest raw materials to final consumers. This market is, say, the primary processing plant: sawmill, veneer mill, pulpmill, or whatever. Is there a tendency for such plants to be built near raw materials? near final consumers? in between the two?

Suppose that manufacturing costs don't differ greatly from one location to another, and, indeed, that the prime costs that do differ are those of transportation. Wherever it locates, the firm incurs a cost of shipping inputs to its plant and outputs from its plant to consumers. If it puts the plant close to the source of inputs, the **procurement cost** of bringing inputs to the plant will be small, but the **distribution cost** of forwarding outputs to consumers will be large. Or if the plant is built near consumers, raw-material procurement cost will be the big cost. The most efficient location is the one where the sum of the procurement and distribution costs is least.

The concavity of the tariff function militates against locations intermediate between materials and consumers. Imagine, for just a moment, that the mill's raw materials weigh the same number of tons as the products into which they are converted, and that the tariff functions for raw materials and end products are the same. Then it's a toss-up whether to put the mill near materials or near consumers (Fig. 44-3).

But as a matter of fact, it's seldom true that the roundwood that goes into a mill weighs the same as the product that comes out. Most commonly, timber processing involves weight loss: Water is removed and residues are left behind. Consequently, there are more tons of material to procure than of product to distribute: The procurement-cost curve is raised and steepened; and total cost is least if the mill is located where the raw materials are: in or near the forest. The firm is able to save expenses by minimizing the distance that the comparatively heavy roundwood has to be shipped. The phenomenon is termed **materials orientation.**

Redraw procurement cost in Fig. 44-3 to show a raised and steepened function. Then redraw total transportation cost and observe that the firm is now materials oriented.

Among primary wood-processing plants, there are exceptions to materials orientation. A railroad may, for various reasons, set advantageous tariffs for the firm that will build a plant at a designated intermediate point between material and consumer centers. **Transshipment points,** where goods must be transferred from one type of carrier to another (examples: rail heads, seaports), are magnets for all sorts of processing industry, for firms can save the cost of one unloading and one reloading by locating there.

In Fig. 44-3, assume that there is a point somewhere between materials and consumers where goods must be transshipped from narrow-gauge to standard-gauge railroad at substantial unloading and reloading costs. Draw the three cost functions and show that total transportation is least if the firm locates at the transshipment point.

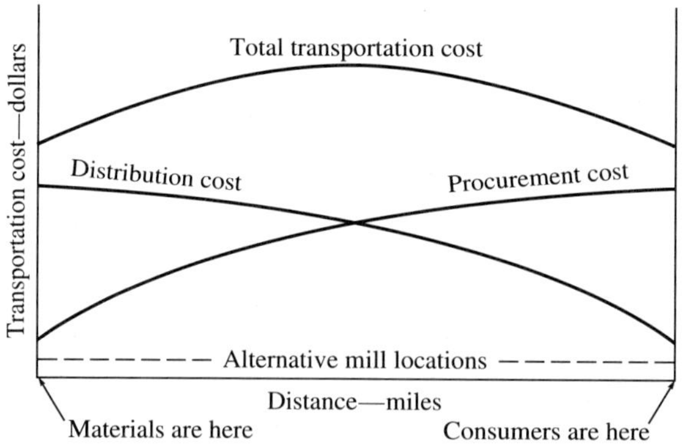

FIGURE 44-3
The firm's procurement, distribution, and total transportation costs in relation to its choice of mill location, assuming raw materials and products weigh the same.

If consumers are concentrated but raw materials are widely scattered, **consumer orientation** may result. At one time, a sawmill was located within the city limits of Columbus, Ohio. Its raw material came from farm woodlands within a wide circle: So far as procurement cost went, one location was as good as another. But Columbus was the consumer, the rent of land was not notably greater there, and this tipped the scales in the city's favor as a mill site.

An unusual example of orientation to *both* materials and consumers was a sawmill that operated for a time in downtown Manhattan, New York City. The mill's materials came from old, but still sound, piling and timbers, by-products of the demolition or remodeling of wharfs.

Of course, costs other than transportation may vary with location and prove decisive. Community A has an abundant supply of clean water to attract a pulpmill. B has certain labor skills. C is a nice place to live. D is eager for industry and holds out incentives, while E shrinks from noise, smoke, and influx of strangers.

The principal examples of consumer orientation are secondary wood manufacturers (whose raw material is the output of the primary industry) and also wholesalers and retailers. Consider furniture manufacture, a secondary wood use. Like sawmilling, it is a weight-losing process, yet furniture is more expensive to ship than the lumber required to make it, because it is comparatively fragile and bulky. So the firm's distribution-cost curve is not the mirror image of procurement cost, as in Fig. 44-3, but is relatively high and steep, and least total cost is found where the consumers are. Comparatively high, steep distribution cost is characteristic, too, of the construction industry. It also holds for paper making unintegrated with pulping. The paper maker receives dried pulp, which transports more cheaply than the end product. The firm wants to be as close to consumers as costs will allow and still have access to freshwater. But when paper making is integrated with pulping, it is drawn to the woods-oriented pulpmill.

Retailers are consumer oriented. Their services are highly perishable with increasing distance from buyers. Wholesalers who carry stocks (termed **yard wholesalers** in the lumber and panel trade, **merchants** in the case of paper) locate at transportation hubs. Some of the biggest lumber wholesalers carry no stocks, but serve as **brokers,** bringing sellers and buyers together for a commission. These brokers, sometimes called **office wholesalers** or **hat-rack wholesalers,** are located at forest-product market centers such as San Francisco, Portland, Denver, Chicago, New Orleans, St. Louis, Cincinnati, New York, and Boston.

Reminder: Firms tend to grow, diversify, and integrate vertically. More and more, when one speaks of the successive phases of the market through which wood flows—roundwood transactions, processing, wholesaling, and retailing—one is *not* necessarily referring to separate *firms.* Some individual firms have their hands in all activities from forest management to serving the ultimate consumers. This is so with lines such as recreation resources—and some timber resources, too.

How is the foregoing related to earlier ideas about location? Chapter 29 proposes that location is a function of land use, and that land is occupied by the firm that can pay the highest price or rent for it. In the present chapter, location is seen as an outcome of transportation cost. You can interpret this view as explaining the earlier theory. Why is the firm able to pay for the use of a certain tract? For one thing, because the firm saves transportation cost by using the tract.

WOOD CONCENTRATION FOR A SOUTHERN PINE PULPMILL

Much of the forestry economy, then, is material oriented. Material and consumer are often widely separated, not only in distance, but also in the number of functions to be carried out between genetic production and ultimate consumption. It is easy to understand, therefore, that the forestry economy has a complex marketing system.

For an illustration, consider marketing in the southeastern U.S. pulp and paper industry. The system is most readily understood if described as three subsystems: (1) **Concentration** of wood raw material at the pulpmill. There are, in general, three sources of wood: first, roundwood from the pulp-paper firm's forests; second, roundwood from the open market; third, chips produced from residue in sawmills and veneer-plywood plants, some of which are owned or controlled by the firm. (2) Equalization of the pulp flows, out of the pulpmill and into the paper mill. Do you remember the paperboard machine that is under fire from the accountants in Chap. 38? The firm didn't make enough pulp to feed this machine and had to buy (go into the market for) wastepaper. In other cases, a firm's paper machines are not able to use the whole pulping capacity, and the firm sells the excess as **market pulp.** Some installations produce *only* market pulp. Others comprise only a paper mill, which is run on market pulp. (3) **Dispersion** of the firm's product lines to the ultimate markets.

This chapter looks at Subsystem 1 and takes marketing in the southeastern U.S. pulp-paper industry as an example. The chapter asks about the procedure followed for bringing wood from a pulpmill's procurement territory (its **timbershed,** or **pulpwood shed**) into the mill yard. It asks how the prices paid for pulpwood and the quantities of wood procured vary from place to place within the shed. It asks how the sheds of mills within a region fit together and how the levels of prices vary from area to area within a region.

We begin with the organization of the wood-procurement function at an individual pulpmill. The form of organization has evolved over the years within the industry, and rarely in the same from firm to firm. Nevertheless, all organizations have features in common. Under the mill manager, either directly or within a woodlands department, is a wood-procurement department. The mill manager gives the department a procurement quota for the year ahead, and it is the department's responsibility to see that the quota is met in a reasonably even flow from least-cost sources. The department's work is vital to the firm because the cost of wood is a big part of pulp value and because in normal times the pulpmill must not be allowed to run out of wood lest the paper firm incur the expense of shutting down the mill. Pulpmills in the southeast used to carry a couple of months' wood supply in the yard, but this inventory, a source of expense, is now often targeted to a 3- or 4-weeks' supply. If the backlog dwindles dangerously, the local druggist is called upon for antacid and neighboring mills are called upon for a loan of wood.

The wood department maintains a staff of district managers, each assigned a fraction of the mill's total timbershed. Each district manager is given a share of the total quota. To meet the assigned share of the quota, the district manager draws on the three sources of wood mentioned earlier. The least expensive wood is residues. Refer again to Fig. 40-3. Some of the smallest mills get by on residues alone, but let's continue with the case in which all three sources are drawn upon. The next source is the firm's own land. At headquarters, the woodlands staff and the wood department agree on the amount of the total wood quota that will be taken from the firm's land. Let's assume that the agreed amount is parceled among the district managers.

To meet the rest of the district's quota, the manager calls on operators distinguished from those discussed up to this point in that they are not employees of the firm. They are contractors. The distinction is, traditionally at least, of great importance in the industry. We'll see why.

The contractors that the district manager works with are of two sorts. One, known in the trade as the **independent contractor,** with emphasis on the adjective, is a logger whose contract with the firm ordinarily runs for three months at a time. The contract states the quantity of wood to be delivered at the concentration point, the price the firm will pay for it, product specifications, and perhaps evidence that the contractor is indeed independent and is abiding by the law, paying insurance premiums, and so on. The contractor employs a crew of three or more workers. The contractor's responsibility is to find the stumpage, negotiate a price with the owner, log the timber and haul the wood to the point of concentration, buy and operate efficient equipment, and maintain friendly relations with all concerned. The pulp firm may specify the results it expects the contractor to achieve, such as the condition of the forest after the cut. It may help the contractor with financing, though ordinarily not beyond acting as a co-signer in borrowing from the bank. The firm strives to display the contractor as an independent business manager, who is told what the firm wants done, but not how to do it.

The second variety of contractor that the district manager works with is the **wood dealer.** The dealer, unlike the independent contractor, may serve a number of pulp firms. Quite commonly, the dealer obtains wood by subcontracting for it with loggers who resemble independent contractors in their procedure. However, the district manager has no direct contact with these loggers. That is the dealer's responsibility.

To supplement the wood-buying that the district manager delegates to contractors and dealers, the manager may himself buy stumpage in the open market. He then turns to independent contractors for the logging and delivery.

Presumably, the contractor system has grown up because it has proved to be the most efficient wood-procurement system for pulp firms. Among advantages claimed for it are the following: (1) The system provides savings to the firm. Loggers are not paid at union rates or allowed union fringe benefits, as they would be as employees. The firm's bookkeeping is greatly simplified. Whether the logger keeps adequate books is the logger's responsibility. (2) The contractor system relieves the firm of legal involvement of many sorts. (3) The system reduces the number of persons with whom the district manager must work. (4) The ultimate responsibility for wood procurement is put into the hands of workers who have the best opportunity and incentives to succeed. They are local people—neighbors who can win the confidence of timber owners. They can operate under a piecework system of pay that offers rewards for hard work.

But the fact that the contractor system has survived the test of time does not prove that it is best, either socially or for private profit. Some pulp firms have tried, on a small scale, using only employees for the wood-procurement tasks. They have found, in some cases, that they can put wood into the mill yard at lower cost than the contractor system is capable of.

What are some of the advantages, in terms of immediate wood cost and in other terms, that a firm might gain by using only its own employees in wood procurement?

GEOGRAPHIC PRICE STRUCTURE AND QUANTITY STRUCTURE

In general, pulpwood prices at the concentration points within a timbershed fall into one of two geographic patterns, or **geographic price structures.**

In a region where wood users have enough elbow room in their procurement so that they do not find themselves obliged to bid against one another very often, a pulp company is able to follow a policy of **delivered pricing.** This means that the pulp firm pays the same price for all wood arriving at the mill, regardless of its origin. The price received by the dealer or contractor at the local concentration point is the mill price minus the cost of transportation to the mill. In effect, the seller pays the freight to the mill. Looking at the **local-price profile** in a shed with delivered pricing, one sees a peak at the mill, and then, in every direction away from the mill, prices that at first decline very steeply and then less and less steeply. The local-price profile (moving to the right away from the mill) is the tariff profile of Fig. 44-2 turned upside-down. The pulp firm is able thus to discriminate against distant sellers because, out there where these sellers are, there are few alternative buyers. Sellers can take the firm's offer or leave it.

Look at the map in Fig. 44-4 (a). The map shows the location of pulpmills in Virginia and North Carolina that used appreciable quantities of pine. Mill capacity per 24 hours is symbolized, and the approximate boundaries of each mill's timbershed are marked by the ends of the lines radiating from the mill symbol. *The year is 1940.* There are eight mills. Their mean capacity is 280 tons per day (range 60 to 560 tons). The average timbershed radius is 75 miles. Four of the eight mills, the four largest, draw half or more of all their pine pulpwood from counties where they have no competition from other pulpmills for pine. The biggest mill of all draws 100 percent of its pine from such counties. *The price structure is delivered pricing.*

Now turn to the map in Fig. 44-4 (b). *The year is 1981.* The number of mills has increased to 13. Their mean capacity is 1,038 tons (range, 250 to 2,050). The average timbershed radius is 100 miles, and some wood is moving as far as 300 miles as the crow flies. One mill out of the 13 has some small freedom from rivalry: It derives 19 percent of its pine from counties where it confronts no other pulpmills. Percentages for the remaining 12 mills are all less than 4 percent.

A glance at the picture in 1987 (not mapped) reveals that the industry had reached a plateau and was adjusting its competitiveness. The cost of new construction had become prohibitively high. As a result, no new mills had appeared. Greater capacity had been built by enlarging existing mills—though slightly, about 100 tons per day on the average. The older plants were being modernized. Firms were selling their far-out woodlands and buying, if at all, more near-in lands. Timbersheds and competition were shrinking.

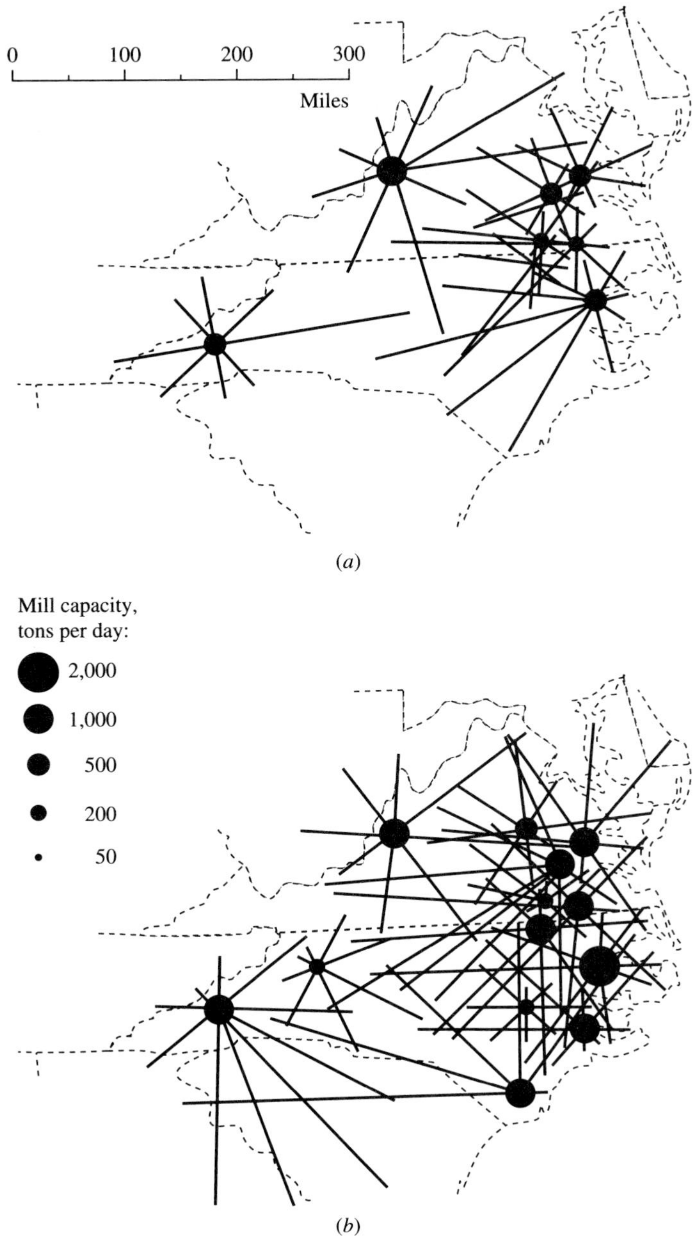

FIGURE 44-4
Timbersheds of Virginia and North Carolina pulpmills that used pine wood as raw material, and daily capacities of the mills: (a) 1940, (b) 1981. Ends of radiating lines mark approximate boundaries of timbersheds. (*Source:* Southeastern Forest Experiment Station, Asheville, N.C.).

Well before 1981, *the price structure had changed to local pricing,* and it remained so thereafter. In **local pricing,** the prices paid for all wood delivered to concentration points throughout a timbershed are the same. The pulp firm picks up the tab for transportation of wood from local points to the mill. It is obligated to perform in this way because sellers have sufficient alternatives. They cannot be obliged to pay the transportation. The local-price profile across a pulpwood shed and its neighboring sheds is a horizontal line. The replacement of delivered pricing by local pricing was a phenomenon of the post–World War II era, commencing about 1950.

With a local-pricing policy, how is price determined? Observe that where pulpmills are packed into a region cheek by jowl, as in Fig. 44-4 (b), their position in the input markets resembles the paper mill's position in the product markets. The pulp firm as a wood buyer is somewhere between a perfect monopsonist and a perfect competitor. The term is **oligopsonist.** Compare the discussion of the oligopolist, beginning on page 329. Under oligopsony as under oligopoly, price leadership is the rule. The leader takes responsibility for finding a (low) price that will bring the wood to the industry.

> Which raw-material market is more competitive, the one with a delivered-price structure or the one with a local-price structure? If the pulpwood buyer in a shed with local-price structure is an oligopsonist, what is the appropriate term for the buyer in a shed having delivered-price structure?

With either delivered or local pricing, a firm wants to buy as much wood as possible per square mile of forest from the square miles closest to the mill. In this way, it offers either sellers or itself the maximum freight-cost saving. Commonly, as one moves away from the mill across successive distance zones of equal width, one finds that the firm is procuring increasing quantities of wood per zone, for although its takings per square mile are diminishing, there are more and more square miles per zone. But then comes the point where the drop-off in acquisition per square mile outweighs the growing number of square miles, and beyond this point, the per-zone acquisition fades away to the boundary of the shed.

The foregoing neat **quantity structure** within a timbershed can be disturbed by various forces. One is the arrival of newcomer mills. Some newcomers, especially small mills, fit readily into the pattern of acquisition, provided the resource is not already overtaxed. But big newcomers, especially if they come into the game with their elbows out, set up repercussions that travel in waves for hundreds of miles, each mill shifting the focus of its operations to accommodate its neighbors.

The proximity of a pulpmill to sawmills and veneer plants brings special forces into play. Through rivalry, these wood users may force local pricing upon one another. Through cooperation, the pulpmill may feed upon the others' residues, while the others benefit from the flow of larger logs diverted to them by the pulpmill. If the pulp firm gets all the wood it wants from small-pole forests, it may not interact with other wood users except when it determines what forest stands it will leave to grow larger timber.

Often pulpmill wood departments are conscious of their part in timber-resource conservation. They are aware of the drain they and other users put on open-market supplies. Some big firms are in close touch with the Forest Service's Renewable Resources Evaluation Project and help devise plans for spreading procurement activities to avoid creating deficit areas.

A WIDE VIEW OF PRICE STRUCTURE

In conclusion, we return to the price topic and take a wide view of price structure, benefiting from the work of T. P. Hunter [1]. He studied the records published in *Timber Mart-South* [2] and worked up a map of the southern states, each state divided into sections. He showed the average price paid for pine pulpwood within each section.

Generally speaking, pulpwood prices are highest along the coast, falling off to the west and north. Why so? Hunter found one significant variable that interests us. The greater the pulpmill capacity within a 100-mile radius, the higher the pulpwood price. This doesn't necessarily signify a departure from local pricing within individual sheds, but it may suggest a softening of the policy in some cases, and it clearly indicates a downward drift in price in those sheds that are farther and farther from the most tightly packed mill belt along the coast.

> But why the tight packing of mills along the coast? Can you venture an explanation?

Wherever there are pockets of lesser rivalry among wood buyers, where sellers have comparatively few outlets for their wood, pulpmills may find that their interest lies in magnifying any geographic price variations that result. This is because sellers in a low-rivalry zone, with its limited alternatives, exhibit a low price-elasticity of supply. (Bear in mind that this refers to dealers and contractors, whose output is sensitive to price; it does not refer to genetic producers.) By contrast, sellers in the high-rivalry areas have a price-elastic supply. The wood department will certainly want to investigate the possibility of raising price just a little where rivalry is high, thus obtaining more wood than before from the elastic suppliers. To compensate itself, the firm can cut prices to the inelastic suppliers; the loss in wood will be small.

The gains obtainable from price discrimination against inelastic suppliers and in favor of elastic suppliers illustrate a widely pervasive **principle of incidence.** Favorable economic changes are most apt to benefit those firms that are elastic, flexible, and able to shift from one strategy to another; while they pass by the inflexible firm. Unfavorable economic changes are most apt to damage those firms that are inelastic, inflexible, and unable to change strategies; but they can be escaped by the flexible firm.

Be all that as it may, the fact remains that the procurement object of the pulp firm is to get an assured flow of enough wood. Under a local-pricing policy, the firm must distinguish itself in the eyes of contractors (and contractors in the eyes of timber owners) through buyer differentiation. Sellers will deal with buyers who are friendly, helpful, constructive, and reliable and who treat them with deference, as equals. To manipulate the price may not be in the buyer's interest.

SUMMARY

Transportation costs vary with distance in such a way that market activity tends to be located either near raw materials or near consumers. Much of the forestry economy is material oriented, and thus, transferring forest resources to consumers is a complex system. We look at just a small part of the marketing system, the part concerned with transferring southern pine from forest to pulpmill.

Each mill draws its wood from a timbershed with the help of a contractor organization. Because of rivalry among wood users, the pulp firm follows a local-pricing policy within its timbersheds and relies on leadership to set the price level. Within the South as a whole, pine pulpwood prices are highest along the coast, where mills are most concentrated. Inland pulpwood sheds operate at lower price levels.

REFERENCES

1 Hunter, T. Parkin: "Regional Price Variations of Southern Pine Sawtimber and Pulpwood," *Forest Products Journal,* September 1982, pp. 23-26.

2 *Timber Mart-South.* F. W. Norris, Inc., Highland, N.C.

CHAPTER **45**

PRICE TRENDS IN THE TIMBER ECONOMY

PRICE-TREND PROBLEM

The question is this: Why have prices followed the trend that they have in the timber economy? Particularly, why has the real price of lumber worked its way relentlessly upward at an annual rate averaging almost 2 percent since the times of earliest record (Fig. 45-1; note logarithmic vertical scale)? And why the contrast between the price behavior of lumber and that of paper (including paperboard and pulp)? The question is readily phrased, but not quite so readily responded to. Nevertheless, let's persevere as straightforwardly as we can: First, clarify the question by examining the data. Second, find a model for explaining the trends. Third, put information into the model and work out a response. Fourth, follow up with a few observations and tests.

THE Z-FORMULA

Foresters have long been concerned over the fact that lumber prices were rising relative to the prices of other things. They saw lumber as the noblest outcome of forestry and feared that it was pricing itself out of the market [5].

Among early students of lumber-price trends were Wilson Compton, executive secretary of the National Lumber Manufacturers Association, Ralph Bryant of the Yale School of Forestry, and Robert Goodman, also of Yale [2, 1, 4]. Compton and Bryant published their ideas in the years immediately before 1920, when all prices were rising, but lumber's was skyrocketing and getting a lot of attention. The U.S. was in the midst of the softwood lumbering era. Compton and Bryant looked at the long trend in lumber price and charged it to rising transportation costs from western production centers to eastern centers of consumption.

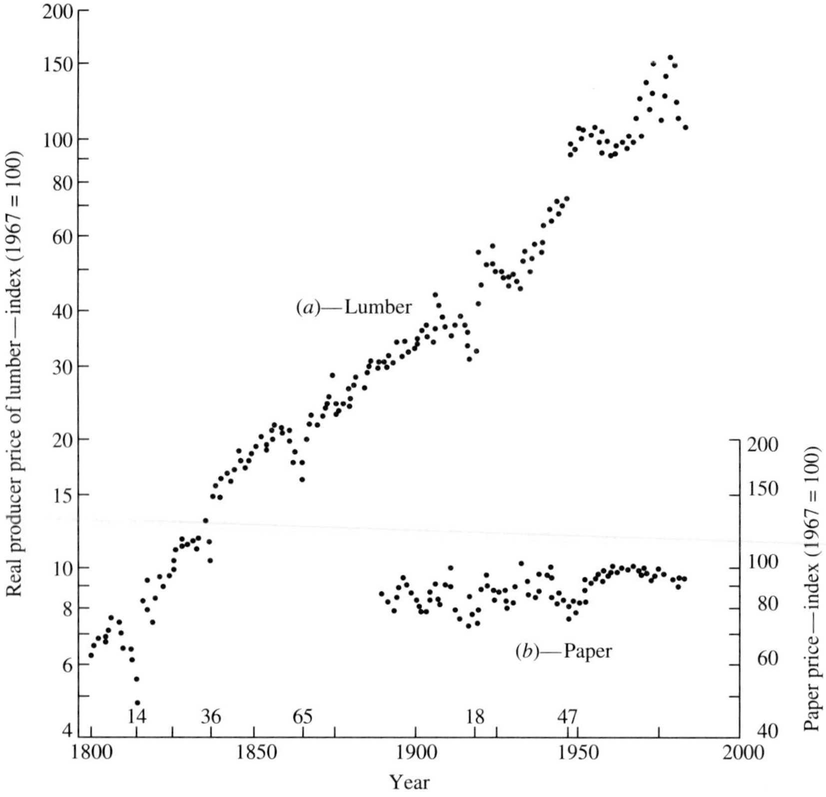

FIGURE 45-1
U.S. average annual real producer price index of (a) lumber, 1800–1982; (b) paper, paperboard, and pulp, 1890–1985. (*Sources:* Warren, G. F., and F. A. Pearson: *Wholesale Prices for 213 Years, 1720–1932;* Cornell University Agricultural Experiment Station Mem. 142, Ithaca, N.Y., 1932; and U.S. Department of Labor, Bureau of Labor Statistics, *Monthly Labor Review.*)

How do we know today that the Compton–Bryant explanation was unacceptable?

Note that both men said that price was going up because cost was going up. They lighted on one particular source of cost; but, in general, their explanation was the same as that offered by almost everyone before and since: the rising cost of bringing lumber from the tree into consumers' hands.

True enough, a few commentators pointed out that price is the upshot of interaction between demand and supply. But what they were talking about were the day-to-day price movements—or the year-to-year movements that are plotted in Fig. 45-1. Over the long haul, price reflects cost. In a surviving industry, it cannot be less than cost. And in a somewhat competitive industry, it cannot be more than cost, counting profit as the cost of retaining entrepreneurial talent.

Just as World War I provoked a flutter of comment about lumber prices, so did World War II. When price controls were lifted after the war and most prices jumped up to fill the vacuum, lumber prices executed a near-record leap. Ellery Foster, a representative of the labor movement in lumbering, decried high relative prices and attributed them to the low efficiency of the industry [3]. He was writing at the close of the peckerwood-mill era (Chap. 33) and was stressing the structural change that the industry had been undergoing. That is, he was stressing *cost* —not, in this case, transportation, but logging and milling. He didn't specify the source of high cost, though he did remark that workers in lumbering were poorly paid.

It was a decade later that an attempt was made at a tight explanation of lumber-price behavior. Joseph Zaremba, in the *Journal of Forestry* [7], pointed out that the trend in gross labor productivity in the lumber industry had long lagged behind the productivity trend in the economy at large. He then argued that high productivity at large had led to generally high wage rates—rates that lumber firms were obliged to pay if they were to stay in business. They couldn't pay these rates out of productivity gains, as most industries were doing. They paid in the only way possible: by raising prices. In a word, a general upward trend in real per-capita income causes real prices to rise in industries that lag technologically. If average annual percentage changes (expressed in decimals) are represented as follows:

In productivity:	
In the economy at large	Q
In an industry	q
In prices:	
Of all commodities	P
Of the industry's product	p

then

$$\frac{1+p}{1+P} = \frac{1+Q}{1+q} \qquad (1)$$

The Zaremba formula says that the rate of change in the real price of an industry's product equals (is explained by) the general rate of productivity change in relation to the industry rate.

If one stations oneself at the point of time when Zaremba was writing and looks back on a Q of approximately 2 percent per year and a q of zero, one finds that the Z formula works fairly well: The upward trend in lumber's real price *was* about 2 percent per year.

> Look up Zaremba's article in the March 1958 *Journal of Forestry*. Try your hand at interpreting his argument, using terms appropriate to the present day.

Again there was a flurry of discussion [6, 8–10], One correspondent asked: What about sawtimber stumpage prices? They have been rising, he said, relative to other prices, just as lumber prices have. (True, indeed, and the trend has continued, as Fig. 45-2 attests.) Isn't raw material one of the mounting costs that lumbermen face? Zaremba had passed lightly over stumpage cost, saying that it was only about 5 percent of the total and had been less than this in the early years.

Zaremba's price data were questioned, just as those in Fig. 45-1 may be questioned. Such long-continued price series have the fault that "lumber" is changing in the meantime, and the mix of "all commodities" is changing. And the technology of lumber consumption is changing—for example, knots may be a defect in one era, but not in another. Thus the base (=100) for computing a price index must be updated from time to time, and this means "splicing" the successive series end to end, with results that may be hard to interpret. Nevertheless, with such a radical performer as lumber price, the gist of the story is unmistakable.

Another correspondent argued that the Z-formula allowed for the effect of only labor cost on product price. The formula assigned no weight to the costs of other inputs such as land and capital, the latter including equipment and working capital as well as raw material.

Let's look at the Z-formula in the following way: Gross labor productivity, as explained at the outset of Chap. 39, is not a measure of labor's contribution to product. It measures total product, however attributable, and expresses it per unit of labor input. In this respect, productivity tells the same story as per-capita income. It tells, for an industry or group, what the output was per unit of input. This unit of input is one worker plus land, capital, and entrepreneurship that are used per worker. Therefore, if I turn to the Z-formula to explain real-price trends in an industry that is average for the whole economy with respect to land, capital, entrepreneurship, *and labor* input costs per worker, the formula should prove out very well. But if the industry is atypical in its labor, land, capital, or entrepreneurship cost per worker, the Z-formula alone will not explain the price trend. I will have to supplement it by identifying the atypical sources of cost. Indeed, turning, now, to two sets of real prices so far unexplored, I find that this kind of supplementing is needed. The two sets are (1) lumber prices in the postwar era of the new technology and (2) prices of paper, including paperboard and pulp.

A TEST OF THE FORMULA

Have a look at the solid and dotted lines in Fig. 45-3, which show labor productivity in the economy at large and in the lumber industry during the years 1947–1980, approximating the era of new technology. For the first time in history, productivity in the lumber industry was rising about as fast as that in all industry, though neither series showed much gain during the final decade. The Z-formula predicts a flat prices trend

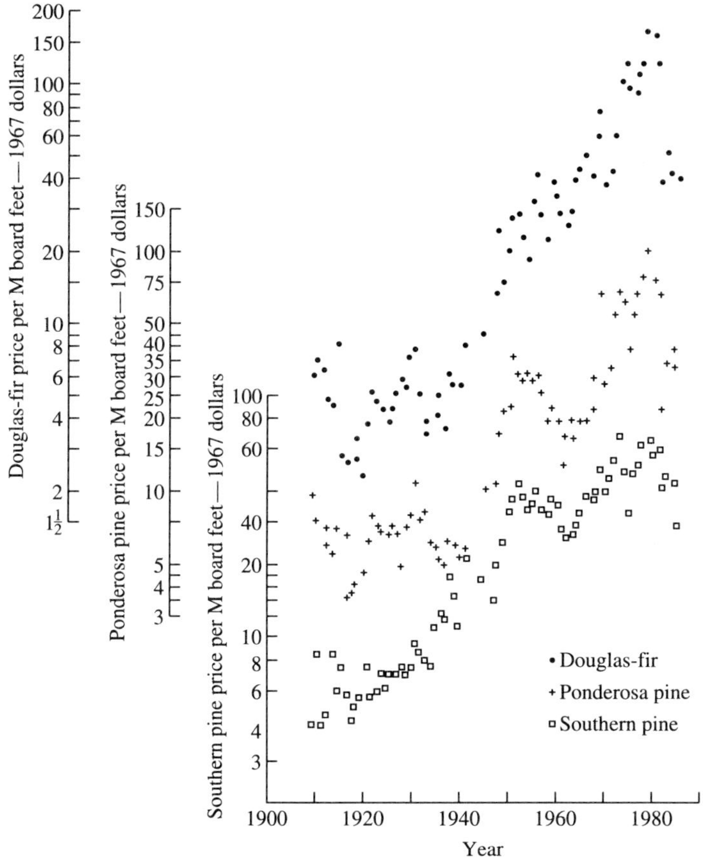

FIGURE 45-2
Real prices of sawtimber stumpage sold from national forests, 1910–1985. *(Sources 1910–1949:* U.S. Department of Agriculture, Forest Service, Forest Resource Rep. 20, 1973, p. 332, adjusted by author to conform to more recent data; *1950–1985:* Ulrich, Alice H.: *U.S. Timber Production, Trade, Consumption, and Price Statistics 1950–1985,* U.S. Department of Agriculture, Forest Service, Misc. Publ. 1453, 1987, p. 28.)

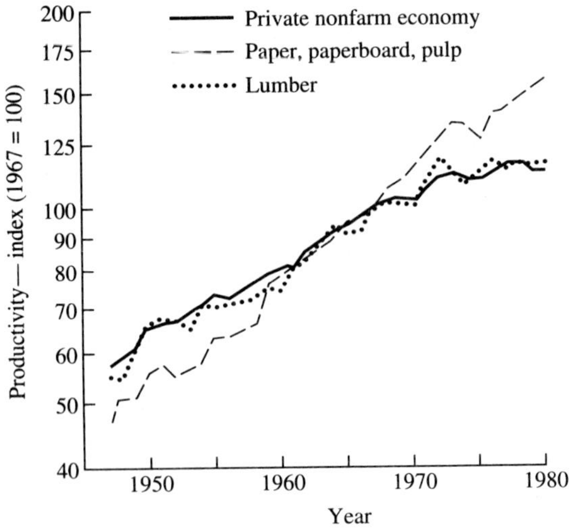

FIGURE 45-3
Gross labor productivity in the private nonfarm economy and in the manufacture of paper and lumber, 1947–1980 (*Sources:* U.S. Department of Labor, Bureau of Labor Statistics, *Monthly Labor Review;* Backman, Jules, ed.: *Labor, Technology, and Productivity in the Seventies,* New York University Press, 1974; Kendrick, John W.: *Postwar Productivity Trends in the United States, 1948–1969;* National Bureau of Economic Research, New York, 1973.)

for lumber, 1947–1980. What does Fig. 45-1 say? At best, it is inconclusive, and one can wish for a longer price record in times of rising lumber productivity before coming to a judgment. Overlooking the four or five boom years in the 1970s, one finds real lumber prices clinging to the 100 index during much of the post–World War II period.

If, in decades to come, lumber productivity holds the national average line while real price resumes its climb, then for an explanation you can look to lumber's uniqueness with respect to some of its inputs, such as stumpage. Douglas-fir stumpage, for example (Fig. 45-2), has shown an upward real-price trend of more than 5 percent for longer than half a century.

> Under what circumstances could you use the Z-formula successfully to explain horizontal trends in *both* lumber productivity and lumber real price?

PULP-PAPER TRENDS

Now, turning to paper, paperboard, and pulp, one steps into a quite different world. The trend in real prices—Fig. 45-1(b)—is amazingly flat. And in recent decades, the annual averages have shown little of the volatility that characterizes the lumber market and that was found even in the paper market before 1950.

In the decades before World War II, paper's price trend conformed to the Z-formula, just as lumber's did. Productivity in the pulp-paper industry and productivity in the economy as a whole were both rising, on the average, about 2 percent per

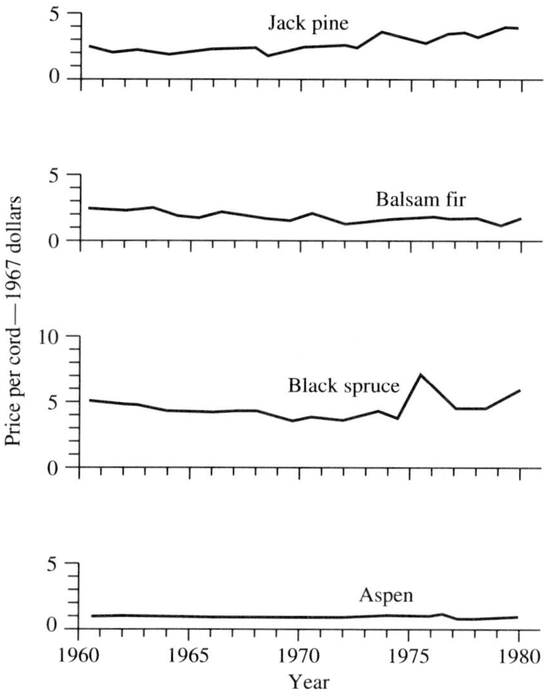

FIGURE 45-4
Real prices of pulpwood stumpage in Minnesota, four tree species, 1960–1980. (*Source:* Lothner, David C., Edwin Kallio, and David T. Davis: *Minnesota and Wisconsin Forest Product Prices, A Historical Review, 1950–1980.* U.S. Department of Agriculture, Forest Service, North Central Forest Experiment Station, Duluth, Minn. 1982.)

year, and the industry held the line on real price. But the war was a turning point. Figure 45-3 says unmistakenly that productivity has been rising faster in the industry than in the economy at large. The Z-formula, then, says that real price has been going down. But it has *not* been going down. In fact, since about 1950, it has taken a firmer grip on the 100 index than it ever did previously.

In partial explanation of lumber's maverick price behavior, we turned to the sawtimber stumpage-cost record. For pulpwood, the record for a long period comes less readily to hand. Nevertheless, some valuable shreds of evidence are available.

Figure 45-4 affords a glimpse of pulpwood stumpage-price behavior in Minnesota in the period 1960–1980. Compare these flat, and in some cases slightly downward trends for the raw material of pulp-paper with the upward trends in Fig. 45-2 for lumber's raw material. Another glimpse of paper raw material is the series of monthly real prices of southern pine pulpwood delivered to the concentration point for shipment to the mill (Fig. 45-5). The data are averages for an 11-state region. They attest to the tight rein held by the industry on its raw-material suppliers. The month-to-month fluctuations are almost entirely the result of changes in the proportions of total wood supply originating in the various sections of the southern states.

The fact of the matter appears to be that the orderly price policies pictured in Figs. 45-4 and 45-5 are those of oligopsonists guided by skillful price leaders. And the policies revealed in Fig. 45-1(b), especially after 1950, are those of an oligopoly with

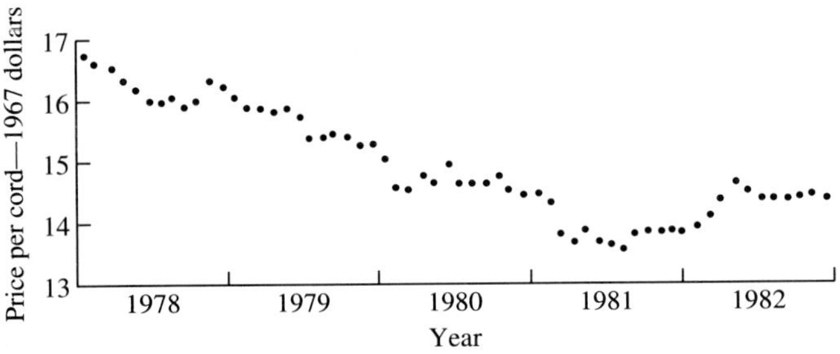

FIGURE 45-5
Monthly average real prices of pine pulpwood delivered to concentration points in 11 southern states, 1978–1982. (*Source: Timber Mart-South,* F. W. Norris, Highlands, N.C. Deflators from U.S. Department of Labor, Bureau of Labor Statistics, *Monthly Labor Review.*)

similarly skillful leadership. The market consequences are understandably a far cry from those of lumber's regional and national auction markets.

What constitutes "skill" in price leadership by an oligopsonist?

SUMMARY

This chapter asks why U.S. lumber prices have shown the trend they have and why the trend differs from the price trend for paper, paperboard, and pulp. The model brought out and discussed is that an industry's product price, in the long term, is a function of the cost of all inputs to production. Cost trends are affected by productivity trends plus the circumstances of industry input that differ greatly from the typical.

REFERENCES

1 Bryant, Ralph C.: *Prices of Lumber,* U.S. War Industries Board Price Bull. 43, 1919.
2 Compton, Wilson: "The Organization of the Lumber Industry," *American Lumberman,* Chicago, 1916.
3 Foster, Ellery: "Lumber Snafu," *Journal of Forestry,* June 1946, pp. 393–400.
4 Goodman, Robert B: *The Price of Lumber,* Yale University School of Forestry, Lumber Industry Series I, 1921.
5 James, Lee M.: "The Trend of Lumber Prices," *Journal of Forestry,* September 1947, pp. 646–649.
6 Settel, Lee S.: "Comments on Stumpage Costs," *Journal of Forestry,* July 1958, p. 517.

7 Zaremba, Joseph: "The Trend in Lumber Prices," *Journal of Forestry,* March 1958, pp. 179–181.
8 Zaremba, Joseph: "Reply to Lee S. Settel," *Journal of Forestry,* July 1958, p. 517.
9 Zaremba, Joseph: "Reply to John A. Zivnuska," *Journal of Forestry,* July 1958, p. 519.
10 Zivnuska, John A.: "Lumber Prices and Labor Productivity," *Journal of Forestry,* July 1958, pp. 517–519.

CHAPTER 46

FLUCTUATIONS IN THE TIMBER-PRODUCTS MARKETS

CYCLES DEFINED

For us as foresters, the idea of fluctuation is altogether ordinary, and the fact of fluctuation is self-evident. In the forest, we observe fluctuations every 24 hours in air temperature, humidity, and movement; light intensity; and biological activity. These are **diurnal** fluctuations. Over slightly longer stretches of time, we observe precipitation alternating with dry weather, windy spells with calm. These are **irregular** fluctuations. Then there are the **annual** cycles that see the coming and going of seasons: warmth and cold, deciduous leaves, bird populations, human visitors, and the height growth and successive rings of the trees. Through such changes as these, the forest extends its waves of influence into the world outside, as when skiers are attracted to the slopes or loggers held off by bad ground conditions or fire fighters called in to meet emergencies during the warm, dry season. And then there are the widely influential fluctuations that reach down through multiples of years—the timber cutting cycles, the rotations, and the long-term shifts in land use.

Influences run, not only from the forest outward, but also from the world or region at large inward through the forestry economy to the forest. Wars are a major influence, and we have already had a glimpse of this. The ups and downs of general economic activity, or **business cycles,** are influential, as are the somewhat longer **building cycles,** which are the ups and downs of the construction industry.

The purpose of this chapter is to pose these questions: (1) What are some of the economic cycles that are keenly felt in forestry and forest industry? (2) What is the cause of the cycles? (3) What is their relation to the forestry economy?

A note about the term **cycle.** A cycle is an up followed by a down, an on by an off, a fast by a slow—or vice versa. Cycles have **amplitude** —that is, severity, or degree of change over time. And cycles have **duration:** the length of time between a given stage in one cycle (say, its **peak** or **trough**) and the analogous state in the next. Some kinds of cycles, such as diurnal or annual, are **periodic:** They have a fixed duration. But many economic cycles are *not* periodic. They are *recurrent but not periodic.*

INTERWAR CYCLES

For the most part, economic cycles are not objective phenomena like rainfall and sunlight. They are, in a sense, creatures of the human mind. They exist by virtue of being discovered. Look again at Fig. 45-1(a). Do you see cycles there? I see a series of six episodes, which I might as well call cycles. Each begins with an explosion in lumber's real price that sets it apart from the preceding cycle and ends, with one exception, in a period of sagging real prices. The ending and beginning dates, are

Cycle Number	Begins	Ends
1	—	1814
2	1815	1836
3	1837	1865
4	1866	1918
5	1919	1946
6	1947	—

Now that I have written these dates (and located them along the horizontal axis in Fig. 45-1), I have a theory about my cycles. With one exception, each ended in the aftermath of a war. The general price level was inflating, but the price of lumber, not a scarce military essential, lagged, so that lumber's real price fell off. When the economy returned to normal, lumber's real-price behavior did likewise. The cycle that ended with the Civil War illustrates the pattern, here somewhat enlarged beyond the 45-1 scale:

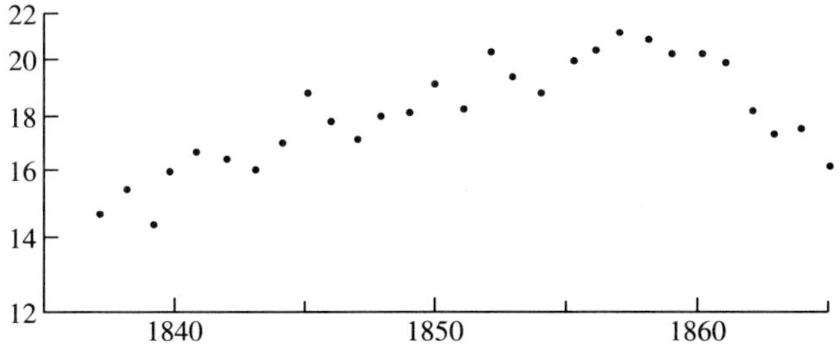

In the cycle that came before this one, the 1834–1836 business boom played the part of a war in depressing the real price of lumber. In the cycle that ended in 1946, wartime price controls prevented a lumber downturn.

How could a business boom depress the real price of lumber? Why did price controls, intended to prevent general inflation, prevent a downturn in lumber's real price?

As a group, the interwar cycles provided the means for implementing lumber's long-term price runaway. The runaway was accomplished in a series of episodes, each separated from the next by a pause and then a headlong leap to catch up.

CYCLICAL BEHAVIOR

Of course, if I think I see a cycle somewhere in the statistics, this alone doesn't make it a cycle. There must be a rationale, a recognizable and recurring sequence of events, an explanation at least implied, and some usefulness toward understanding the ways of the world.

Consider the business cycle. Of all economic cycles, it has been given by far the most attention, for it affects everybody. It has been studied by a multitude of scholars, who have worked out careful definitions of it. For example,

> Business cycles are a type of fluctuation found in the aggregate economic activity of nations that organize their work mainly in business enterprises: a cycle consists of expansions occurring at about the same time in many economic activities, followed by similarly general recessions, contractions, and revivals which merge into the expansion phase of the next cycle; this sequence of changes is recurrent but not periodic; in duration business cycles vary from more than one year to ten or twelve years; they are not divisible into shorter cycles of similar character with amplitudes approximating their own [1].

By following that definition, scholars have developed a business-cycle chronology dating from 1843, and there is amazing unanimity on the dates. The average cycle in these 150 years had a duration of 51 months [2].

All economic cycles of variable duration have, at bottom, the same causes: growth of the economy, requiring firms to take new decisions over time, lags between taking decisions and bringing them to fruit, deficiencies in human foresight and wisdom, the tendency of individuals to follow the crowd, and the crowd's tendency to overreact. Consider the behavior of a yard wholesaler of lumber. Early in Chap. 40, the firm is seen as managing its stock so as to equate marginal revenue with its guiding rate of return.

One can well believe that this is the firm's aim. But look at what may happen to disrupt the smooth flow of the firm's business over time. A developer-builder starts a project in the community and chooses the firm as its lumber source. Foreseeing, after more or less delay, a heavy drawdown of stock, the firm places orders with its supplier, a big office wholesaler, who, let's say, fills this and other new orders by calling upon mills that must acquire and log the necessary stumpage. Aware of the lags in the system and foreseeing that the heightened business activity will continue (a "persistence forecast," page 240), the yard wholesaler enlarges its orders. Meanwhile, its stocks are falling. To avoid the risk of losing the developer's confidence, the firm buys lumber from various local sources.

After a while, the enlarged orders from the central sources are being filled. The developer stops buying lumber for the project and doesn't start another. Now the wholesaler's inventory hits a peak, and so does the carrying cost. Orders come in regularly, but they are comparatively small and times are hard. The firm stops buying lumber and seeks to reduce stock. In all likelihood, it may carry the reduction to a point well beyond the optimum. One **inventory cycle** has been completed. If there had been no conspiracy of circumstances such as those mentioned above, there would have been no cycle.

> If firms had perfect information about current events and if they had perfect foresight, would economic cycles be eliminated? Would they all be eliminated, or only certain kinds? What kinds?

Two points about the cycle just described hold particular interest. One is the possibility that a small fluctuation in activity at the consumer end of the forestry economy will make large fluctuations at and near the forest end. Thus a 5-percent increase in wholesaler activity can easily lead to building new mills and opening up timberlands hitherto inactive—both very steep increases. And decreases work on the same scale. Such magnification of changes in consumer spending illustrates the operation of the **acceleration principle.** The tendency is for genetic and primary manufacturing production to feel the brunt of fluctuations in ultimate demands. The demand for the output of genetic and primary activity is derived from these ultimate demands.

The second point about the inventory cycle just described is that under some circumstances a change in investment produces a change in employment and income that changes consumer spending more than proportionately. Thus the workers laid on in the woods and mills by wholesalers' new orders may want new housing that will sustain wholesale activity and many other forms of activity in support of ultimate demands. Here one sees the working of the **multiplier principle** and sees that cyclical expansion and recession both tend to be self-perpetuating.

> Can you draw a clear distinction between the acceleration principle and the multiplier principle? Imagine that you are at work writing a textbook; have a try at explaining the distinction in writing.

All in all, inventory cycles in lumber and other materials are characteristically short—say, a couple of years at most from peak to peak or trough to trough. This is

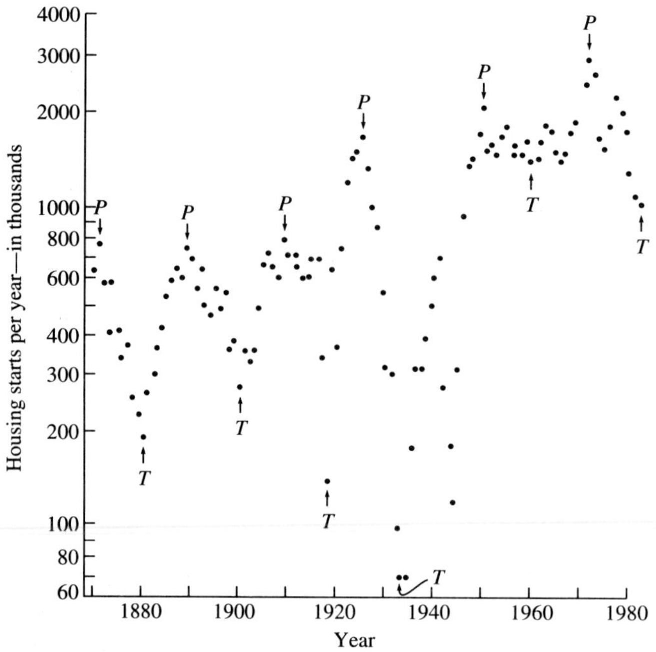

FIGURE 46-1
Annual numbers of housing starts in the U.S., 1870–1982, identify the peaks (P) and troughs (T) of $5\frac{1}{2}$ building cycles. (*Sources:* Long, Clarence D., Jr.: *Building Cycles and the Theory of Investment,* Princeton University Press, 1940; Marcin, Thomas C.: *Population Change as Related to Long-Term Cycles in Residential Construction in the United States,* U.S. Department of Agriculture, Forest Service, Res. Pap. FPL 392, 1981; U.S. Department of Commerce, Bureau of Economic Analysis: *Survey of Current Business.*)

because lags in inventory acquisition are commonly minimal, and because inventory itself is not durable: It can be turned over and liquidated.

A longer cycle in which timber resources are closely involved is the building cycle. The $5\frac{1}{2}$ cycles that occurred in the United States between 1871 and 1981 (Fig. 46-1) had an average duration of just 20 years. Actually, the cycles represented in the figure are *housing* cycles, involving single and multifamily dwellings, but not other kinds of buildings. They are long cycles because houses are durable: Once the demand is filled (overfilled), the supply is there to stay for quite a while. Beyond this, during the cycle's long recession, the building industry becomes disabled, especially in its force of skilled workers, and requires time to get under way again. Recent examples of housing recessions are the precipitous declines to 1933 and 1943 and the continuation of the last drop-off shown in Fig. 46-1 to another record low in the early 1990s [3].

The longest economic cycle in forestry is little understood and is not usually thought of as a cycle. It is the fluctuation from forest resource abundance, through depletion and rebuilding, back to forest prosperity. We have almost no examples of a full cycle in the United States (the southern pine region being perhaps the exception) and must turn to Europe. One thinks of some of the Scandinavian nations as illustrative. The cycle can measure well over 100 years, and it is highly variable from case to case.

FORESTRY ILLUSTRATIONS

In every region of the developed world, each sector of the economy is fluctuating at a pace that is partly its own and partly enforced upon it by other sectors. The whole economy is in motion. One may think of the business cycle as embodying the aggregate of fluctuations. If the chronologies of the components differ greatly, the aggregate is a mild cycle, such as were many in the United States after 1950. If, on the other hand, major elements of the cycle reach peaks or troughs at the same time, the result is a severe cycle. The downswings that all struck bottom in the early 1930s provided our most extreme example. Every cycle is unique—in timing, and in the identity of its dominating elements.

Illustrative of fluctuations in which forestry is heavily involved are those of the years 1978 through 1982. The period ended at or near the trough of a cycle in general business and in housing construction. Some indices are plotted month by month in Fig. 46-2. It was a period marked by some of the highest interest rates in U.S. history. Rates on bills (bonds) sold at auction by the U.S. Treasury for 3-month financing are plotted in Fig. 46-2(a). Rates on home mortgages, which are used in the purchase of almost all private housing, were higher yet, hitting 20 percent in 1980 and 1981 and deepening the depression in the construction industry; see Fig. 46-2(b). However, toward the end of the period, interest rates were declining far enough to produce what appeared to be a trough in the building cycle in October 1981, at a **seasonally adjusted annual rate** of 854,000 starts.

(A seasonally adjusted annual rate of output works like this: If an industry exhibits no annual cycle in production, so that with a horizontal trend, each month's output is $\frac{1}{12}$ of the yearly, any monthly figure can be converted to an annual basis by dividing it by .0833. . . . But suppose there's an annual cycle, in which, say, July, output averages 20 percent of the year's total. Then the seasonally adjusted annual rate in July is the monthly rate divided by .2.)

The downswing in the building cycle forced down the real price of a principal building material, Douglas-fir lumber [Fig. 46-2(c)]. [Here one gets a little insight into the monthly behavior of lumber price in some of the more recent years plotted in Fig. 45-1(a).] Douglas-fir lumber production fell away [Fig. 46-2(d)].

Worsening times worsened the situation of labor in the forestry economy. In the Eugene-Springfield area of Lane County, Oreg., in the heart of Douglas-fir country, where lumber governs economic life, unemployment rose from less than 6 percent of the labor force to more than 14 percent [Fig. 46-2(e)]. (The **labor force,** in U.S. parlance, is everyone 14 years or older who has or is actively seeking paid employment.)

> Trace the flow of cause and effect—in your own mind or on paper—from the national or world economy to the economy of the Eugene-Springfield area in the years 1978–1982.

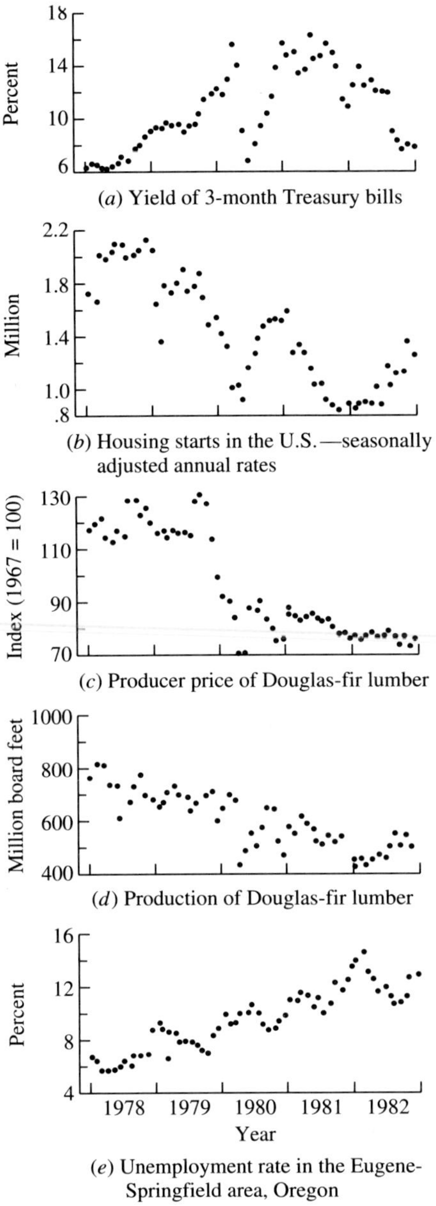

FIGURE 46-2
Some impacts of recession on the forestry economy of the Douglas-fir region, monthly data, 1978 through 1982: Unemployment rate in the Eugene-Springfield area, Oregon. [*Source* (a), (b), (c), (d): U.S. Department of Commerce, Bureau of Economic Analysis: *Survey of Current Business;* (e): U.S. Department of Labor, Bureau of Labor Statistics: *Employment and Earnings* (monthly).]

SUMMARY

A cycle is an up followed by a down. It is recurrent, but many economic cycles are of variable duration. The discovery of an economic cycle is useful, because it increases understanding of economic life. Among economic cycles related to forestry are a very long exploitation-conservation cycle, a long interwar cycle in lumber price, a moderately long building cycle, a moderately short business cycle, and a short inventory cycle.

REFERENCES

1 *The Business Cycle and Public Policy, 1929–80,* a compendium of papers submitted to the Joint Economic Committee, Congress of the United States, November 28, 1980. 96th Cong., 2d Sess., Joint Comm. Print, Washington, 1980, p. 20.

2 Moore, Geoffrey H.: "A Long-Run Look at the Business Cycle," *In The Business Cycle and Public Policy,* op. cit., p. 21.

3 *The Wall Street Journal,* Eastern ed,. January 27, 1992, p. A1.

CHAPTER **47**

APPRAISING FOREST RESOURCE VALUE

THIS CHAPTER'S QUESTION

The problem to be solved is this: Here is a forest resource; how much is it worth? The chapter mentions some different situations in which an estimate of value may be wanted and describes some models. It aims to acquaint you with appraisals, not to make you an appraiser.

The worth, or value, sought is market value or some counterpart of market value. In other words, it is objective value (**value in exchange,** to use the classical term), as distinguished from the subjective value (**value in use**) that is a topic in Part Two. There, the emphasis is on demand. Here, as it were, supply also is introduced. The outcome is the flame ignited in the meeting of these two.

Value, remember, is the quality of an object or action or idea that makes it a resource, exchangeable with other resources in the realm of possession. Here *possession* includes not only ownership in the ordinary sense, but also access to enjoyment—for example, enjoyment of scenery (page 79). The chapter inquires about the magnitude of a resource value: the quantity of the other resource that may be exchanged for this one.

Recall how we view the market and market value, at times narrowly, at times broadly. In the last analysis, markets are circumstances in which choices are made: the stock market, the log market, the legislature, the courtroom, the manager's desk. . . . In each circumstance, alternatives are weighed, decisions made, and values determined or implied. The assignment in this chapter is to look to the market in whatever sense is appropriate to the question. We look to the market to find the value of the resource in question.

Remember that the magnitude of a resource value depends on the *predictions* of the persons who are making the market decisions. It depends on their *anticipations* of benefits and penalties from the resource. The value of a forest depends on the net benefits that the market decision makers *expect* from the forest. And the value of any output from the forest is determined by *forecasts* of its net benefits. One cannot remind oneself too often that the past is relevant only as a guide to the future and that the present is only a dimensionless point of time.

WHY RESOURCES ARE APPRAISED

Why appraise the value of a forest resource? Appraisals are made to guide market decisions—that is, choices among alternatives. Indeed, all decisions involving alternative values—*all* decisions—are based on appraisals, even though often highly informal. If I plan to buy or sell a resource, or if I plan to pay or be paid for its use, I want an appraisal as a guide to a fair (or unfair) deal. The tax assessor who has been instructed to estimate the "fair market value" of property has the same want. The alternatives in such cases are other comparable resources and other comparable values.

Again, if a resource is lost, I seek an appraisal so as to repay the loss or be repaid. Or if a resource is damaged and a decision is to be made about it, I want an appraisal. If I contemplate a resource investment and have choices of the kind, size, location, or other characteristics of the investment, I estimate costs and revenues to try to identify the preferable resources and values. Which factory site or forest site is best? What uses of a forest are best? Which management procedures are best? Where there are alternative resources and values—alternative objects or actions or ideas that have market value or some counterpart of it—appraisal is wanted.

The alternatives in question are simultaneous. Or if not simultaneous, the appraiser is obliged to make them so, perhaps by methods such as are sketched in Chap. 19. Otherwise the resources and values are not comparable. For instance, the value of a loss is *not* the resource value before the loss minus its value afterward. This is a *change* in value—over time. The appraiser wants a *difference,* Δ: what the resource would have been worth now in the absence of loss minus its actual worth now. Refer to page 199. Or again, what is the value of this water source, wildlife population, or other subset of a forest property? It is the value of the property including the subset minus the value without the subset, both estimated as of the same point in time.

> To view appraisals as guides to choice may seem to involve a contradiction. For if markets are circumstances in which choices are made, and appraisals are guides to such choices, of what use are markets? Why not just let the appraisers make the decisions?

NEED FOR CLEAR DEFINITIONS

Since resource appraisal derives from and leads into the comparison of alternatives, a common unit of measure is wanted for appraisals. There is little question that the unit sought is the dollar. The value we seek is money value. Some markets determine an **explicit value** in money terms. Examples: the markets for land, stumpage, pulpwood, forage, and the like. (Explicit money value per unit of a resource is the **price** of the resource.) Other markets determine only an **implicit value,** expressed in terms of some counterpart of money. Value, in this case, is implicit in that it is implied in peoples' behavior as barterers. Our task then is to search for the money equivalent that interprets the barter process. (Implicit money value per unit of a resource is the **shadow price** of the resource.) Example of implicit value: The legislature sets aside a tract of public timber as wilderness. In doing so, the legislature implies that the value of the wilderness resource is no less than the explicit value of the timber forgone in the set-aside.

(Obviously there may be a dilemma here. Suppose the legislative committee, before reporting out a wilderness bill, turns to you as a consultant. They ask you what the value of the proposed wilderness is and how it compares with the values that would be sacrificed. You could tell them that this is precisely the decision they are making. Or you could search for an alternative approach, one that uses "objective" evidence and escapes circularity.)

Again, since resource appraisal involves comparing alternatives, there is clear need to specify precisely what resource or combination one is trying to appraise. It is, for example, the timber on a forest tract? All, or just part of it? The substance of it, or some idea about it, such as silvicultural system? Or, is it the land alone? Or, perhaps the forest as a whole, with all its potential uses, its road network, and an adjacent wood-processing plant? This depends, of course, on the alternatives. If you are the tax assessor, you want the value of the entire property. However, if the owner asks you as a consultant what is to be gained by selling hunting rights in season, then you want to estimate the value of such rights and the value of other resources that will have to be given up if hunting rights are marketed.

Further, you want to specify precisely from whose viewpoint an appraisal is to be made. A question often asked concerns the value of a visitor-day of recreation. Who is asking the question, and what purpose will the reply be put to? Is the question being asked by a private landowner who wants a shadow price that will help set an admission fee? Or, is the questioner a public agency that seeks to learn what it can justifiably pay for the tract as a park? In the latter case, a social value is sought. The question falls in a realm sometimes termed **welfare economics.** Most of the models for answering questions about social value differ from those applicable to private value, though the underlying theory may be the same.

The public and the private viewpoints commonly differ with respect to the range of resources they encompass. A public agency planning a timber sale may be obliged to predict its influence on forest accessibility, community income and well-being, recreation values, and water and air quality. A private firm, on the other hand, may shrug off the side effects as **externalities,** considering them only as they may bear on public relations and thus net revenues. Yet while the public firm and the private firm are looking at quite different sets of values, both can have the same general aim of efficiency in the sense of maximizing the excess of positive over negative results.

Accounting for social benefit and cost involves specifying the public in question, and this can be a perplexing assignment. Who are these people, and in what dimensions are resource benefits and costs distributed to them—in the geographic dimension (and, if so, across how large a region)? among interest groups? over time?

IMPLICIT VERSUS EXPLICIT VALUES

When one considers the wide variety of resources subject to appraisal, one is impressed by the number of items that have implicit, but not explicit, value. Whenever people who are interested in the economic aspects of forestry get together to discuss what topics they are eager to learn more about, they almost invariably give implicit values prominent billing. This has long been so [7], and one is tempted to add that it may always be so. The argument is that social value information is wanted for efficient, equitable resource management: for budgeting, planning, and resource allocation. Indeed, social value appraisals for forest and rangeland resources are now required by federal law. The acts in question are the subject of a later chapter, on social planning for forestry. How do we know what timberland acreage to set aside as wildlife preserve or how much old-growth Douglas-fir to give over to the spotted owl or how far to go in creating hardwood cutovers for deer habitat or how many den trees to leave unless we can express the value of the animal and timber resources in commensurable terms? The question has at least two sides.

1 Ours is a number-loving culture, striving to express ideas in numerical terms. The essence of sound resource allocation, in our culture, is to quantify the alternatives. We reject the idea that appraising implicit value is either hopeless or useless. Certainly we delight in its intellectual challenge. Interest in the subject is growing. This is partly because of the vogue among public executives and legislators to demand evidence (predictions) of the cost-effectiveness of proposed government projects. The federal Forest Service used to have a task force for appraising forest resources other than wood and forage so as to have guides to multiple-use forestry. Now there is a team at the Rocky Mountain Forest and Range Experiment Station, Fort Collins Colo., designated to study methods for appraising implicit values. (References 3, 12, and 18 cite some of the output of this team.) The University of Arizona used to have a center for the Assessment of Noncommodity Natural Resource Values. The Center's first workshop, attended by academic and federal personnel and reported on in 1980 [17], took up the topic of wildlife values.

2 The other side of the matter is that even in our society, the vast majority of resource-allocation decisions are made without objective measures of the values at stake. Indeed, comparatively few of our social markets, and certainly few of our private markets, use financial units to weigh the things exchanged. Each of us as an individual transacts most everyday affairs in a currency of smiles, pleas, threats, and kicks in the shins. Or we may trade in the political market, where the medium of exchange is an appeal to loyalty, patriotism, generosity, and other human virtues.

> Stand back and look, for instance, at U.S. wildlife resources and then at timber resources. Has our precise knowledge of timber values put our timber affairs in better shape than our wildlife affairs? Explain.

THREE KINDS OF UTILITY

The value of a resource arises in at least three kinds of utility, or service:

a-service: the opportunity to own the resource or use it directly: to own the house built of wood products or to live in it as a tenant, to use a recreation site, to shoot birds and take possession of them or simply to hunt them.

b-service: the opportunity to sense the resource: to admire the house, to watch and listen to birds.

c-service: the opportunity to think about the resource: to find satisfaction in the existence of timber and wood products, birds, and other wildlife.

The wildlife literature has a common expression. The use of a-service is referred to as the "consumptive use" of wildlife, and the enjoyment of b- and c-services is called "nonconsumptive use." These are terms to be avoided. Using any of the three categories of service is an act of consumption. The misunderstanding arises in thinking of consumption as using *up,* an issue reviewed on page 56.

The value of resource *R,* then, is the power of its a-, b-, and c-services to substitute for those of resource *S* for which it exchanges. It follows that in order to appraise a resource value, one observes what amounts of other things people exchange for the resource.

Let's emphasize that it is indeed actual instances of exchange that are to be monitored, not someone's view of what "ought" to be. "Oh, the privilege of photographing wild zebras ought to be worth *much* more than that!" Or again, "In a competitive market, the pulpmills would be paying a *much* higher price for wood!" This chapter's interest is in objective value, a positive (nonnormative) concept. The chapter looks at the market, not to weigh its faults and virtues as Chap. 43 does, but to study its value outcome.

TRANSACTIONS EVIDENCE VERSUS ANALYTICAL APPRAISAL

What is the value of a tract of stumpage? It is the sum that the buyer is paying the seller, interpreted in view of all the constraints that buyer and seller have agreed to accept. Or it is the sum per unit of measure, a price expressed either as a single average or as averages for categories of timber.

But suppose the stumpage hasn't been exchanged yet. An appraisal is wanted to guide buyer and seller. In this event, two methods are common. Both are used for private and for public stumpage. One is to observe what prices were actually consummated in recent sales of stumpage like that at hand. This is certainly the straight forward approach. It is called appraisal from **transactions evidence** and is detailed in Chap. 49.

Transactions-evidence appraisals are used for forest resources other than stumpage—indeed, for any resources that have explicit value. Real estate is the most common case: land and vegetative cover and buildings. The real value of a property is appraised as the sum—say, per acre—for which it and comparable properties changed hands recently, with adjustments for departures from comparability. Such departures are inevitable if there are differences in land use. They may be occasioned, too, by the passage of time and by differences in land productivity (for instance, site quality), accessibility, distance from markets, topography, and other circumstances.

We can find still other examples of transactions-evidence appraisals in forestry.

When a loss, as of a structure by fire, is estimated as the cost of replacement, transactions evidence is called upon. It has been used, too, in appraising urban forests and trees [1], the views observable from urban residential lots [13], and also water in some situations [3]. Range forage may be appraised from transactions evidence. The U.S. Department of Agriculture reports annually the unit prices of forage sold for grazing on private land [19].

In some cases, stumpage is so variable in ways that affect its value—species, stocking, quality, total quantity for sale, physiography, roading, distance from market, and so on—that finding a counterpart or simulating one by regression analysis was, in the past, unreliable. And it may be shunned today by force of custom. In such cases, stumpage value is simulated indirectly, in an **analytical appraisal.** The appraisal begins with a forecast of the "selling price" (market price) of the product into which the stumpage will be converted—say, lumber. It then subtracts a prediction of "conversion cost," mainly logging, transportation, milling, and selling. The remaining "conversion return" is divided into two parts, one for the seller as stumpage, the other for the buyer as a profit margin. The simplest procedure is to divide the conversion return 50–50. A fuller description is given in Chap. 49.

(Models resembling the analytical have been used for appraising resources other than stumpage—for example, water for agricultural irrigation in Arizona [3].)

> Why not appraise stumpage at the cost of producing it? After all, sellers have nurtured their crop and paid taxes and other costs. If people want the timber, should they not reimburse the sellers?

The prime cause of stumpage supply's price inelasticity is the overwhelming dominance of the stock-supply condition, so that quantity is insensitive to price. For stumpage, short-run responses, which *do* heed price, take a long time to materialize, since they must be generated by such inputs as fertilizer and fire-control effort. And as for cost of production, there is no reason to suppose that it could be effective in regulating the flow of stumpage into the world the way the flow of processed products depends on whether prices are expected to cover costs. Stumpage growers as such don't retrench or go bankrupt if the business isn't profitable (Chap. 25). Public forests stay in business willy-nilly, the stumpage factory humming along merrily and undaunted, and so, for the most part, do private forests.

In recent years, federal agencies, notably the Forest Service, have been scolded by environmentalists and others for selling stumpage at less than cost [1]. *Cost,* here, is not production cost, but the cost only of sale preparation, especially road building. The issues, though murky, are clearly not those touched upon in the preceding paragraphs.

Whether a transactions or analytical appraisal is made, the value of stumpage marketed is the **liquidation value** of the stumpage. Before stumpage is sold from the tract, the whole forest may be appraised by capitalizing anticipated net revenues to estimate forest expectation value. The difference between this value and the forecast expectation value of the forest with stumpage removed, or of bare land, is the **stumpage expectation value.** Comparison between stumpage's expectation value and its liquidation value helps reveal if the stumpage is financially mature. This is the same financial-maturity model as in Part Three, only with different terminology.

TRAVEL-COST MODEL

Now we move into a field that has been termed "amenity resource valuation." The field concerns placing money values on resource that do not pass through markets in the usual sense [14].

Consider a case where a resource is exchanged for money, but not directly as in the foregoing seller–buyer examples. The money passes partly or entirely into the hands of third parties. The case concerns the value of forest recreation. The model is widely usable with other resources—for example, wildlife and even timber. It was introduced by Harold Hotelling [11] in 1947 and has been reviewed, improved, and elaborated by a host of scholarly successors [6, 8, 20], notably Marion Clawson [5]. It's called the **travel-cost model.** We will follow a path traced out by Cesario and Knetch [4] in 1976.

Picture a state park, Forest Park, where people from surrounding territory may come to enjoy, without charge, **day recreation,** notably picnicking, scenic walking and hiking, photography, bird-watching, and the like. They travel from home and return on the same day.

You are commissioned to estimate Forest Park values. You station yourself at the park entrance. As each carload of recreationists arrives, you learn how many adults and children are in the party, where they live, and other information to sharpen your analysis. After a sampling of the hours of the day and the days of the year, you are in a position to construct a year's estimate of park attendance: numbers of persons by zones of round-trip travel distance, the same sort of function as that produced by King on hunting small game in Missouri (Fig. 13-4). With information on cost of automobile travel per mile and marginal outlays per person for supplies and equipment, you convert the vertical scale of your graph from miles to out-of-pocket dollars. You now have a first insight into the amounts that visitors have stood ready to pay for the park-recreation experience—that is, the subjective values of the experience.

But the scholars emphasize that there is another cost of travel besides the out-of-pocket cost: the cost of time spent in traveling. One school of thought relates this cost to the wage rate in the community where the traveler lives. From studies of commuters' time costs, they estimate that adults incur costs equal to perhaps half the wage rate in their home community; children a fraction of this. You adopt in your analysis a time-cost allowance of 4 cents per minute for adults, 1 cent for children.

Now you can estimate the demand for the entire recreation experience: cumulative number of visitor-days in relation to travel cost consisting of out-of-pocket plus time cost (Fig. 47-1). For simplicity, the demand curve is shown as a straight line. The point at which the line strikes the horizontal axis is your estimate of total park attendance for the year. The intersection with the vertical axis marks the highest travel cost that anyone has stood ready to incur for recreation at Forest Park.

What is the social value of the entire recreation experience involved in the use of Forest Park? Observe that the market for this use is not a single market, but a multitude of markets. Indeed, each carload of visitors (carload C) has entered a separate market, conceivably different from all the others. The demand in the market is found on the function in Fig. 47-1. The supply is a perfectly elastic horizontal line signifying that the state park agency has stood ready to furnish an unlimited number of visitor-days of recreation opportunity at Forest Park at the carload-C price. Value in carload-C's market is read at the point on the demand curve where supply intersects it. Total value is the sum of carload-C values—that is, the area under the demand curve. This is an annual value, or rental, subject to capitalization as discussed in Chap. 19.

The foregoing total value is the social value of the entire recreation experience involving Forest Park: anticipating the experience, traveling to and from the park, enjoying the park itself, and reliving the experience afterward. It calls for amendment by (1) subtracting the social marginal cost of providing special park services such as upkeep and trash collection and (2) adding the marginal benefit of park use to nonusers.

Observe that the unamended social value of the entire recreation experience is numerically the same as the total consumer surplus in Fig. 47-1. But the two *concepts* are different. Social value as we define it is the sum of all carload-C values, each determined by supply and demand in its own market. Consumer surplus is what consumers

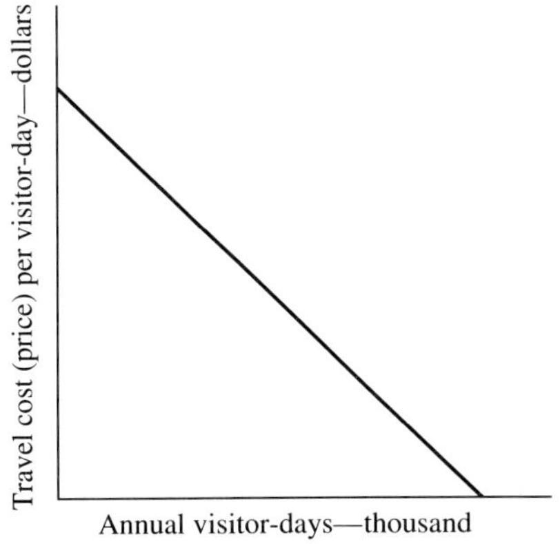

FIGURE 47-1
Aggregate annual price-demand for the entire recreation experience involved in the use of Forest Park (idealized draft).

stand ready to pay for the recreation opportunity, over and above what the state charges them for it.

> Do you agree that the two concepts are different? Explain why or why not.

FURTHER USES OF THE MODEL

At this point, we can move readily into the other question answerable with travel-cost analysis: How much *can* the state charge for the use of Forest Park? Or better, what is the social value of Forest Park? How much is it worth to the state to have this park rather than not have it? This is the question that the state might want answered in order to compare the park use of this or comparable land with alternative uses that would yield money revenue. What site should be selected for a park? What fee should be charged for its use? What would be the related effects of new highways?

To get at the foregoing questions, you are asked to assume that if the state should charge a park admission fee, people would respond to it in their attendance just as they respond to the cost of travel. In other words, you are asked to assume that if the state charges $1 per person *admission cost* to Forest Park, visitors from each travel-cost zone around the park will alter (reduce) their attendance per hundred population in the zone to the same number as the no-fee attendance from the zone where *travel cost* is $1 greater per person.

The calculations are in Table 47-1. Six $1 travel-cost zones are recognized (column 1). For each, your estimate of annual visitor-days such as might be read in Fig. 47-1 is set down in column (2). Column (3) is the total population of each zone, and (4) is visitor-days in percent of population—that is, frequency of use of the park. The remaining columns estimate the number of visitor-days demanded (attendance) in relation to entrance fee charged per visitor. Column (5) repeats column (2) and says that if price (entrance fee) is zero, the quantity demanded is 15,000 visitor-days per year.

We can think of each column (5) figure as the product of columns (3) and (4). By the same token, if we want to estimate quantity demanded at a $1 price, we assume that the 10,000 people in zone (a) will attend the park in the same proportions as zone (b) people when there is no fee—that is, 10,000 × 13.3 percent = 1,330 [column (6)]; 30,000 ×. 06 = 1,800; and so on in column (6), which shows total annual attendance dropping to 5,350 visitor-days per year on account of the $1 price (entrance fee). We make this calculation under the assumption, stated earlier, that people respond to a fee just as they do to travel cost.

> Trace for yourself the calculation of the numbers in columns (7) through (10).

TABLE 47-1
CALCULATION OF AGGREGATE PRICE-DEMAND FOR DAY USE OF FOREST PARK [LINE (G) IN COLUMNS (5)–(10)] FROM HYPOTHETICAL DATA INFERRED IN FIG. 47-1 [COLUMN (2)]

Round-trip travel-cost zone, dollars per visitor (1)	No. of annual visitor-days (2)	Total population, in thousands (3)	Ratio of visitor-days to population, % (4)
(a) 0.01–1.00	5,000	10	50.0
(b) 1.01–2.00	4,000	30	13.3
(c) 2.01–3.00	3,000	50	6.0
(d) 3.01–4.00	2,000	70	2.9
(e) 4.01–5.00	1,000	90	1.1
(f) 5.01 or more	0		0.0

Round-trip travel-cost zone, dollars per visitor (1)	No. of annual visitor-days when entrance fee per visitor in dollars is					
	0 (5)	1 (6)	2 (7)	3 (8)	4 (9)	5 (10)
(a) 0.01–1.00	5,000	1,330	600	290	110	0
(b) 1.01–2.00	4,000	1,800	870	330	0	0
(c) 2.01–3.00	3,000	1,450	550	0	0	0
(d) 3.01–4.00	2,000	770	0	0	0	0
(e) 4.01–5.00	1,000	0	0	0	0	0
(f) 5.01 or more	0	0	0	0	0	0
(g) All zones	15,000	5,350	2,020	620	110	0

The demand function represented in line (g) is graphed in Fig. 47-2. This is the annual aggregate price-demand for day admission to Forest Park. The social *value* of the park, by the same reasoning as for Fig. 47-1, is the area under the curve. This, like the other, is an annual value, or rental, subject to capitalization by the methods of Chap. 19. Again, like the value in Fig. 47-1, it calls for amendment by subtraction of park maintenance costs and addition of marginal revenues to nonusers.

> Suppose that you calculate the data for Table 47-1 and Fig. 47-2 on the assumption that children are charged an admission fee one-half of that for adults. Will you expect to come out with the same or a different result? Why?

A travel-cost approach may be taken to situations other than the flow of scattered visitors to a concentrated attraction. For instance, refer again to Fig. 13-4, concerned

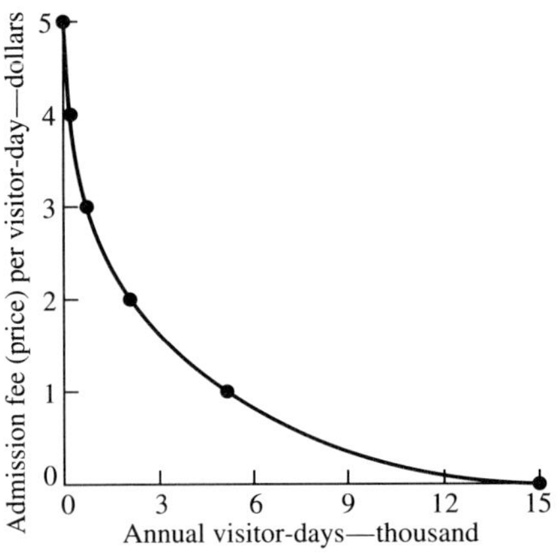

FIGURE 47-2
Aggregate annual price-demand for admission to Forest Park. (Basis: Table 47-1.)

with small-game hunting in Missouri. Further examples are big-game hunting [12] and both cold-and warm-water fishing [18] in Idaho.

A number of weaknesses mar the beauty of the travel-cost model. Here are a few of them:

1 The model accounts for the values of a-service and b-service, but cannot appraise c-service to nonvisitors.

2 There is no clear way to account for the *benefit* that may accrue to the visitor in the travel between home and tract. Travel is part of the whole experience. Even some of the *cost* (benefit?) of travel, such as that of time, is hard to estimate.

3 The model disregards the fact that a visit to the resource area may be incidental to travel for other purposes.

4 The resource consumed usually can't be measured. Often one must take refuge in a proxy such as the visitor-day. Added to the difficulty of measurement is the fact that the character of the resource—of the visitor-day—is a function of demand. It changes with the degree of crowding.

5 It may not be realistic to assume that visitors have the same reaction to a dollar admission fee as to a dollar travel cost. Visitors may behave unpredictably if they are asked to pay a fee when admission previously was gratis.

6 Visitors may not, on the average, be alike from zone to zone respecting income, culture, consumption technology, and other circumstances that bear on demand. Almost inevitably, visitors from distant zones have access to more substitutes for the resource being appraised than nearby visitors do. Allowing for such biases presents special difficulties.

As the list is further lengthened, the items become more technical. One dare not lose sight of the fact that minor flaws are negligible in a procedure that as a whole is rough at best.

CONSUMER-SURVEY MODEL

Another model that looks to the consumer for value appraisal is the **consumer-survey model,** known alternatively as the **contingent-value** [2] or **bidding-game model** [16]. The model is applicable to evaluating either a-service or b-service, and hope is looming that it can be used for c-service, too. The appraiser uses questionnaires or personal interviews in an effort to learn how much consumers would be willing to pay for their current experience and for an alternative experience—for example, hiking that involves more encounters with wildlife. In this way, the appraiser can try to discover the extra revenue associated with a change, so as to compare it with the extra cost.

The consumer-survey model has the advantage over the travel-cost model of addressing itself directly to the weighing of alternatives. And yet, in the common case where the survey model calls for consumers to respond to a hypothetical situation (what would you do if . . .?), the model may hit a snag. Consumers' estimates of how they would respond *if,* even when thoroughly honest, are not thoroughly reliable. For consumers to envision imaginary circumstances that they have never encountered may be out of the question.

When consumers are interviewed to discover how much more they would be ready to pay for a given experience than they are paying in the case at hand, some will respond with a zero bid. Such a bid may be, not in fact a zero valuation, but rather a protest by the respondent against the bidding game itself or some aspect of its terms. To check results for the presence of zero bids thus motivated is desirable.

THREE MORE MODELS

Let's turn, now, to three models for appraising implicit value that lean on the decision maker for help. A resource is appraised by observing the firm's actions or judgments that imply resource value. This is the same sort of approach as the one mentioned on page 144 for estimating a forest owner's guiding rate of return.

The **multiple-use model** is based on the procedure represented in Table 28-1. The table specifies the production functions of two rival forest products. It is used for finding the best combination of the products when their unit values are known. The multiple-use model starts with the same production functions. It calls for finding a forest or forests where the manager has arrived at a combination of two products that is satisfactory and successful. If this can be taken as the best combination and if the unit value of one product is known, the value of the other can be inferred by working the Chap. 28 procedure backward. This model was proposed by G. R. Gregory in 1955 [10].

The **input-output model** was described as an appraisal device by Warren Flick in 1975 [9]. Input-output is a system for analyzing an economy. It starts with a description of the flows of payments among all the industries or economic sectors in a nation or region and draws conclusions about the effect of a unit change in each sector on each of the others. Used for appraisal purposes, input-output analyzes money and product flows within a firm, where the sectors are the firm's enterprises. The model can respond to such questions as this: At the present level of fish output from the firm's forest streams, what is the marginal cost of raising output by one unit? If the firm can be assumed to have found its best combination, this marginal cost equals marginal revenue—that is, the value of a fish. Flick found that the Bureau of Land Management program on the Eugene District in Oregon implied a value of 4 cents each for smolts, young salmon ready to head downstream and out to sea. Some details of input-output analysis are given in Chap. 48.

In the **market-conference model,** a value consensus is sought from a group of decision makers or other knowledgeable persons. Refer again to pages 211 and 212, centering on Table 28-2. A Cranberry Wilderness is proposed on the Monongahela National Forest. One analyst has calculated the annual **direct** (variable) social **cost** of the wilderness: the opportunity costs consisting of coal, oil, gas, and timber output forgone; the wilderness management cost; and the private land acquisition and other costs. The total cost comes to about $30 per anticipated visitor-day in 1978, about $20 in 2000. The members of the congressional subcommittee ask themselves if the Cranberry wilderness is worth this much. They go over the affirmative and negative arguments and take a vote. As it turns out in this particular case, the ayes have it, and since January 1983, the Cranberry is wilderness by law.

A variant of the market-conference model is found where the conferees are members of a single public agency or other firm. They generate values that may be termed "administratively assigned." We have had occasion already to observe the use of implicit values administratively assigned by the federal Forest Service.

Where it can be made reasonably objective, the market-conference model comes about as close as any to being universally applicable to implicit-value appraisal. Yet even it, like all the others, is poorly suited to appraising c-service. The consumer of this service may never have laid eyes on a porcupine or ears on a loon's maniacal mirth. Yet this person is aware of the loon's and the porcupine's existence and takes pleasure therein. What is such pleasure worth? Who knows?

SUMMARY

It boils down to this: There are two kinds of market value (objective value): explicit, as for timber, and implicit, as for wildlife. Each may be appraised either from the individual or from the social viewpoint. If the resource to be appraised is plainly a subsystem of a larger system, then, strictly speaking, the appraisal should be arrived at as the difference between the value of the entire system and its value in the absence of the subsystem. Resource values arise out of three kinds of utility, consisting of opportunities to (1) use, (2) sense, and (3) think about the resource.

Explicit value can be appraised from market records. Implicit value must be appraised indirectly, by inferring value from the behavior of consumers, producers, or others whose judgment is believed to be reliable. There are several different models for appraising implicit value.

REFERENCES

1 Anderson, L. M., and H. K. Cordell: "Residential Property Values Improved by Landscaping with Trees," *Southern Journal of Applied Forestry,* March 1985, pp. 162–166.

2 Bishop, R., T. Heberlein, and M. Kealy: "Contingent Valuation of Environmental Assets: Comparisons with a Simulated Market," *Natural Resources Journal,* 1983, pp. 619–633.
3 Brown, Thomas C: *Monetary Valuation of Timber, Forage, and Water Yields from Public Forest Lands,* U.S. Department of Agriculture, Forest Service Gen. Tech. Rep. RM-95, 1982, pp. 25–26.
4 Cesario, F. J., and Knetch, J. L.: "A Recreation Site Demand and Benefit Estimation Model," *Regional Studies,* October 1976, pp. 97–104.
5 Clawson, Marion: *Methods of Measuring the Demand for and Value of Outdoor Recreation,* Resources for the Future, Washington. Reprint 10, 1959 (5th printing, 1981).
6 Clawson, Marion, and Jack L. Knetch: *Economics of Outdoor Recreation.* The Johns Hopkins Press, Baltimore, 1966 (4th printing, 1978).
7 Duerr, William A., Henry J. Vaux, and Charles H. Stoddard: "Some Research Needs in the Economics of Forestry," *Proceedings, Society of American Foresters Meeting, 1954,* Washington, 1955, pp. 161–162.
8 Dwyer, John F., John R. Kelly, and Michael Bowes: *Improved Procedures for Valuation of the Contribution of Recreation to National Economic Development,* University of Illinois, Water Resources Center Rep. 128, Urbana and Champaign, 1977.
9 Flick, Warren A.: *Resource Flows and Values,* U.S. Department of the Interior, Bureau of Land Management Tech. Note 276, 1975.
10 Gregory, G. Robinson: "An Economic Approach to Multiple Use," *Forest Science,* March 1955, p. 13.
11 Hotelling, Harold: Letter to the director, National Park Service, U.S. Department of Interior, 1947. Quoted in Outdoor Recreation Resources Review Comm. Rep. 24, *Economic Studies of Outdoor Recreation,* Washington, 1962.
12 Loomis, John B., Dennis M. Donnelly, Cindy E. Sorg, and Lloyd Oldenburg: *Net Economic Value of Hunting Unique Species in Idaho: Bighorn Sheep, Mountain Goat, Moose, and Antelope,* U.S. Department of Agriculture, Forest Service Resource Bull. RM-10, 1985.
13 Magill, Arthur W., and Charles F. Schwartz: *Searching for the Value of a View,* U.S. Department of Agriculture, Forest Service Research Pap. PSW-193, 1989.
14 Peterson, George L., B. L. Driver, and Robin Gregory: *Amenity Resource Valuation,* Venture Publishing, State College, Pa., 1988.
15 Popovich, Luke: "How Do You Account for Deficit Sales?" *Journal of Forestry,* October 1984, pp. 594–599.
16 Randall, Alan, Berry Ives, and Clyde Eastman: "Bidding Games for Valuation of Aesthetic Environmental Improvements," *Journal of Environmental Economics and Management,* January 1974, pp. 132–149.
17 Shaw, William W., and Ervin H. Zube, eds.: *Wildlife Values,* University of Arizona, School of Renewable Natural Resources, Center for Assessment of Noncommodity Natural Resource Values, Institutional Series Rep. 1, November 1980.
18 Sorg, Cindy F., John B. Loomis, Dennis M. Donnelly, George L. Peterson, and Louis J. Nelson: *Net Economic Value of Cold and Warm Water Fishing in Idaho,* U.S. Department of Agriculture, Forest Service Resource Bull. RM-11, 1985.
19 U.S. Department of Agriculture, Economics, Statistics, and Cooperative Service: *Farm Real Estate Market Developments* (annual).
20 Various authors: "Estimating Recreation Benefits, a Discussion," *Journal of Forestry,* January 1980, pp. 21–28.

CHAPTER **48**

INPUT-OUTPUT ANALYSIS

In Chap. 47, one of our three last models for appraising implicit resource value was the input-output model. It behooves us now to elaborate a bit on input-output analysis, especially since this is a useful method for analyzing an entire economy by bringing out the interdependencies among its sectors—for example, the dependence of the forestry sector on others and vice versa. We are happily reminded of one of the earliest propositions in this book, that everything depends on everything else.

Our model is the **input-output** model for **interindustry** analysis developed by Wassily Leontief [2] in the 1930s and first given a major forestry application by Hays Gamble [1] in the 1960s.

AN ECONOMY MEASURED IN PHYSICAL UNITS

Our illustrative economy and data are highly simplified. We assume a rural forestry economy, a community of 800 inhabitants occupying 30,000 acres of land. The land is given over mainly to two activities: forestry (tree growing and some simple wood processing) and farming. We take these two activities as the **sectors** of our economy and set their **transactions** in a matrix [Table 48-1(a)]. The top row of figures shows the output of the *forestry* sector in a recent year in thousand cubic feet of wood. Each figure is forestry's input into the sector named at the head of the column. Thus, of forestry's total 1,000 M-cubic-foot output during the year, firms (families) in the sector itself used 120 M—for sawmill sheds, roads, and other construction. Farming used 270 M—for buildings, fences, and the like. The rest of the wood output went to meet the **final (export) demand** of the outside world for stumpage, logs, bolts, and lumber.

The *farming* row we can interpret similarly.

TABLE 48-1
DEVELOPMENT OF INTERDEPENDENCY COEFFICIENTS IN *PHYSICAL UNITS* FOR OUR HYPOTHETICAL ECONOMY

a. Annual transactions, or flows

Output from	Input into: Forestry	Input into: Farming	Unit of measure	Final (export) demand	Total output
Forestry	120	270	M cu. ft. of wood	610	1,000
Farming	40	60	Tons of crops	2,900	3,000
Labor	35	60	Numbers of workers	—	95

b. Technical coefficients

	Forestry	Farming	Unit	
Forestry	0.120	0.090	M cu. ft.	Per unit of product of input sector
Farming	0.040	0.020	Ton	
Labor	0.035	0.020	Worker	

c. Interdependency coefficients

A unit change in final demand for output of	Forestry	Farming	
Produces this change in total output of			
Forestry	1.1411	0.1048	M cu. ft.
Farming	0.0466	1.0247	Tons

The *labor* row tells us that 35 workers were employed as an input to the forestry sector, and 60 were occupied in farming.

> Use your imagination to develop a plan of fieldwork for deriving the input-output matrix on the left side of Table 48-1(a).

From the transactions data, we derive **technical (technological) coefficients.** These tell us, for each unit of a sector's total output, the number of units of input that the sector requires from each of the others. Thus, the leftmost column of coefficients in Table 48-1(b) says that the forestry sector requires 120 ÷ 1,000, or .12 M cubic feet per M, from itself; 40 ÷ 1,000 = .04 ton of crops per M cubic feet from farming; and 35 ÷ 1,000, or .035 worker per M cubic feet, from the labor, or household, sector.

Now we work from technical coefficients to **interdependency coefficients,** which tell us the effect of a unit change in final (export) demand for each sector's product on the total output of each sector. Let

X_1= Forestry sector's total output
X_2– Farming sector's total output
Y_1= Final demand for wood
Y_2= Final demand for farm crops

Then

$$X_1 = .120X_1 + .090X_2 + Y_1$$
$$X_2 = .040X_1 + .020X_2 + Y_2$$

and when we solve for X in terms of Y, we get

$$X_1 = 1.1411Y_1 + .1048Y_2$$
$$X_2 = .0466Y_1 + 1.0247Y_2$$

That is, we get the matrix of interdependency coefficients displayed in Table 48-1(c). This matrix says that a unit change in final demand for wood causes a change of 1.1411 units in total wood output and .0466 unit in total farming output. A unit change in final demand for farm crops produces a change of .1048 unit in total forestry output and 1.0247 units in farming output.

How so? How, for example, can a unit change in the final demand for wood lead to more than a unit change in wood output? It comes about thus:

First, we observe in the matrix of technical coefficients that the forestry sector takes .12 unit of its own output for every unit of its total production. That is, only 1 - .12 or .88, of any forestry output can be released outside the sector. Thus, to export a unit of wood requires that 1 ÷ .88, or 1.1364, units be produced. But this isn't all.

Second, producing 1.1364 wood units places a demand on the farming sector measured by technical coefficient .04. And we are reminded that farming, like forestry, consumes some of its own output: indeed, .02 of it, so that farms must produce 1 ÷ .98 = 1.0204 units for every unit they supply to others. Thus the multiplier .04 is raised to .04 × 1.0204 = *.0408,* and the new wood units stimulate the output of 1.1364 × .0408—that is, .0464 crop unit. And now the forestry sector feels a new impact, because farming, with .0464 new crop unit to produce, calls upon forestry for .09 wood unit per crop unit, which means that forestry must product .09 × 1.1364, or *.1023,* unit per crop unit (the reason for italicizing two figures in this paragraph will be evident very soon). Hence the total new impact on forestry is .0464 × .1023 = .0047 wood unit, which must be added to the 1.1364 that we already calculated. In this addition, we have already reached the interdependency coefficient 1.1411.

But clearly the economic echoing and reechoing that we have traced through two rounds continues through endless further rounds, and if we carried out all the decimals far enough, we could see that we have a rapidly decaying geometric series with a ratio of .0408 × .1023 = .0042, the sum of which equals the interdependency coefficient. So it is with the other three coefficients: The first terms in the series are unique, but the ratios are all .0042 so long as we assume no change in technology (technological coefficients).

Why are the ratios all the same?

DERIVING SHADOW PRICES

The matrices that we have set in physical terms could just as well be built with financial data. Indeed, the latter is by far the more common way, permitting the input-output information to be gathered in the form of money expenditures and receipts by a sample of households and other firms.

However, the physical approach is useful to us because it permits us to make an interesting and somewhat realistic assumption about the market and thereupon *derive* prices **(shadow prices)** from which to generate financial data. The shadow-price approach will come in handy for the solution of other problems. The assumption: Output prices are just sufficient to cover all input costs. Let

$$P_1 = \text{Unit price of forestry output}$$
$$P_2 = \text{Unit price of farming output}$$
$$W = \text{Wage per worker per year}$$

Then

$$P_1 = .12P_1 + .04P_2 + .035W$$
$$P_2 = .09P_1 + .02P_2 + .02W$$

In other words, prices are assumed to be proportional to the labor embodied in each unit of output: a **labor theory of value** [3]. This can be seen more readily when the foregoing relationships are rephrased:

$$P_1 = 1.1411 \times .035W + .0466 \times .02W$$
$$P_2 = .1048 \times .035W + 1.0247 \times .02W$$

In either case, it turns out that

$$P_1 = .0409W$$
$$P_2 = .0242W$$

so that if W is \$12,000 per year, the price of wood is \$490.45 per thousand cubic feet, and the farm-crop price is \$289.94 per ton.

The labor theory of value is a convenient oversimplification. Explain.

THE ECONOMY MEASURED IN FINANCIAL TERMS

Now we can set up a financial transactions matrix, Table 48-2(a), deriving it from the physical matrix and the dollar values per physical unit. Note that in this truncated little economy of ours, the forestry and farming sector together are self-sufficient in their wood and farm-crop requirements; the total wage bill is just paid by the two sectors' exports; and we are saying nothing about imports—that is, the next-to-bottom row in Table 48-2(a) is blank, and we pass over the question of how households spend their income of $1,140,000.

From the transactions data, we derive technical coefficients as before [Table 48-2(b)]. The starred coefficients are, as we should expect, the same as the physical coefficients. The other two forestry-farming coefficients differ from their physical counterparts by a ratio equal to the price ratio, $P_1 \div P_2$, about 1.7. The labor coefficients are greater than their physical opposite numbers by the ratios of the annual wage to the wood and farm-crop price, respectively, about 24.5 and 41.5.

TABLE 48-2
DEVELOPMENT OF INTERDEPENDENCY COEFFICIENTS IN *FINANCIAL TERMS* FOR OUR HYPOTHETICAL ECONOMY

	Buyers				
Sellers	**Forestry**	**Farming**	**Total internal**	**External (exports)**	**Total sales**
a. Annual transactions, in thousand dollars					
Forestry	58.85	132.42	191.27	299.18	490.45
Farming	11.60	17.40	29.00	840.82	869.82
Labor	420.00	720.00	1,140.00	—	1,140.00
Total internal	490.45	869.82	1,360.27	1,140.00	2,500.27
External (imports)	—	—	—	—	—
Total purchases	490.45	869.82	1,360.27	1,140.00	2,500.27
b. Technical coefficients: Dollars paid by buyers per dollar of product					
Forestry	0.120*	0.152			
Farming	0.024	0.020*			
Labor	0.856	0.828			
c. Interdependency coefficients					
A dollar change in final demand for output of	Forestry	Farming			
Produces this change, in dollars, in total output of					
Forestry	1.1411*	0.1773			
Farming	0.0275	1.0247*			
Total export demand multiplier	1.1686	1.2020			

> Explain the foregoing relationships.

Finally, we can derive interdependency coefficients from the technical coefficients by the same arithmetic as before. They are given in Table 48-2(c). The starred coefficients, again, are the same as their physical counterparts, while the other two differ in proportion to the price ratio, about 1.7. Taking the left column of figures in Table 48-2(c) for illustration, we learn that a $1 change in final (export) demand for wood will produce a $1.14 change in forestry activity and a 2.75-cent change in farming. The total effect on the economy is the sum, $1.17, or a ratio of 1.17. This ratio we can term the **export demand multiplier** for wood. Counting all the economic reverberations within the economy, it measures the stimulation that comes from purveying a particular consumable to the outside world and getting new money in return, to be passed from pocket to pocket.

FAULTS AND VIRTUES OF INPUT-OUTPUT

Input-output analysis has its weaknesses. It is a "data hog," and in real life calls for expensive surveys to feed its gluttonous appetite. Its functional relationships are linear and therefore cannot predict very reliably the outcome of large changes in the economy. Its predictive capabilities are limited on other counts, too: It has no built-in constraints such as in real life are imposed by resource availability. And again, its technology (matrix of technical coefficients) is fixed at a past period and does not readily lend itself to revision to suit the future outlook.

> How do you suppose it might be possible to amend a matrix of technical coefficients so as to introduce forecasting? Use our oversimplified illustrative economy as your example.

Faults or no, the input-output model has outstanding capabilities in regional analysis. It can identify the economic sectors whose exports have large multipliers and whose growth may therefore be beneficial to the region. It can help predict the impact of alternative developments being contemplated, and thus help guide a choice. A public agency, or a private industry, or even a large firm, can devise an input-output analysis in which the entity itself constitutes one sector, so that its role in the economy—its "importance" and its sensitivity—can be readily grasped.

After the admittedly large task of assembling data is finished, the arithmetic necessary for deriving interdependency coefficients can be carried out in something of a jiffy on the computer with the use of matrix theory. Thus Hays Gamble was able to recognize 29 sectors in the Sullivan County economy—something arithmetically akin to calculating the phases of the moon. And U.S. Department of Commerce input-output analyses of the U.S. economy [4] work with more than 500 sectors!

SUMMARY

The input-output model is a device for interpreting data on the flows of activity in an economy. The economy is divided into sectors, and data are gathered on flows, ordinarily in money, among sectors and between them and the outside world. The model is then used to generate coefficients that measure interdependencies and thus serve both to describe and to predict the economic impacts of change.

REFERENCES

1 Gamble, Hays B.: *The Economic Structure of Sullivan County, Pennsylvania,* Pennsylvania Agricultural Experiment Station Bull. 743, June 1967.

2 Leontief, W. W.: *Structure of American Economy,* Oxford University Press, New York, rev. ed., 1951.

3 This approach was suggested to me by Professor Warren A. Flick, Auburn University.

4 U.S. Department of Commerce, Economics and Statistics Administration, Bureau of Economic Analysis: "Benchmark Input-Output Accounts for the U.S. Economy, 1982," *Survey of Current Business,* July 1991, pp. 30–71.

CHAPTER 49

STUMPAGE APPRAISAL

SIGNIFICANCE OF STUMPAGE APPRAISAL

Now we move from valuation in general to the estimation of *explicit market value*—that is, the forecasting of it. We take the value of stumpage as a case in point. Stumpage is standing timber destined soon for harvest. Its value is derived from the value of the serviceable goods that will be made from the harvest. These goods are logs, bolts, and other round products; split, hewn, and sawn wood; veneer, chips, particles, and panels; pulp, paper, and other things derived by pulverization or chemical treatment; and all the multitude of products of further manufacture.

Stumpage appraisal is one of the oldest conventions in forest-products marketing. In some circles today, as in the sale of public timber, the convention is one of the most elaborate, occupying bushels of pages in the rule books.

The value of stumpage can be estimated by either prospective buyers or prospective sellers or both, as a guide to their bargaining in the market. Many public timber managers are required to sell stumpage at no less than its appraised value. A stumpage-value estimate can serve, too, in assessment for property taxation, in calculating losses, in insuring against loss, and so on. Here, we'll think mainly about appraising stumpage for the market.

Note that the subject is the *appraisal* of value. Appraised value and market value as actually consummated may turn out to be equal. But commonly they do not. And even when they do, as in the case of public-timber sales where appraised value is announced, the identity may result from imperfect competition among buyers rather than excellence or good luck on the appraiser's part. Many forces enter into market value.

What are some of the forces behind stumpage's market value? First, there are the characteristics of the tract to be logged, the **logging chance:** accessibility, area, physiography, roading, timber quantity, quality, stocking, species, size, and the like. Second are the prospective characteristics of the market: end products in view, places where their values will be determined, costs of transportation, degree of competition among stumpage sellers and buyers, terms of sale, characteristics of loggers, especially their technology, bearing of other sales on this one, proposed time and duration of the sale, general economic conditions, and so forth.

All stumpage appraisals take at least a stab at foreseeing the characteristics of the logging chance and the market.

Apparently the independent variables in stumpage value are essentially all on the demand side of the market. Why is this? *Hint:* Picture a tract of national forest stumpage to be appraised and sold to the highest bidder. Sketch a graph of price over quantity. Draw the supply function, then the demand.

LUMP-SUM TRANSACTIONS

If you want some insight into the market value of anything, you look to the market for the purpose. The stock market will tell you what the price of your favorite wood-products stock has been; the lumber market will tell you about lumber; and presumably the stumpage market will tell you about stumpage. What could be more obvious?

In fact, it *is* obvious in the case of securities and even for lumber. But it is not quite obvious for stumpage. Our federal Forest Service, which has been appraising national forest stumpage for going on a century [3], only recently started to move away from the customary analytical approach and to look right at the stumpage market itself, taking a **transactions-evidence,** or **direct,** approach to appraisal. The trouble with the historical record as a source of information on value is that each stumpage transaction is unique in respect to both the logging chance and the market.

When timber is appraised for sale, by whatever method, one of the terms of sale in particular must be stated: how the stumpage is to be paid for. Two methods are common. In one, the parties to the transaction will agree in advance on a single, **lump sum** total of money to be paid for all the harvested timber within the sale **boundary.** In the other, the parties will agree on a price per unit of measure for each category of roundwood. Payment is then made on the basis of the **scale** (quantity) of each category cut and loaded for removal. The lump-sum approach has long been in disrepute because it has so often resulted in an inordinately low return to the nonindustrial private timber owner. Yet lump-sum sales backed up by good timber cruises carry two really significant advantages over scale sales. The lump-sum sales encourage the timber buyer to practice close utilization, for the marginal stumpage cost of any tree or section is zero. The stick that the firm couldn't afford to utilize if it had to pay extra stumpage may be merchantable if there's no additional charge. Furthermore, the seller is spared the cost of supervising the sale for utilization practices and of scaling the harvest. The Bureau of Land Management sells timber in lump-sum transactions.

It has been argued that the utilization advantage of lump-sum sales holds largely in the shorter term: that in the longer term, scale transactions may produce full use of the timber resource. Can you explain this argument?

TRANSACTIONS EVIDENCE

Not surprisingly, pioneer work on transactions evidence was done on relatively uniform material: sales of national forest sawtimber in the coastal plain of Texas, Louisiana, and Mississippi. In the early 1950s, Sam Guttenberg [2] collected data on 243 sales made during a $4\frac{1}{2}$-year period. The scatter of prices looked like this, with a mean of $32.70 and a standard deviation of $6.90 (the chart illustrates the uncertainty about price discussed at the opening of Chap. 31):

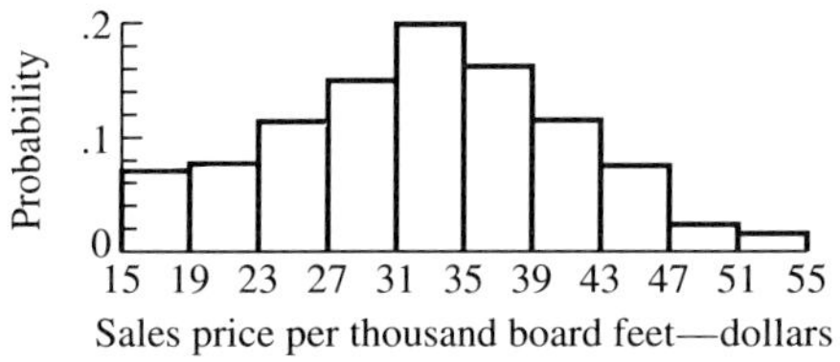

Of course, there is transactions evidence right here; a stumpage value of $32.70 per thousand board feet. This is a valid figure, but Guttenberg wanted more precision. He resorted to regression analysis to account for some of the variables responsible for the price scatter and thus reduce the standard error. He arrived at this list of variables:

M board feet of pine in the sale *(Q)*
M board feet of pine per acre *(C)*
M board feet of hardwood per M board feet of pine *(H)*
Board feet of all timber per square inch of basal area of all trees *(W)*
Wholesale price of No. 2 common pine boards, dollars per M board feet, 1 month before sale *(L)*

Representing stumpage price per M bd ft as *P*, Guttenberg came up with this equation:

$$P = .18Q + 2.28C - 4.08H + .17W + .54L - 24.97 \qquad (1)$$

You will observe that four of Guttenberg's independent variables describe the logging chance; the other one describes the market. Evident in the coefficients is the fact that pine was the key to stumpage value in the region studied. Hardwood was a liability. The standard error of *P* is $5, not bad for so far-flung a territory.

Since the early work, other studies have followed in growing numbers. For a time, only the research agencies made such studies. In recent years, with the gradual adoption of transactions evidence in public sales, administrative groups such as National Forest Administration within the Forest Service have taken a hand. Today all national forests east of the Pacific Coast states use transactions evidence in their stumpage appraisals, and soon only Alaska will be left clinging to analytical appraisals. The independent variables in the transactions-evidence equations may number in the twenties, covering a good range of logging-chance and market characteristics. The procedure varies somewhat from region to region, but in general it is to subdivide a state into zones and build equations from recent sales within each zone.

But one should hasten to recognize that transactions-evidence appraisals don't all involve a lot of mathematics. One can turn, for example, to pulpwood roundwood markets in regions of the U.S. where oligopsony and price leadership result in price uniformity, not only geographically, but also over time. And uniformity finds its way from roundwood price to stumpage price. Many local residents, just from common knowledge, can recite the going price or range. Refer again to Fig. 45-4, especially Fig. 45-4(d). Here, in the case of Lake States aspen, is a nearly perfect example of uniformity.

If stumpage prices in a region are only moderately variable, persons who work with informal stumpage appraisals in the field develop the ability to recognize the causes of the variation and allow for them in judging the value of any tract. A common method is to start with a going price per unit of measure and amend it to allow for the tract's departures from the norm.

Most forested states give local property-tax assessors guidance in their timber assessments, usually in the form of transactions evidence of market value. A good example comes from western Oregon [1]. The state Department of Revenue each year would gather information on stumpage sales, both public and private. For each sale in each county, the Department would report to assessors the price per unit of measure, along with information on "species, quality, volume, age, the presence of disease and of defect and breakage, stand density, costs of removal, accessibility to point of conversion, topography of the site and surrounding country, and other relevant factors."

ANALYTICAL APPRAISALS

Turning now to the analytical, or **residual,** method of stumpage appraisal, we see a model of great historical significance. Though it has been supplanted progressively by the more straightforward direct method as the years have moved along, the analytical method is still in use on some federal forests, most state forests, and widely in industry. Take a simple, old-time example of the method. This example has all the essentials that we want to be aware of.

In the days of the softwood lumbering era (page 247), lumbermen had "landlookers" and appraisers to work out ahead of the logging. The landlooker estimated the quantity of timber on the next section, and the appraiser translated quantity into value. The thinking ran something like this:

1 We can saw easily 20 million ft of lumber from the section.

2 The lumber will be worth $10 per thousand board feet at the mill: $200,000 total.

3 Logging, hauling, and milling will cost us $7.50 per M ($150,000 total), leaving a margin of $50,000 in all.

4 For profit, we should allow 10 percent of our investment in railroad and mill allocated to the section: $35,000.

5 Therefore we can afford to pay $15,000 for the stumpage, which amounts to 75 cents per M board feet.

Let's take these old data, which come from the past century, and apply today's terminology to them:

	Per M bd ft		Total	
Selling price		$10.00		$200,000
Logging cost	$2.00		$40,000	
Hauling cost	1.00		20,000	
Milling cost	4.50		90,000	
Conversion cost		7.50		150,000
Conversion return		2.50		50,000
Profit margin		1.75		35,000
Appraised stumpage value		$ .75		$ 15,000

In simplest terms, an analytical stumpage appraisal starts with the forecast selling price of an end product and works backward, subtracting all the forecast costs that intervene between the product and the standing timber. These intervening costs fall into two categories: first, the aggregate costs of "conversion" (standing timber to product), both fixed and variable with output; second, an allowance for profit (or, if you like, for profit and uninsured risk). What's left over is for stumpage. Hence the term "residual" appraisal. If there's nothing left over, it means that this particular stumpage is economically inaccessible unless the owner wants to pay to get rid of it.

Stumpage value, then, is designed to answer this question: How much can a buyer afford to pay for timber and still cover all costs and make a satisfactory profit? Or stumpage value answers this question: How much can a seller expect to receive for timber, assuming that any buyer would want to cover all costs and make a profit too? If the stumpage appraisal is to respond to the buyer's question, then it uses the buyer's cost rates, based on the buyer's conversion technology. If the appraisal is to reply to the seller's question and there are buyers competing for the timber, then the appraisal ordinarily assumes the most efficient technology likely to be available among buyers.

SOME DETAILS OF ANALYTICAL APPRAISALS

Glancing back at the tabulation in the next-to-last paragraph, you see that there are three parameters that determine appraised stumpage value: (1) selling price, (2) conversion cost, and (3) profit margin. (Conversion return is just 1 minus 2.) Consider, now, each of these three.

1 The selling price that provides the starting point for analytical appraisals is customarily the price in the first well-defined market that the logged timber will pass through on the way to ultimate consumers. This is a market where there are real transactions between sellers and buyers that are open enough to public view so that prices are known. For a seller-oriented appraisal, a competitive market is to be desired as a starting point, but often cannot be used or even found.

The appraiser's job is to study the historical data and foresee what the price will be at the time to which the appraisal applies. A "persistence forecast" (page 240) is commonly resorted to. But if a significant period is to elapse between appraisal and cutting, or in the course of cutting, the parties may agree to **indexing** the selling price or the appraisal by adjusting it to changes in the product market.

Sawlog stumpage appraisal has a good starting point, the lumber market. The markets for veneer, plywood, chips, and other residues are often well defined. In parts of the West Coast, sawlogs and veneer logs used to go in large quantity into well-defined markets, where they were graded and sold to processors at prices that were available publicly. These prices could provide a happy starting point for analytical stumpage appraisals because they let the appraiser escape the task of estimating milling cost. To a limited extent, such use is still made of log prices.

> Analytical appraisals of pulpwood stumpage never, so far as I am aware, start with paper or pulp selling price, even though this is well documented. They start with pulpwood prices. Why?

2 Without doubt, the most arduous and arguable estimates in analytical appraisals are the estimates of conversion cost. Think of the painstaking data collection and grocery-store arithmetic that may go into defensible forecasts of the costs of site preparation, felling, bucking, limbing, slash disposal, yarding, loading and unloading, road building and maintenance, logging camp construction and maintenance, processing plant construction, maintenance and operation, office operations, and administration! Furthermore, these costs, both the fixed and the variable in relation to output, are commonly allocated to units of output. Such allocation of the *fixed costs* is ticklish business, as mentioned in passing on page 293 and explained on pages 394 and 395. Following is a glimpse of the cost items included in estimating just the *logging cost* on one Forest Service sale in the Lake States in the time before transactions-evidence appraisals:

I "Development" cost, fixed and variable with output
  A Camp construction, maintenance, moving, salvage

- **1** Headquarters camp
- **2** Portable logging camps
  - **a** 25 one-man shacks
  - **b** 25 two-man shacks
  - **c** 6 tractor sheds
  - **d** 25 outhouses
  - **e** 15 wells
- **3** Payroll taxes and compensation insurance

II "Overhead" cost, fixed with output
- **A** Supervision
  - **1** Superintendent
  - **2** Clerk checker
- **B** General overhead
  - **1** Trucks: purchase, operation, salvage
    - **a** Pickup
    - **b** Carry-all
    - **c** General supply
  - **2** Operation of camp utilities
  - **3** General office and miscellaneous expense
- **C** Fire equipment: purchase, maintenance, salvage
- **D** Fire fighting
- **E** Fire insurance
- **F** Property taxes
- **G** Payroll taxes and compensation insurance

III "Direct" cost, variable with output
- **A** Felling, bucking, skidding
- **B** Brush disposal
- **C** Loading, hauling, decking

No wonder the Forest Service is abandoning analytical appraisals!

3 Profit margin is the third parameter leading to appraised stumpage value. To this point in the story, the appraiser has arrived at a conversion return. This question is how to divide this return between a profit margin (for the buyer-operator) and stumpage value (for the seller). What allowance shall be made for a profit margin? Given the profit margin and having estimated the conversion return, the appraiser has the stumpage value in hand:

$$\text{(STMP VALU)} = \text{(CONV RETN)} - \text{(PROF MARG)} \qquad (2)$$

The old-time appraiser who was introduced earlier used the **investment method** to derive a profit margin. The appraiser estimated the margin as a given percent of the investment in mill and railroad, allocated to the unit of stumpage. What percent? Ten, because this was the traditional norm for the lumber industry:

$$\text{(PROF MARG)} = .10\text{(CAPL IVST)} \qquad (3)$$

Another method for deriving a profit margin is to figure it at a given percentage of the conversion return. This method is known as the **profit factor** method, the profit factor being the given percentage, expressed in decimals:

$$\text{(PROF MARG)} = \text{(CONV RETN)} \times \text{(PROF FACT)} \tag{4}$$

The Bureau of Land Management has used the profit factor method, taking a factor of 0.50, which divides the conversion return 50-50 between buyer and seller. Using 0.50 has the appeal of simplicity and "fairness."

But the philosophy underlying the now outmoded investment method has proved to be the enduring philosophy: If there is to be an allowance for profit (and uninsured risk), the amount of it is most logically hinged to the buyer's prospective costs of operation. The costs in question, so the argument runs today, are not just the capital investment, as in the investment method. They include the entire conversion cost. And not only that: They include stumpage value, which is another cost borne by the buyer. The upshot of this argument is the **profit ratio** method, in which the profit ratio is the profit margin expressed as a percentage (in decimals) of conversion cost plus stumpage value:

$$\text{(PROF MARG)} = [\text{(CONV COST)} + \text{(STMP VALU)}] \times \text{(PROF RATO)} \tag{5}$$

Of course, the relationship is unusable in this form, since the appraiser won't yet know the stumpage value. So we rearrange:

$$\text{(PROF MARG)} = \text{(SELL PRIC)} \times \frac{\text{(PROF RATO)}}{1 + \text{(PROF RATO)}} \tag{6}$$

Thus one finds profit margin hinged to selling price—that is, all costs plus profit margin to boot. The profit ratio method has been preferred by the Forest Service and many other users of analytical appraisals.

Assure yourself of the consistency of formula (6) by deriving it from formula (5).

All well and good, but where does the profit ratio come from? It comes from analysis of past timber sales within the territory. For these sales, selling price, conversion cost, and stumpage value are known, and hence conversion return and profit margin are known. One need only solve expression (5) for the profit ratio. The profit ratio is derived from transactions evidence!

And when we reflect on the matter, don't we come to the conclusion that every parameter used in analytical appraisal comes from transactions evidence? The analytical appraisal is a form of the transactions-evidence appraisal. The "regression equation" in an analytical appraisal is expression (2) combined with formula (6)!

SUMMARY

In brief, two general methods are available for appraising the market value of stumpage. The first method, which appears to be in the ascendant today in the United States, is the transactions-evidence, or direct, method. It consists of analyzing recent stumpage sales in the region so as to identify—possibly with the help of regression analysis—the prices of timber most nearly like the stumpage at hand. The second is the analytical, or residual, method. This figures stumpage value as the selling price of the marketed wood product minus the costs that intervene between the product and the standing timber. The chief question here is how the residual after out-of-pocket costs, the "conversion return," should be divided between buyer ("profit margin") and seller ("stumpage value"). A number of procedures are in use. You may well be impressed by the similarity, at bottom, of the transactions-evidence and analytical-appraisal methods.

REFERENCES

1 Carlan, Samuel A.: "Property Tax Assessment Procedures on Timberlands in the United States, 1966-1976, M.S. thesis, Virginia Polytechnic Institute and State University, Blacksburg, 1978.

2 Guttenberg, Sam: *Influence of Timber Characteristics upon Stumpage Prices,* U.S. Department of Agriculture, Forest Service, Southern Forest Experiment Station Occasional Pap. 146, 1956.

3 Weiner, Alfred A.: *The Forest Service Timber Appraisal System, A Historical Perspective, 1891-1981,* U.S. Department of Agriculture, Forest Service, FS-381, 1982.

CHAPTER **50**

CONVERSION SURPLUS AS A MEASURE OF VALUE

CONVERSION SURPLUS CHARACTERIZED

From near the beginning of this book, you have been bombarded with two propositions in particular. First is the proposition that *What are the alternatives?* is the central question in forestry economics as problem solving (page xiii). Second is the proposition that if you want to weigh alternatives, you look only at the features that differ from one alternative to another. This is the *principle of relevance* (page 147). For example, only the things that vary with output are relevant to deciding on an output; the fixed items don't matter.

The present chapter keeps up the bombardment by introducing a measure of value, **conversion surplus,** that conforms to both foregoing propositions. Thus the measure is well designed for choosing among alternatives.

Conversion surplus is a useful measure of the value of any raw material or good-in-process that is to be converted into a product. The material's conversion surplus is the market value of the product minus the variable cost of the conversion. Conversion surplus of stumpage is the same "residual value" as in Chap. 49, except that it is net, not of all costs, but only of variable costs. The trick is to identify the costs that are variable and those that are fixed for the alternatives at hand.

> Try defining *variable* and *fixed.* Then take for illustration the problem of choosing a location for a visitor center at Forest Park. Name some of the benefits or costs that the problem solver will consider because they vary with location. Name some that are disregarded because they are fixed.

What are the alternatives that conversion-surplus value helps weigh? Among those that this book talks about are level of timber stocking, time of harvest of a tree or timber stand, and intensity of fertilizing or fire management. Where these topics concern forest or tree value, we can read "value" as "conversion surplus."

Conversion surplus is a useful measure of value, too, for the firm that is deciding on the closeness of material utilization or the products to be made from given material.

PARADOX OF THE EXPENSIVE PROFITS

We begin with the "Paradox of the Expensive Profits." This is how Sam Guttenberg began when he authored conversion surplus in 1949 [2]. Look for the analogy between this story and the story of the paperboard machine in Chap. 38.

Lumber Company K saws 10,000 M bd ft per year, drawing all its raw material from its own timberlands. Of its total cut, 7,000 M bd ft come from butt and middle logs with an average conversion return of $40 per M board feet. Top logs provide the other 3,000 M bd ft; their conversion return is a *negative* $2 per M. Recall that the conversion return (page 387) is the difference between lumber selling price and the aggregate costs of logging and milling, including "overhead," which in Company K's case is $7 per M, or $70,000 annual total. The yearly net conversion return is $280,000 from the butt and middle logs minus $6,000 loss from the top logs, or $274,000 altogether: $27.40 per M board feet.

Dissatisfied with its use of top logs, the company sells them to a pulpwood contractor for $3 per M in the woods, attached to the tree tops. This looks like a profitable arrangement: changing a $2-per-M loss to a $3-per-M revenue on 3,000 M. But the company *does not obtain timber to offset what it diverts to pulpwood.* And so its total lumber production drops from 10,000 M bd ft to 7,000, and its overhead of $70,000, on a per-M basis, rises $3, from $7 to $10. Here is the new annual account compared with the previous one:

Previous account		New account	
7,000 M @ $40 conv. return	$ 280,000	7,000 M @ $37	$ 259,000
3,000 M @ $2 conv. loss	− 6,000	3,000 M @ $3	9,000
Net conv. return, total	$ 274,000	Return, total	$ 268,000
per M	$ 27.40	per M	$ 26.80

In giving up a $2 loss to make a $3 gain, the company has cut its net return per M by 60 cents and its total net return by $6,000!

What is wrong with the company's calculations? It fails to recognize that since it will not replace the diverted top logs with other timber, the $70,000 overhead is fixed with reference to the question of the logs' use. Therefore, overhead is irrelevant to the choice of a use for the logs, and the inclusion of overhead in the valuation is an error. If the company's original accounting ran like the following, there would be no misunderstanding.

Return before overhead cost:	
Butt and middle logs:	
7,000 M at $47 per M	$329,000
Top logs:	
3,000 M at $5 per M	15,000
All logs	$344,000
Overhead cost	− 70,000
Net return	$274,000

Shown the immediately foregoing figures, the company would recognize that by selling its top logs for $3 per M, it would be choosing an inferior alternative and losing, thereby, $2 per M on 3,000 M: $6,000 altogether. That is, the company would recognize that the top logs are worth $5 per M if sent to the mill, and that to sell them for less would mean a loss. For the question and the circumstances at hand, $5 is the value of the logs per M. This measure of the logs' value is their *conversion surplus: the difference between selling price and variable costs of conversion.* Negative $2 per M is the top logs' conversion *return,* the difference between selling price and all conversion costs (except profit), including those that, for the question at hand (what to do with the top logs), are fixed.

> In the accounting for Company K, why is it a crucial matter whether the company obtains timber to offset what it diverts to pulpwood? If the company doesn't obtain offsetting timber, what price must it get for the top logs as pulpwood in order to increase its net revenue? If the company can replace the top logs with other material of equal volume and value without affecting its silvicultural program, what price must it get for the logs as pulpwood in order to increase its net revenue?

Note that although the $70,000 overhead cost is fixed with reference to the utilization question, it is *not* fixed, but *variable,* for the question of the firm's net revenue. Therefore, one is right to include it in net-revenue accounting. This accounting is addressed to an implied question: How much can the company earn if it continues in business, compared to what it could earn if it ceased operations? The overhead cost is variable with these alternatives.

(It's useful to emphasize a rule that is mentioned on page 293 and again on 388: a **rule about the allocation of fixed costs.** The rule is, *don't try to allocate fixed costs to units of output* if it's with reference to output that the costs are fixed, and it commonly is. Such allocation is necessarily arbitrary, and the economic implications of it are necessarily invalid. The allocation is arbitrary because if costs are unrelated to output, they cannot be attributed to output or assigned to units of output. The economic implications are invalid because, whatever they are, they could just as acceptably be reversed by choosing a different allocation rule—for instance, per foot of log length instead of per board foot of log volume. In a buyer's stumpage appraisal, overhead costs are seen as variable with output, and their allocation to such a unit as the board foot is justified on the grounds that this is the unit specified or implied in the purchase.)

CONVERSION-SURPLUS VALUE OF FORESTS

To fortify understanding of conversion surplus, let's examine the function of this value measure. What can it do to help one respond intelligently to professional questions? And what *can't* it do? Let's begin by understanding the difference between (1) conversion surplus and (2) "stumpage value" in the sense it is used in Chap. 48. Some of the material comes, again, from Guttenberg, writing, this time, nearly 20 years later [1].

One difference between stumpage's conversion surplus and "stumpage value" is that conversion surplus must almost invariably be estimated by an analytical approach, whereas there are various ways of estimating stumpage value. So, in drawing a comparison, let's think about analytical appraisals.

Let's also recognize that the two measures of value differ in viewpoint. Stumpage value can be seen from the view of either the stumpage's current owner or its prospective owner; it is useful to both seller and buyer. Conversion surplus, on the other hand, is designed for the use of the firm that contemplates converting some material—stumpage, logs, or whatever—into a product. It is a guide to the firm's channeling of this material.

The conversion surplus of any material is defined as the difference between the selling price of the material's prospective product and the variable costs of converting the material into this product. The product is what the firm sells—what leaves the firm's hands. Of course, it's true of *any* valid measure of value that it is product price minus variable costs of conversion. It's certainly true of stumpage value. The difference in magnitude, if any, between the conversion surplus of given stumpage and its stumpage value is in the identity of the costs that are regarded as variable. With conversion surplus, some of the costs may be fixed with reference to the question at hand. Go back to Guttenberg's "paradox" for a moment. The question is how to channel the top logs, considering that if the company doesn't put them through the sawmill, it won't find other logs to saw instead. Given this question, all the overhead costs of milling, and of logging, too, are fixed: the costs of owning the mill and logging equipment, the salaries of permanent employees—all the costs that would persist no matter how the company channeled its logs: \$70,000 per year altogether.

Is it conceivable that stumpage's conversion surplus and stumpage value could be numerically identical? Certainly. What is the conversion surplus of stumpage that the owner will sell, incurring no special cost of sale? It is the selling price of the end product as it leaves the owner's hands (stumpage value) minus variable costs of converting the stumpage to this end product (zero). Conversion surplus in this case, as in any case, is serviceable for making the channeling decision: To which buyer should the stumpage be sold? Reply: Other things equal, to the buyer who will give the stumpage the largest conversion surplus: the buyer who will pay the most for it, above the cost of sale.

> To the owner-seller of stumpage, what is the material? What is the product? What is the process of conversion? What does the conversion cost consist of? Is it variable or fixed? How does the conversion process resemble the process of stock supply?

CONVERSION-SURPLUS VALUE OF TREES

Now turn to another numerical example, this time an individual tree. The tree is a Douglas-fir, diameter at breast height 40 in, net volume 2,700 bd ft by the Scribner Decimal C log rule. The tree is owned by a firm that operates a logging show and manufactures plywood, structural lumber, and pulp chips. What is the tree's conversion surplus in 1985 dollars?

Net sales value of end products		$1,146.72
Variable costs (mostly labor) of converting the tree to end products:		
Felling tree	$ 7.86	
Bucking stem into logs	7.85	
Yarding logs	49.36	
Loading, trucking, unloading logs	65.97	
Manufacturing end products	396.40	
Selling end products	4.42	
Total variable conversion cost		531.86
Conversion surplus of the tree		$ 614.86

Let's analyze this $614.86 conversion surplus so as to understand its meaning and appreciate its usefulness to the firm. Let's imagine that the firm has extensive logging operations under way in stands that include the tree at hand. These operations and the steps of hauling, processing, and selling that succeed them are hoped to be profitable in the sense that *all* the firm's costs—of forest ownership and management and of owning and managing the conversion facilities—will be covered by the after-tax sales value of end products, with enough to spare to pay a return to the owners.

The manager, let's suppose, approaches the tree and asks himself whether the tree is worth felling. (Other considerations may lead him to have the tree felled even if it's a cull, but let's keep the story simple.) Suppose, the manager says to himself, that we fell this tree. We will incur a $7.86 cost that we would not incur otherwise. Are there products in the tree that will make felling this tree worthwhile? The manager turns to his data and finds that the butt log alone will pay for felling this tree. If just the butt log is separated from the felled tree and is then yarded, taken to the mill, and converted to end products, the net return from these products will pay all the variable costs of these steps from felling onward, and there will be *something* left over as a contribution toward fixed costs and profit. In other words, the tree has positive conversion surplus and is worth felling.

While the tree is being felled, let's review the notions of variable and fixed felling costs. The variable cost of felling is all the cost incurred if the tree is felled but not incurred if it isn't felled. It includes the crew's wages and all the firm's related outlays in the crew's behalf during the period between their arrival at this tree and (bucking aside) their arrival at the next tree. It includes these wages and related outlays because the crew would have something else to do if this tree were not felled. The crew would not stand idle. Compare the contrasting *fixed*-cost case in the "Paradox" where the company does not obtain timber to offset what it diverts to pulpwood. The variable cost of felling also includes fueling and maintaining the crew's chain saw, so far as this tree is responsible, and strictly speaking, it also includes attributable depreciation. And that's all. It does not include the cost of supervision: The supervisor has to be around anyhow. It does not include any part of the cost of getting workers and equipment into the woods and out again. This cost is incurred in any event. All such costs are fixed with reference to felling this tree.

Now the tree has been felled, and the next question the manager puts to himself (let's imagine) is how many logs to take from the stem. Suppose that the manager starts at the butt and appraises the logs successively, asking, for each one, whether it will pay its way through the conversion process. Will the products from this log yield a net return great enough to pay the variable costs of bucking, yarding, hauling, manufacturing, and selling, with something *(anything)* left over to help pay fixed costs and profit? If so, the log has positive conversion surplus and should be bucked and forwarded up the line.

When the point on the stem is reached where the next log length has no positive conversion surplus, this is where utilization, through this particular conversion sequence anyhow, ceases. The total conversion surplus of the tree has been maximized—at $614.86. This is the tree's contribution to paying the firm's fixed costs and profit. It is on the strength of such contributions, from this and the other trees on its harvest areas, that the firm stays in business with net returns as its goal.

Now it becomes evident that $614.86 is the *value* of the tree to the firm. It is the tree's exchange value: The firm gives up the possession of the tree and receives, in return, $614.86 that it would not receive in the absence of the transaction. Now, also, the character and origin of the values used in Part Three to solve problems of stocking, thinning, financial maturity, and so on become evident. All those values are conversion-surplus values! Conversion-surplus values are used in such problems because the problems are of value maximization, and conversion surplus is the value that the rational tree-owning firm seeks to maximize.

The conversion surplus of a tree is not its stumpage value to a buyer in the sense of Chap. 49. If one were to calculate the stumpage value of the 40-in Douglas-fir to a buyer, allocating overhead costs and profit allowance on the basis, say, of Scribner board feet, one would find that it is $340.75. This value, remember (selling cost aside), is the conversion surplus to the stumpage seller. Here again one sees conversion surplus working as a guide to raw-material channeling: The firm chooses the channel in which the tree's conversion surplus is maximized.

In the case where conversion surplus is $614.86, the tree's end-products sales value can be likened to the firm's marginal revenue per unit of output; and the variable cost of conversion, to marginal cost per unit of output. Conversion surplus, then, is the difference between marginal revenue and marginal cost. Economic theory teaches that the firm's best combination of resources is the combination that *equates* marginal revenue and cost, so that the difference between them, for the very last unit (of output, in this case), approaches zero. The best combination is here because it is here that the *total* excess of revenue over variable cost is maximized:

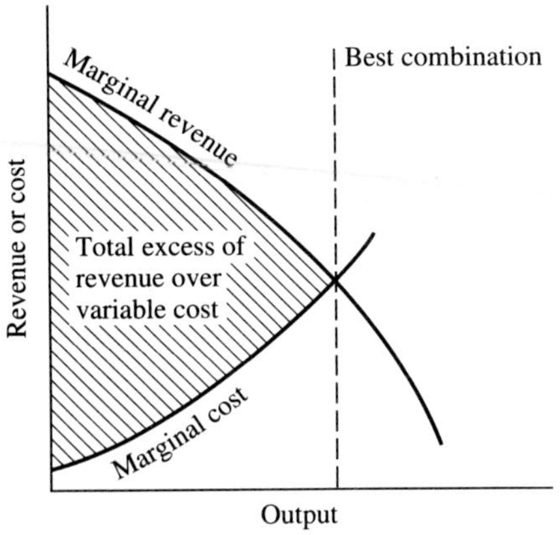

The best combination carries no guarantee of profit. All the firm knows is that at this point, its total net revenue is greatest or total net loss the least. Which one it is depends on fixed costs.

So it is with conversion surplus. The firm plans to proceed up the tree stem with its log-making until the conversion surplus of the next log is zero or less. In this way, it maximizes the conversion surplus of the tree. In the same way (though we didn't go into it), the rational firm proceeds down the ranks of trees, stopping its utilization at the trees too poor to have any positive conversion surplus. Thus the stand's conversion surplus is maximized: The fund is maximized from which fixed costs—and profit, if any—can be paid.

Is it inconsistent to state the firm's financial aim as maximizing the excess of marginal revenue over cost, and then to restate it as equating marginal revenue and cost? Why not? What "run" is assumed in the statements? long run? short run?

CONVERSION SURPLUS AS COMPARATIVE VALUE

This chapter is entitled Conversion Surplus as a Measure of Value.

"Oh ho!" someone may exclaim. "A measure of value, is it? Will you pay $614.86 for that Douglas-fir tree?"

"No," you are obliged to concede, "I'll confess I won't. Some of a buyer's costs are left out."

"Then it isn't a measure of value after all, is it?"

Conversion surplus is a measure of **comparative value.** It is a measure for comparing alternatives. The $614.86 is the current value, to the tree's owner, of cutting and processing the tree, when the value of not cutting it is zero. The owner may forecast the tree's conversion surplus as of the end of the next cutting cycle and find that it so greatly exceeds $614.86 that the tree is not now financially mature. In this case, cutting the tree now is a good alternative, but cutting it later is a better one.

If you are thinking of buying the tree (as opposed to not buying it), you can calculate its conversion surplus to *you* as a buyer. You can start with the $1,146.72 net sales value of end products and subtract *your* variable costs of conversion and an allowance for profit, amounting to a total of $805.97. The difference, the conversion surplus, is the stumpage value by analytical appraisal, $340.75. Would you pay this much for the tree? Presumably so.

SUMMARY

To put the story of conversion surplus in a few words, this measure of value is a guide to the channeling of raw material or goods-in-process. Calculated as the difference between product sales value and the variable costs of converting the material to the product, conversion surplus states the value of the material to the owner firm. That channel of utilization is preferable which offers to maximize conversion surplus.

Like the total excess of revenue over variable cost, conversion surplus is the fund for paying fixed cost and profit. The financially rational firm aims to carry each of its conversion processes to the point where surplus is just extinguished for the last units of output. In this way, total conversion surplus is maximized.

REFERENCES

1 Duerr, William A., and Sam Guttenberg: "Conversion Surplus as a Measure of Value," *Forestry Chronicle,* August 1968, pp. 24–27.

2 Guttenberg, Sam, and William A. Duerr: *A Guide to Profitable Tree Utilization,* Southern Forest Experiment Station, Occasional Pap. 114, New Orleans, November 1949.

CHAPTER 51

HIGH POINTS IN PART FIVE

1 Marketing, in a narrow sense, is the exchange of information and of title and the performance of complementary services that would take place in a perfectly competitive economy. In a broad sense, marketing is the foregoing plus market planning, product or buyer differentiation, and market administration such as are essential for the imperfectly competitive firm.

2 A market is the context of location, time, persons, and activities in which marketing takes place. Any circumstances in which choices are made may be viewed as a market.

3 Market organization comprises three characteristics of an industry. First is its structure: degree of concentration, product or buyer differentiation, barriers to entry, and so on. Second is its conduct: firms' policies regarding price, product, and rivals. Third is its performance: contribution to attainment of national economic goals. Forest resource management and forest-products production, manufacture, and marketing can be analyzed helpfully in terms of structure, conduct, and performance.

4 Market geography is a study of where marketing activity is located and why it is so located. It includes a study of timbersheds and their price and quantity structure and also an analogous study of sales areas, though in this book such areas are recognized only in passing.

5 Real prices of lumber and of sawtimber stumpage have long shown a strongly upward trend, while those for pulp-paper and for pulpwood stumpage have long held steady or drifted downward. Differences between the industries in gross labor productivity help explain the price-trend differences. So do differences in competitiveness.

6 In the midst of its trends, the forestry economy undergoes cycles of activity. The business and building cycles are prominent. Besides these, a medium-length interwar cycle and a long conservation cycle may be recognized.

7 Appraisal of forest resources, which is to estimate (*not* determine) market value, may be carried out by judging

a Explicit value (as of timber) by means of
- **(1)** Transactions evidence
- **(2)** Analytical appraisal
- **(3)** Conversion-surplus model

b Implicit value (as of park resources) by means of
- **(1)** Travel-cost model
- **(2)** Consumer-survey model
- **(3)** Multiple-use model
- **(4)** Input-output model
- **(5)** Market-conference model

Any value appraised in money units may be used as appraised, changed to allow for uncertainty, adjusted to estimate loss as a difference between alternatives, or moved in the time dimension with the help of compound interest.

8 The value of a resource rests on three kinds of attributes: services that the resource performs. These are a-service, in which the resource is used directly; b-service, in which it is seen or heard; and c-service, in which it is consumed (whether pleasurably or not) by knowing of its existence. Generally speaking, the appraisal methods are effective for a- and b-services, but not for c-service.

PART **SIX**

SOCIAL INSTITUTIONS

CHAPTER 52

PREAMBLE TO PART SIX

Let's take **social institutions** to mean the arrangements in the forestry economy at large that vary from one culture to another. They are arrangements that can't be validated in mathematical language. You and I have been taught to accept some particular arrangement as normal, though we recognize that there are others who hold a different view. We take it for granted that we are right or at least not strange. And the others take it for granted that they are right or at least not strange.

INSTITUTIONAL ECONOMICS

To be sure, our whole economy is at root institutional. And economics—ours or someone else's—is dialectical. Yet on the surface are some apparently undeniable principles: Net benefit is maximized if, as marginal return diminishes, marginal benefit and cost are equated. The features of a project that are fixed with reference to the alternatives are irrelevant to the choice. Many of the 51 chapters that precede Part Six rest on such principles as these. The many that don't, though clearly institutional, are admitted into Parts One through Five because they demand to be recognized as belonging to the geographic setting, the consumer, the genetic or manufacturing production, or the market.

Now, however, comes a group of chapters that has its own institutional identity. If the chapters relate most closely to any of the foregoing groups, it is to *setting*—though not to *geographic* setting. They close the circle of subject matter as they return to context in a new sense.

> How do you explain the fact that Culture A accepts Procedure A and thinks of other procedures as unacceptable, while Culture B does the same respecting Procedure B? Is there any way to find which culture (which procedure) is right? If there is, how do you define *right?* Can you straighten out the apparently confused thinking involved in these questions?

Examples of social institutions in the current meaning are legislation, custom, and articles of faith about the working of the forestry economy. Six sets of such laws, customs, and articles of faith are taken up. First are those concerning labor and workers. Second comes land, its ownership and use. Third are public fiscal and monetary policies. Fourth is income taxation. Fifth is property taxation. Sixth, social planning, completes the list.

The space devoted to taxation reflects the dominance of this institution in the working of our forestry economy.

SUMMARY

This preamble makes two points. First, it tries to justify segregating certain social institutions into a separate part of the book. The institutions discussed don't fit in earlier parts. And as they relate to forestry's social context other than geographic, they close the book on an appropriate note. Second, the preamble sketches the subject matter of Part Six: labor, land, financial policy, taxation, and social planning.

CHAPTER 53

LABOR IN FORESTRY AND WOOD INDUSTRY

Who are the people that work in the forestry economy? What kinds of jobs do they have? Are these attractive jobs? What are the working conditions? the rates of pay? the hazards of work? What are the relations between employees and employers? Are employees unionized? Do they change jobs often? We can look at highlights.

In the forestry economy, where capital is characteristically the dominant agent of production, the topic of labor is nonetheless of great moment. The human beehive of production and supervisory workers and consumer-producers creates forest resources out of what is otherwise neutral stuff. The individuals come in all degrees of talent, productivity, ambition, fortune, and control over their own and other's actions.

INDUSTRIAL REVOLUTION

Today's labor economy dates from the industrial revolution of the latter part of the eighteenth century in England. Before then, goods were commonly made by families in their homes. Then about 1770, James Watt invented the modern condensing steam engine, and "the earth became pregnant with new earths." The engine could drive machines installed in factories governed by capitalists. The capitalist families also operated mines to feed coal to the engine boilers. They hired adults and children to work from predawn to dark for starvation wages. These workers carried economic development on their backs, and for many years the law, like their masters, largely shrugged off responsibility for their circumstances. A publicist of these troubled times was Charles Dickens. His novels spoke gently of poverty, exploitation, greed, and injustice.

The London of those frightful days was the London also of Karl Marx. This man studied the industrial revolution. The message he read was not of want for loving human kindness and generosity. It was a strident insistence on one more class struggle. "The proletarians have nothing to lose but their chains." Meanwhile, the capitalists' coachmen were rounding the street corner below Marx's sordid room. The cracking of their whips aggravated the man's stomach ulcer pains.

In the U.S., our industrial revolution dated from Civil War times. Our cities and our capitalist factories grew and flourished as the nineteenth century careened on into the twentieth. Our labor publicists were the "muckrakers," polemicists such as Upton Sinclair, whose most famous novel cries out against the inhuman treatment of a family of Lithuanian immigrants employed in the Chicago stockyards.

LABOR UNIONS

U.S. labor never suffered grinding mistreatment on a scale comparable to England's, because labor in the critical years had an escape route to the rural west, where our liberal land laws encouraged settlement and proprietorship. Nevertheless, there was plenty of strife between labor and employers, and the law for long years looked the other way. In the wood-using industries, the United Brotherhood of Carpenters and Joiners of America was formed in 1881. Carpenters were traditionally the trade responsible for tree cutting and processing, and the Brotherhood early proclaimed its jurisdiction over "all that is made of wood or that was ever made of wood"[7]. In 1990, it listed a membership of 700,000 in 1,568 locals across the country [1]. Another early union (1905) was the Industrial Workers of the World (IWW), active in the logging camps and sawmills of the West Coast near the turn of the century. Radical and militant, the IWW came to be feared by the general public and by the moderates within the labor movement. It drifted into oblivion as its former sympathizers grew fat and contented.

Altogether, five labor unions with a reputed total of well over a million members listed themselves in 1990 as active in forest products: logging, lumbering, panel manufacture, construction, furniture, and pulp-paper. Among their activities are lobbying for favorable legislation, supporting friendly political candidates, collective bargaining, educating members, supporting research in their behalf, and publishing periodical house organs. These unions are found in the big concentrations of industry: pulp-paper nationwide and other lines of production chiefly in the West and South. The national total union membership in three branches of wood industry in 1980 was:

	Thousand members	Percentage of all workers
Paper and allied products	455	65
Furniture and fixtures	215	47
Lumber and wood products	215	33
	855	49

Even nowadays, much of the forestry economy is too fragmented and scattered and fiercely conservative and paternalistic to be unionized. As a union organizer, what appeal and what pressure can you exert upon the employees in a little sawmill whose owner serves as head sawyer and hires a half-dozen of his neighbors to help him run the outfit? Chances are, you will get cold shoulders all around and be told to go back to the city where you belong.

However, a union is a union, traditionally the natural-born enemy of management. When the great peak association of management, the National Association of Manufacturers (NAM), was formed in 1895, its overriding aim soon became to suppress labor organization. It used all conceivable tactics to fight unions, from gentle persuasion and injunction to the billy and gun. An early NAM leader proclaimed that "organized labor knows but one law, and that is the law of physical force—the law of Huns and Vandals, the law of the savage" [3]. Within wood industry, many a goon squad was formed to fight union action.

How do you explain the traditional antagonism between management and labor?

Now we have a new world. Is it more civilized? Certainly it is more regulated. A series of laws commencing in the 1930s with the New Deal controls wages, hours, working conditions, and a host of benefits; provides for orderly disputation and arbitration; and strikes a balance of sorts between the powers of the contending parties. Their commando posture has been replaced by sound bureaucracy. And meanwhile the environmentalists have come on stage to provide a common, even a uniting, adversary.

However, from its peak in the mid-1970s, union membership in most branches of the U.S. economy has drifted downward. Unions in general were jolted by the U.S. president's 1981 decision to fire striking members of the Professional Air Traffic Controllers Organization. An AFL–CIO spokesperson commented that "other employers, public and private, interpreted this as a declaration of open season on unions and went all-out to block, weaken, or be rid of them" [4]. By 1990, unionization had sunk to 15 percent of the labor force, down from roughly a third at the peak [8]. Forest-related employment followed a parallel trend. For example, in the three years following the traffic controllers episode, membership in the carpenters union was said to have fallen from about 680,000 to 620,000 [6], and the latter figure may be an overstatement.

FORESTRY WORKERS

Employees in our forestry economy number some 2.6 million, more than $2\frac{1}{2}$ percent of the U.S. total. Most of the story is in Table 53-1. Besides, there are perhaps 80,000 persons (20 percent female) employed in managing forests and related lands. Manufacturing paper and allied products and marketing all wood products are the principal sources of employment in the forestry economy, together accounting for about two-thirds of all forestry-related employees.

TABLE 53-1
AVERAGE NUMBER OF EMPLOYEES ON NONAGRICULTURAL PAYROLLS, BOTH SEXES AND PERCENTAGE OF WOMEN, AND AVERAGE HOURLY EARNINGS, BOTH SEXES, 1986

	Employees		
Industry	**Total, in thousands**	**Women, %**	**Average hourly earnings, in dollars**
Manufacturing wood products:			
Logging	81.5	15.7 (Logging through Other "lumber and wood products")	11.11
Sawmills, planing mills	199.2		8.59
Veneer, panels	256.0		8.23
Wood buildings	70.9		7.63
Other "lumber and wood products"	119.2		6.76
Wood household furniture	132.2	31.2	6.36
Pulp, paper, paper products	689.9	24.5	11.14
Wholesale trade in wood products	351.4	29.1	9.17
Retail trade in wood products	643.3	52.4	7.24
All wood products	2,543.6	30.0	8.84
All nonfarm	100,167.0	45.7	8.75

Source: U.S. Department of Labor, Bureau of Labor Statistics: *Employment and Earnings,* March 1987, pp. 44–55, 83–95.

Table 53-1 also reports on average hourly earnings in each of the forest industries. The data illustrate some of the weaknesses of public census statistics on forestry. Note, for example, the first figure in the earnings column, $11.11 per hour in logging, nearly the highest figure in the column. This is not realistic information. There are two causes. First, the Bureau of Labor Statistics (BLS) census is taken by mail and therefore requires mailing lists. Where are loggers' names and addresses to be found? Mainly among the big, stable, unionized outfits in the West, and this is where wage rates are highest. Eastern loggers, by and large, are too numerous, insignificant, and mobile to be listed. The problem that a census-taker has in pinning down loggers is stressed on page 299. Second, the BLS census concerns employees, an excellent base of information in most industries. But in logging, much of the work is done by the self-employed. See the comment on poverty among the southern shortwood loggers, page 261.

The deficiencies of data found in logging extend also into lumber and other fragmented lines of activity. You can well bear the deficiencies in mind as you look through the statistics in this chapter and read about them. Thoughtfully interpreted, the relative standings of the occupations are useful information. But by and large, the absolute annual values vary a good bit over time: For either depressed or strongly prosperous times, current data such as the Census provides may be misleading [10].

> Study the three columns of figures in Table 53-1, disregarding the last two lines. Which of these figures would you expect to fluctuate from year to year by the greatest percentages? Why? Which figures would you expect to fluctuate the least? Why?

Pay rates vary significantly by geographic region [2]. Let us recognize the South as the 12 states from Virginia to Texas, the North as all else east of the hundredth meridian, and the West as the remainder of the U.S. Then we come upon the interesting finding that for the past 80 years, there has been a nearly constant relationship among regions with respect to hourly pay rates in *all manufacturing,* including wood products. Let the average pay rate in the North be represented by an index of 100. Then the three indexes are

West	107
North	100
South	80

In the lumber and wood products industry, including logging, the South shows an index of 75 that tends to rise; the West, an index of 140 that tends to move downward. In paper and allied products (logging excluded), the southern index rose above 100 in the late 1950s and continues slowly upward; the western index fluctuates between 115 and 120.

The general tendency of southern wage rates to be depressed is perhaps attributable to capital shortage and to the region's focus on land as the prime agent of production. Low wage rates are chargeable also to the South's deficient **human capital:** the amount of education and work experience embodied in the labor force. The region's good showing in pulp and paper is evidence of this industry's leadership in the "New South." Among southern industries, only petroleum and publishing pay higher wage rates than pulp-paper.

The West's extra wage-rate advantage in timber products is evidently a reflection of timber-resource quality: large tree diameters and heavy per-acre stocking.

One can throw more light on the question of human capital. Some branches of the forestry economy show up poorly with respect to human capital. Consider workers' median number of years of schooling:

Land management	12.4
Paper and allied products	12.1
Lumber and wood products, except logging	10.2
Logging	9.0

The median for all U.S. labor is 12.3 years. Land management's good showing, and pulp-paper's also, stems from the relatively high proportion of college-trained professional workers.

> What changes might have to occur in order for the median number of years of schooling to be more nearly equal in all four branches of the forestry economy?

Or consider the percentage of employees on duty for 50 or more weeks per year:

Pulp-paper	81
Land management	71
Lumber, etc.	67
Logging	45

Or labor turnover rates: markedly higher in lumber, wood-products, and logging than elsewhere.

Or injury rates, in which logging stands out as one of the most dangerous of all U.S. industries:

	Frequency	Severity
Logging	42	6,160
Lumber and wood products	34	2,890
Paper and allied products	14	940
Furniture and fixtures	22	910
All manufacturing industries	15	760

Frequency here means number of disabling work injuries per million man-hours worked. *Severity* means number of days of disability per million man-hours worked.

What kinds of jobs do the workers do? If one recognizes seven occupations, the numbers of workers for (1) lumber and wood products, including logging, (2) furniture and fixtures, and (3) paper and allied products look about as given in Table 53-2.

In the forestry economy, as one might expect, workers are rural people to a greater degree than one finds elsewhere, except in agriculture. The percentage of male workers who reside in rural areas is 80 in logging, 60 in other lumber and wood-products manufacture, 45 in furniture and fixtures, and 30 in paper and allied products—as compared with 25 in the rest of the nonfarm economy.

TIMES STILL CHANGING

Thus the hierarchical order—from particles and slices to slabs and chunks—that we have seen so often as manifesting the advantages of pulverization shows up prominently in the field of forestry labor. Workers in logging and lumber manufacture have

TABLE 53-2
EMPLOYMENT IN THREE BRANCHES OF WOOD-USING INDUSTRY, BY OCCUPATION, 1980

Occupation	Lumber and wood products, No.	Furniture and fixtures, No.	Paper and allied products, No.
Management	41,070	24,300	37,340
Professional staff	9,890	10,200	30,190
Technical	5,090	3,610	10,500
Service	13,520	8,510	11,570
Production, material handling, maintenance	534,720	348,250	519,530
Clerical	41,920	50,490	73,240
Sales	11,060	10,470	16,540
All occupations	657,270	455,830	698,910

Source: U.S. Department of Labor, Bureau of Labor Statistics: *Handbook of Labor Statistics,* Bull. 2217, June 1985, p. 225.

long appeared, by almost any criterion, disadvantaged. Traditionally, the work of the little logger and sawmiller was looked down upon as inferior, undesirable; and this social stigma carried over to some extent in all woods work and wood-manufacturing employment.

But, as recognized in Part Four, times have changed and are continuing to change. The character of wood-using industry has been remolded by economic and technological forces, both in the West and in the East. The same thing has been happening in the logging industry, especially in the once deeply disadvantaged eastern industry. Equipment has come in to replace human muscle. Many of the disadvantages of work in the changeable and uncertain outdoors—the heavy, nonuniform raw material, the variable terrain and weather—are being overcome. Attention is being given to the training of woods and mill labor; schools have been established, where the courses cover not just equipment operation and safety, but also planning, recordkeeping, and other facets of the job context. Today's logging-injury frequency of 40 and severity of 6,000 are high, but they are down from 65 and 7,000, respectively, in two decades.

> Why should the degree of subdivision of an industry's raw material or product have a bearing on the status of labor in the industry?

On page 29 in comments about the comparative advantage of the Japanese log-processing industry, a remark is made about the high labor morale and skills in the industry. Japan's system of industrial labor relations has produced some close bonds between management and labor, even in the largest firms. Workers have received training, personal attention, and fringe benefits without the necessity for striking a posture of adversary vis-à-vis the employer. Here were re-created the attitudes of the old American family lumber business, strengthened in a modern sense by concern for workers' safety, convenience, and advancement. Although the Japanese labor system has exhibited faults in recent years [5, 9], it is an interesting and perhaps useful question whether the system has features that would strengthen the American wood industry of the future.

In any event, you and I have a major concern for labor in genetic forestry and wood industry. Our concern is this: What advancement in the well-being of workers has the forestry economy accomplished? And what are the prospects? In the short-term ups and downs, we see, of course, periodic depression for industry and its labor. We see close at hand the slump of the early 1980s, with its inflation, high interest rates, and uprooting, migration, and bankruptcy of industrial activity, amid loss of markets and jobs. And even closer at hand, we see the 1990s recession following a wide swing in activity.

We see prosperity and depression for professionals, too, linked to redefinition and deemphasis of forestry, especially in the federal government. But, reading the longer trend, we are obliged to temper pessimism. We have witnessed truly remarkable improvements in safety, other working conditions, productivity, material rewards, and social status of forest-related labor. And for professionals, we can foresee renewal accompanying orientation to the new world.

SUMMARY

The long and short of it is this: The labor force in forestry and wood industry, especially outside the pulp-paper industry, has traditionally lagged behind most branches of the American economy with respect to social and financial status. The same conclusion is derived from most of the criteria studied: types of jobs, working conditions and safety, turnover, education, pay rates, employment of women, unionization. However, the status of forest and wood-industry workers is improving. There is some reason to believe that the workers are gaining on those in other industries—and that the depression in professional activity, too, is only temporary.

REFERENCES

1 Burek, D. M., ed., Gale Research Company: *Encyclopedia of Associations,* New York, 25th ed., 1991.
2 Kaiser, H. F., Jr.: "Interregional Wage Differentials in Forest Products Industries," *Forest Products Journal,* July 1973, pp. 12–16.
3 National Association of Manufacturers: *Proceedings, 8th Annual Convention,* Washington, April 14–16, 1903.
4 *Time,* March 26, 1990, p. 57.
5 *Time,* March 23, 1992, pp. 44–46.
6 U.S. Department of Commerce, Bureau of Labor Statistics: *Directory of U.S. Labor Organizations,* 1986–1987 ed. The Bureau of National Affairs, Washington, p. 4.

7 U.S. Department of Labor: *Brief History of the American Labor Movement,* Bull. 1000, 1950.

8 *The Wall Street Journal,* Eastern ed., November 30, 1990, p. A14.

9 *The Wall Street Journal,* Eastern ed., November 19, 1991, p. A1.

10 Webb, Roy H., and William Whelpley: "Labor Market Data (including a section on "Suggestions for Further Reading")," *Economic Review* (Federal Reserve Bank of Richmond), November–December 1989, pp. 15–22.

CHAPTER 54

LAND-USE REGULATION

LAND OWNERSHIP AND USE

Something is said in Chaps. 29 and 44 about the economics of land use—about how the ownership or use of a given tract tends to fall to the activity that can occupy it to best advantage: the one that can pay most for it out of the proceeds of occupancy. The principle is that land, like much else in an economy of rivals, goes to the highest bidder.

But circumstances may intervene to thwart land's efficient use. The performance of the land market may not be good. For example, the market may be monopolized or it may be controlled by forces that serve other than social goals. And then again, society may decide in a particular case that economic efficiency is not what it wants. For instance, a community may prefer relief from noise and congestion to the efficient location of transportation routes.

Wherever the play of our social-political-economic system produces unpalatable results—results that society can be led to oppose successfully—deliberate regulation of human behavior is the outcome. If people in the house next door are engaging in gunplay, you and your other neighbors may ask the police to step in. If the owner of a landmark historic mansion plans to tear it down because it costs so much to maintain, the county may prevent the demolition by purchasing the property.

Land is earth surface and its attributes: what lies above and beneath the surface, the physical climate, the social context. **Land ownership** is holding a bundle of rights in the use of land. The bundle is never, in any culture, completely unlimited: There are always quite a few things that owners are prohibited from doing on their own land or others' land. That is, society always retains for its own disposal some of the rights in the land-ownership bundle. Thus society is able to change owners' rights in land use, withdrawing some and perhaps adding some. Such changes in the bundle are acts of public regulation: calling the police, buying the historic property. Public land-use regulation is the subject of the present chapter.

Let's first remind ourselves of the principal means for regulation. In the final analysis, they are the two just used for illustration: (1) resort to the law and (2) the purchase of rights. Then let's take a recent example.

> If land is indeed the earth's surface and all its attributes, then *land use* in the last analysis includes all activity. Try your hand at defining *land use* so that it is a part of forestry economics in a reasonably acceptable sense.

REGULATION BY LAW

Of all the means open to the public for regulating land use, probably most fall under the general heading of resort to law. The law may be statute or common law, or it may be less formal: dictates of the culture that have no immediate legal connection. Under the U.S. system, the federal constitution lays the legal foundations and is the final recourse when the validity of any law is in question. The first ten amendments to the constitution, the Bill of Rights, specify some types of behavior (land uses, if you like) that are to be encouraged. They are not prohibitions, as are so many of our laws. The constitution assigns a set of responsibilities to the federal government and leaves the rest to the states.

Speaking of rights, an important one that the owner of land retains under our law is the right to prohibit trespass on the property. Thus the forest owner may "post" the land boundaries so as to exclude hunters. This is not an instance of *public* regulation, but it's worth passing notice.

One of the powers implicitly left to the states by the federal constitution is the police power. It is under this power in recent decades that the states—and by delegation, the local governments—have promulgated valid land-use regulations. Ever since the late 1930s, the court's view of police power has continually become broader. In the 1950s, the U.S. Supreme Court established the principle that the police power is not "confined to a narrow category," but "extends to all the great public needs," and that states and their subdivisions may legislate to make the community "beautiful as well as healthy." The goals of legislation may be "spiritual as well as physical, aesthetic as well as monetary" [5].

Under the police power, the states may regulate private forest-land practices by legislation. California, Oregon, Washington, New York, and Massachusetts were among the early states to have forest practice acts [6]. In the 1930s and 1940s, there was some sentiment in favor of federally inspired regulation of forest practices, but it did not bear fruit [2]. The states have come through as the principal public regulators of private forestry practice [1].

Perhaps the chief land-use regulatory measures that flow from the police power are **zoning** measures. The states authorize their counties and towns to pass ordinances that map communities into various categories of residential, commercial, and industrial zones and open space. Such segregation of activities is undertaken to promote good living and working conditions, successful business, maintenance of property values, and the like. We think of zoning as an urban practice, but rural zoning has been tried, too. In northern Wisconsin in the 1930s, a time of widespread depression and property-tax delinquency, zoning was tried as a means for clustering habitation and leaving the rest of the land to the forest. A principal purpose was to cut the cost of community service, especially schooling and roads, and thus to ease the tax load. Another example of rural zoning, a current example from northern New York, will enter this chapter shortly.

Note that the function of zoning is, in a sense, negative: to prohibit certain uses of land, not primarily to promote certain uses.

Another law-oriented device for regulating land use is taxation. Property taxation, a function of the states and their subdivisions, is the prime illustration. The property-tax law and its administration can be designed to encourage or discourage the forest use of land or to foster specific management practices. As you see in Chap. 58, taxation achieves such ends by rewarding or not penalizing the taxpayer who carries them out. Income taxes bear upon land use; examples are given in Chaps. 56 and 57.

In recent times, both state and federal governments have taken to statute-writing in defense of environmental conservation. They have passed acts regulating air quality, water quality, disturbance of the soil surface, changes in the landscape, and other matters that bear closely on land use. Chapter 10 touches upon environmental conservation.

REGULATION BY PURCHASE OF RIGHTS

Consider, now, the second broad category of approaches to land-use regulation: the purchase of rights.

Outright purchase of land is a perfectly clear case. One recalls that national forests in eastern U.S. were created by purchase, under the Weeks Law of 1911. Indeed, all public land projects, federal or not, that were not supplied directly from the public domain, took the purchase route. The public, once it becomes owner, can regulate land use by setting the appropriate policies of management.

However, the public needn't purchase all the rights in the land bundle in order to regulate use. Often, it's feasible to buy just the rights at issue: to buy an **easement.** Most easements are for routes of access across property. But there are other types of easement. A **scenic easement** ensures that the landowner's behavior will be governed so as not to impair a given view or landscape that society values. Under the principle of **transferable development rights,** these rights are identified as part of the ownership bundle. They may be purchased from a landowner by local government or a consortium of neighbors in order to prevent unwanted subdivision and residential development of the owner's property. Easements are binding on landowners who follow the original grantor, and they tend to lessen the market value of property.

Why should easements tend to lessen the market value of property? Phrase your reply in recognition of our definition of ownership, that it is the holding of a bundle of rights.

REGULATION IN NEW YORK'S ADIRONDACK PARK

Commonly, each instance of forest-land-use regulation concerns one or a very few of the uses to which forests may be put—that is, the potential *services* of forests. Furthermore, in each instance, allowing or preventing the use in question becomes an issue in the community. Typically there is a principal group or agency defending each side of the issue, and the general public lines up pro and con. So it is in the illustration taken up here. The forest use at issue is residential development. The principal contending parties are developers on one side and state regulatory agencies on the other. The means for regulation is zoning: a zoning project of unprecedented scale, involving some of the most restrictive land-use provisions in the country. The place is New York State's Adirondack Park [4].

The Adirondack Park, occupying 6 million acres in the mountainous northern extension of New York, is the largest park, state or federal, in the U.S. Its area is about that of the state of Vermont. It is an unusual park in that it comprises 40 percent public lands (the Adirondack Forest Preserve), 28 percent lumber and pulp company holdings, and 32 percent other private tracts. The preserve lands within the park were recognized in the last years of the nineteenth century by legislative statute and constitutional amendment as forest to be kept forever wild, where no further sale or removal of timber was to be allowed. The preserve was justified publicly as a backlog of forest resources and as a safeguard for its five major watersheds. Its creation provided protective backcountry for the large estates of railroad and financial notables from New York City. Of all the determined efforts by business, commercial, and conservation interests over the years to reopen the preserve for resource use, none ever succeeded. The "forever-wild" principle is supported by a politically invincible group of preservation organizations and newspaper publishers.

The Adirondack Park is a region of mountains, plateaus, lakes, and streams, heavily forested with young-growth hardwoods. Of the mountain peaks, 46 rise above 4,000 ft. Waterways total 5,000 miles; lakes and ponds number more than 2,000. In winter, some of the coldest spots in the U.S. are found here. Snows are deep, and before the snowmobile, the snow season closed activity over much of the land.

In the park, people live and work as well as play. Major sources of employment are forest products, mining, and tourism. There are $5\frac{1}{2}$ million acres of forest of commercial quality. Within the park are the largest open-pit iron, titanium, and garnet mines in the world, and mines for many other substances. In summer, the park's population increases from 125,000 to more than 9 million. These statistics, while they picture feverish economic activity, also attest to the region's devastating seasonality, with a third of all workers unemployed in winter. Here is a prime example of a wide-amplitude periodic cycle in the forestry economy (Chap. 46). Poverty abounds among the permanent residents, a condition reinforced over many generations by isolation from the rest of the world.

Then, in the 1960s, students of the region became impressed by the mounting changes in its life. The completion of Interstate Highway 87 from Albany to the Canadian border put the park within a day's drive of 55 million people. Isolation was fading and use burgeoning. Out-of-state developers were studying the purchase of huge land tracts. State agencies became alarmed over the park's vulnerability to unplanned, unwanted exploitation. The governor in 1968 appointed a study commission to consider the future of the park and submit management recommendations.

The study commission's 1970 report took the position that immediate action was necessary "if the Adirondacks [are] to be saved." It pointed out that 50 percent of the park's private lands are held by 1 percent of the owners, that 53 percent of these lands are in the possession of 626 owners, 3 of whom own more than 125,000 acres each, and 32 of whom hold more than 10,000 acres each. Of the forest lands, the state and the private owners have approximately equal acreages, which emphasizes the conclusion that protecting the privately held spaces is essential to maintaining the park's character. The commission recommended that an independent Adirondack Park Agency be created, with statutory authority over both private and public lands. Notably, the agency should have both planning and land-use control over the private lands.

Created in 1971, the Adirondack Park Agency produced a "master plan" for *state* lands. The governor approved the plan. This part of the agency's work went without a hitch.

Hitches appeared as the agency moved toward its second responsibility, the privately owned lands within the park. An issue was struck between residents' want of material income and the state's want to preserve the natural beauty of the region for everyone.

Early in 1973, the Adirondack Park Agency completed a plan for the use of private land. The plan classed each privately owned parcel into one of six categories, based on the degree of residential or industrial development that would be permitted. It required agency review of all significant development proposals within the park.

In mid-1973, the state legislature approved the Adirondack Park Agency's plan for the private lands. Thus the plan became law. For simplicity, let's refer to it as *the Act.*

CONTROVERSY OVER REGULATION

The Act came under attack at once. Take one instance as illustrative [2].

A year or so before the Act was passed, a developer, Paparazzo by name, bought an 18,500-acre tract of land near the village of Tupper Lake, population less than 5,000. In late 1972, still before the Act, Mr. Paparazzo proposed a project for his newly purchased land: Ton-Da-Lay, to be a second-home community of 20,000 residents, with camper-trailer space, "ranchettes" around a common range, condominiums, "estate" waterfront lots, and a central 3,000-acre area with shops, resorts, restaurants, and an Adirondack museum.

State approval for Ton-Da-Lay rested with the commissioner of environmental conservation, whose department reviewed requests for water-supply permits within the park. The commissioner, viewing the Paparazzo case as a precedent, held 29 days of hearings.

Mr. Paparazzo testified that he planned to develop only 8 percent of his land, though he admitted that his development plans might change.

The president of the Tupper Lake Chamber of Commerce testified that Mr. Paparazzo was different from most developers: He had hired a local attorney and bought a home in the village for his family. "They showed they wanted to be part of us," he said. He expressed his position in these terms: "We've got a bunch of super-birdwatchers who are against Ton-Da-Lay. They have to let us make a living. We have to survive, and if they put a fence around us and say No to development, we're in trouble. Ton-Da-Lay is going to upgrade our tax base. It will be good for our stores."

The Sierra Club representative, presumably one of the super-birdwatchers to whom the Chamber president alluded, testified that the townspeople were so poor and so pre-occupied with "just getting along" that they didn't have "time to think of the consequences."

An assemblyman testified, "It's a man's own property, and a man can do what he wants with it."

> What are the arguments pro and con the assemblyman's contention?

In mid-1973, just as the Act was being passed, the commissioner rejected Ton-Da-Lay's application. He explained that he was preventing subdivision of the park and destruction of its natural beauty of forests and lakes. He was not against all development, he said, just overdevelopment. He would consider a revised application from Ton-Da-Lay, but "clearly something much less than that proposed here."

Mr. Paparazzo declined to revise his application. Developers view sparse land occupancy with alarm. He decided to take his case to court. In order to test the authority of the Adirondack Park Agency, he laid out a subdivision, Jordan River Estates, with thirty 10-acre lots. He got a water-supply permit from the Department of Environmental Conservation and clearance from the town of Tupper Lake under its zoning and subdivision ordinances.

There followed a tangled series of court actions. The agency brought suit, asking for an injunction against Jordan River Estates. The case went to the state's highest court, the Court of Appeals, which refused to grant an injunction. However, in 1977, the agency did win from the Court a ruling that upheld the Act. Late in 1978, Mr. Paparazzo moved to appeal the ruling. He hoped that the case would reach the U.S. Supreme Court, but soon abandoned the hope. In 1991, the Act stood, though its constitutionality had not been tested. Mr. Paparazzo had sold his land to a wood-using firm. And the agency was embroiled in, among other matters, the question of the composition of its membership [3].

SUMMARY

It comes to this: Land ownership is the holding of a bundle of rights in the use of land. Society can regulate land use by changing the rights in the owner's bundle. Society has two principal means for doing this: legislation and purchase. For illustration, this chapter takes a legislative example, a large, complex, and far-reaching zoning project in northern New York.

REFERENCES

1 Cubbage, Frederick W.: "Public Regulation of Private Forestry," *Journal of Forestry,* December 1991, pp. 31–35.

2 Davis, Richard C., ed.: *Encyclopedia of American Forest and Conservation History,* Macmillan, New York, 1983, pp. 578–580.

3 *Glens Falls* (New York) *Post-Star,* November 17, 1991, p. B1.

4 Harvey, Heather: "A Concept of the Field of Land Use in Relation to Forestry," Ph.D. dissertation, Virginia Polytechnic Institute and State University, Blacksburg, 1979.

5 Honnold, John Otis, Jr.: "Police Power," *In Encyclopaedia Brittanica,* vol. 18, Chicago, Encyclopedia Brittanica, Inc., 1968, p. 156.

6 Jones, John Carter: "The Background of State Forest-Practice Legislation and Its Impact Upon Forest Land Use and Management," M.S. thesis, Virginia Polytechnic Institute and State University, Blacksburg, 1977.

CHAPTER **55**

INFLUENCES OF FISCAL AND MONETARY POLICIES

Monetary policy consists of goals and measures respecting the nation's money supply and interest rates. It is directed toward influencing economic activity as reflected in GNP, employment, prices, and the balance of international trade. **Fiscal policy,** on the other hand, involves government spending, taxing, and borrowing.

The principal public agency in the U.S. monetary policy is the Federal Reserve Bank, though the Treasury Department is also concerned and has partly overlapping responsibilities. In fiscal policy, the president and Congress make the decisions.

Among the dimensions of economic activity that fiscal and monetary policies influence, let's think about six that carry over prominently into the forestry economy:

1 Rates of interest
2 Rates of monetary inflation
3 Economic trends and cycles
4 Public expenditures on specific projects
5 Value of dollar abroad
6 Income taxation

The purpose of this chapter is to ask about the first five of these six dimensions of economic activity: *What is their influence on forestry?* The sixth, income taxation, is studied separately in Chaps. 56 and 57.

RATES OF INTEREST

Rates of interest as an outgrowth of monetary and fiscal policies are the rates on a variety of loans. The percentages vary within a range, depending on the term of the loan, the lender's and borrower's status, the type of security offered, and so on. A rate commonly referred to is the prime rate of interest, the rate charged by commercial banks to their most reliable corporate customers. Another is the rate paid by the U.S. Treasury on its short-term bills and longer-term notes and bonds. Another is the money-market rate, the average on a selected set of loans to high-class borrowers, made by money-managing firms on behalf of their investor-clients. In this chapter, the term **market rate of interest** is used to characterize the general level of interest on sound, short-term transactions.

Rates of interest impinge on forestry through their effect on the cost of stock holding. If you hold $10,000 worth of stock—land or improvements or timber or other forest resources or wood products such as a frame dwelling—you incur an annual cost in the form of interest. This is either **explicit interest** that you pay on the loan that let you acquire the stock or **implicit interest** if you negotiate no loan. Such implicit interest is your opportunity cost of holding stock: the revenue you forgo by having the resource as stock instead of liquidating it and earning interest on the proceeds.

When one ponders the matter, one realizes that stock-holding costs in the forestry economy are enormous, almost beyond belief, and that therefore the rate of interest is the determinant of what emerges as the prime cost of forestry. This leading role of interest is created because forestry is such a **capital-intensive** industry.

> What is your definition of forestry as an industry? How would you measure its capital-intensiveness? Would you measure it as value of capital (including land) per unit of labor? per acre of land? How do you think forestry ranks among industries with respect to capital intensity measured in each of these ways—or in others that you may choose?

Consider a rough, possibly conservative, appraisal of the three principal types of stock in the forestry economy—land resources, forest products, and processing and marketing facilities—as of 1990:

Stock	Value, in billion 1990 dollars
Land, 750 million acres: inherent scenic attributes; cover of timber (715 billion cubic feet), other trees, forage, and other living things; lakes and streams; improvements and underlying resources	400
Forest products: incorporated into dwellings of 82 million households and into other structures; furniture, fixtures, and other resources of forest origin; stocks in process of manufacture and marketing	700
Forest resource processing and marketing facilities: other assets of lumber, plywood, panel, and products manufacturers (36,000 firms); pulp, paper, and products manufacturers (57,000 firms); wholesalers, retailers, shippers	300
All stock	1,400

The total value of about \$1.4 trillion in 1990 was nearly half the amount of the public debt of the U.S. In other words, it was almost as large a stock value as one is likely to encounter in a perusal of the U.S. economy's dimensions. A 1-percent difference in the interest rate would make a \$14 billion difference in the yearly cost of forestry stock holding. Furthermore, a 1-percent *change* in the rate could affect *implicit* interest cost proportionately. Explicit interest cost, taking the form, as it does, of payments on mortgages and other long-term instruments, is less sensitive, as a total, to *current* change.

The interest rates generated by monetary and fiscal policies exert an influence on current forestry and forest-industry practice mainly through one of two avenues.

One avenue is the direct connection between interest rates and mortgage rates on new construction. Market interest rates set an opportunity cost of lending money on mortgages. The mortgage rate therefore tends to lie at or above the market rate. And the higher the mortgage rate, the more costly a dwelling or other structure, since buyers borrow, on the average, about three-fourths of the cost and pay the mortgage rate on the loan. Consequently, the fortunes of a major part of wood-using industry and its suppliers fluctuate counter to interest rates. When rates are high, structures are expensive and sales of construction materials are slow. (Refer again to Fig. 46-2.)

The second avenue is the relationship between market interest rates and the guiding rate of return of producer firms. A firm's guiding rate (Chap. 20) is a function of the firm's supply and demand for capital. Hence the guiding rate, like the mortgage rate, tends to lie at or above the market rate: If the firm is a lender, it is unlikely to lend at any interest rate lower than it can earn elsewhere. And if it is a borrower, it will borrow to invest in projects that promise to pay at least the cost of borrowing. Since the intensity of forestry is inversely proportional to the guiding rate of return, the fortunes of a major part of the forestry economy fluctuate counter to interest rates. (Refer again to Chaps. 24, 29, and 40.)

RATES OF MONETARY INFLATION

An aim of the Federal Reserve Board and of fiscal agencies is to hold the value of the dollar reasonably steady (page 322). Deflation was, until the 1940s, the traditional accompaniment of economic recession. As prices and even wage rates fell, sellers and borrowers with fixed dollar obligations were threatened with impoverishment. Nowadays, with both prices and wage rates heavily resistant to reduction, it's inflation that people fear. As the value of the dollar sinks, customers' real level of living falls, unless their current wage rates are given a compensating boost. And lenders' real repayments from borrowers tend to fall away.

What is the net effect of inflation on an economy? Conceivably, if all sectors in the economy inflate and are expected to inflate at the same pace, the influence of inflation is nil. All real values hold constant, and no one gains or loses as compared to an economy with a perfectly stable unit of value. Theoretically, inflation abroad must march in step with that at home.

The catch, to be sure, is that not all dimensions of an inflating economy inflate at the same rate. As a result, some sectors of the economy are advantaged, while others are disadvantaged.

> Which sectors tend to be advantaged: those in which unit values of output are inflating at a high rate or those in which unit values are inflating at a low rate? Why?

The forestry economy, because of its large base of land and durable capital, enjoys considerable gains in inflationary times—that is, when inflation is in prospect. In such times, people turn away from money and the promissory notes (bonds) of borrowers. They turn toward real estate and, other things equal, would rather borrow than lend.

In inflationary times, then, investments in the forestry economy are, by and large, encouraged. Investments in timber property are particularly attractive because timber values are rising so much faster than values in general (Fig. 45-2).

But inflation is by no means purely helpful. It raises interest rates, and high rates cause damage to the construction industry and other forestry sectors. Again, the majority of those workers who retain employment in the forestry economy probably suffer a reduction in real wages in inflationary times.

It would be helpful if one could characterize without qualification the influence of inflation on forestry. If one were forced to express a categorical judgment, it would be that inflation is harmful [1]. Certainly, price stability is one of our national economic goals, and forestry is, above all, merely part of the national economy.

ECONOMIC TRENDS AND CYCLES

Inflation aside, what can one say about economic trends and cycles?

You remember (Chap. 45) that when foresters and forest-industry people became aware of lumber's real-price trend and of similar trends in sawtimber stumpage, they expressed concern that sawlog products would price themselves out of the market. They saw forestry and wood industry as losing their mainstay resource. Pulp-paper, on the other hand, and its wood raw material appeared to be escaping the close clutch of scarcity. Differences in price behavior between the two contrasting groups of wood products show up clearly in consumption differences (page 96). Surely it is to an industry's disadvantage to experience long-term upward trends in real price of product.

> Why is this last statement not inconsistent with the earlier thought that an inflating sector is advantaged?

What about timber's vaunted renewability? How do you square this with the price and consumption trends? Evidently pulpwood supply is more readily renewable than sawlog supply. Surely pulp-paper takes precedence, in its access to raw material, over industries that use unpulverized wood. Perhaps we are seeing the early stages of an era in which the pulverizers are gradually overwhelming (undercutting) the nonpulverizers.

In any event, real-price trends in forestry have widened the separation between industry segments and promise to continue the process.

Turning to cycles in economic activity, one finds a set of truly major influences on the forestry economy. Let's think about the general business cycle and the building cycle, bearing in mind Chap. 46.

Economic cycles strike the forestry economy as fluctuations in demand. In sectors such as recreation and range, they affect forest use directly; use slackens off in periods of contraction and rebounds during expansion. In the timber sector, the first to feel cyclical change is the wood-using industry. Where oligopoly prevails, notably in pulp-paper, the industry inclines to hold the price line during a recession and ride out the cycle by adjusting output to the changing quantities demanded by the market. Such a policy, though judged wisest for the industry, still puts it under heavy strain because of its inflexible scale of operations. The lumber industry, by contrast, long responded to cycles in somewhat the opposite manner. In the 10 business cycles between 1904 and 1940, southern pine lumber prices fluctuated at least 3½ times as widely as production [3]. Today, policies are the same, though contrasts between price and output behavior are less extreme.

At any rate, the timber cycle, as it moves back into the woods, sees its amplitude enlarged. Activity in some sorts of timber and in some timbered areas can change from brisk to nil in the course of a 10-percent contraction in end-product demand. Activity in nontimber sectors, too, can undergo exaggerated fluctuation. And forestry employment and income ride the same roller coaster.

It may well be contended that forestry, because of its generally long period of production and its resistance to changes in policy, is notable among economic sectors in its sensitivity to cyclical change. Such change generates uncertainty, whereas the forestry firm craves stability. The farmer who cuts a stand of timber in response to today's outlook can't mend the stems back to the stumps when the outlook changes. Thus the highest guiding rate of return that the firm acquires during the ups and downs of the cycle tends to be the effective rate for the period and perhaps for subsequent periods, too. The cycle works against intensive forestry.

PUBLIC EXPENDITURES ON SPECIFIC PROJECTS

The public expenditures on forestry that fall out of our fiscal policy have an obvious direct influence on the forestry economy that scarcely calls for elaboration. Such expenditures can be influential because of the socialization [2] of U.S. forestry: the extent of public forest-land ownership, the traditional role of the public agencies in forestry research, and the attitude toward private nonindustrial forest owners, that they can appropriately be subsidized to encourage socially desirable forestry practice.

One point about public forestry budgets that deserves emphasis is that appropriations for public forest management may change from year to year in response to political winds as well as the economic cycle. Such changes introduce extra uncertainty into the program of the forest manager. Furthermore, to the degree that public and industrial budgets fluctuate in unison, their cycling creates periods of low demand for professional as well as other employees, confronts forestry schools with changeable and sometimes plummeting enrollments and budgets, and causes waste of social capital in the form of professional graduates, faculty, and administrators.

Since the 1960s, an issue has arisen respecting national forest budgets: their allocation between the timber and wilderness uses of federal forest land. Funds have been lost from timber production, not simply by subtraction, but also by changes in forest use.

VALUE OF THE DOLLAR ABROAD

Since the time when gold as a monetary standard was abandoned by world nations, central banks have had the responsibility for overseeing the rates at which their currency is exchanged against others. Under normal circumstances, a central bank can tie its currency to that of some other nation by standing ready to buy or sell its own funds at a fixed rate in terms of the other funds. Or a currency can be allowed to float, in which case it tends to gain in value if the nation's exports exceed imports; or if its money is relatively cheap to begin with, so as to encourage exports; or if it has held inflation down or interest rates up (so as to attract foreign savers); or some favorable combination of these events.

The prime influence of exchange rates on forestry is through their effect on international trade. If, for example, the U.S. dollar weakens (is cheapened) against the British pound, so that a pound exchanges for more dollars than before, the result is to lower U.S. prices to people in Britain and induce them to buy more: to buy more U.S. hardwood lumber and to send more tourists to the U.S.

> When foreigners visit U.S. national parks, does this add to U.S. imports or to U.S. exports?

Trade is an essential feature of modern economic life. In an economy where every family is self-sufficient, people are poor, having an income based on only what they can themselves produce. But if these family firms can specialize, each becoming expert and efficient at a line of production in which it is comparatively advantaged, firms can exchange their products and thus raise their incomes (page 28). With larger firms of optimal efficiency, and with free trade both within and among nations, high prosperity is within reach. The desirable rates of international currency exchange are those that are conducive to such prosperity.

SUMMARY

Five dimensions of U.S. economic activity that are influenced by our fiscal and monetary policies are these: (1) the rate of interest, (2) the rate of monetary inflation, (3) economic trends and cycles, (4) public expenditures, and (5) the value of the dollar abroad. For each dimension, the question is raised, How does it affect forestry?

Very generally speaking, forestry and forest industry are favored by low interest rates, stable current prices, freedom from fluctuation in general and in forestry economic activity, and encouragement of international trade. In this list, one sees a recitation of some major U.S. economic goals.

REFERENCES

1 Black, Robert P.: "Reflections on the Strategy of Monetary Policy," *Economic Review* (Federal Reserve Bank of Richmond), July–August 1990, pp. 3–7.

2 Duerr, William A.: "Timber Supply; Goals, Prospects, Problems," *American Journal of Agricultural Economics,* December 1974.

3 TerBush, Frank: "Southern Pine Prices and Business Cycles," M.S. thesis, State University of New York, College of Forestry, Syracuse, 1951, pp. 43–44.

CHAPTER **56**

FEDERAL TAX ON CORPORATE INCOME

TAXES EXPLAINED

The case at hand concerns one of the principal taxes in the United States: the federal income tax on corporations. The questions are the calculation of the tax, especially where timber-based income is concerned, and the effects of the tax on the behavior of corporate forest-land owners. The model is the Internal Revenue Code and its interpretations, as modified by the Tax Reform Act of 1986.

Taxes may serve at least two purposes: to supply public revenue and to regulate private activity. The income tax is a powerful instrument for both purposes. Its impact on behavior is a particularly notable feature of American life. Americans dedicate a mass of precious talent to tax matters, studying the laws and rulings, examining the records, weighing the alternatives, and arguing over the result.

So far as their revenue function is concerned, taxes are levied by governments on taxpayers in response to the principle that certain types of expenditure are more fruitfully made by public than by private firms. Military and police operations, control of currency, relations with other governments, and the regulation of commerce are simple examples of activities generally conceded to lie in the governmental sphere. Foresters might wish to add to the list the ownership and management of some of our forest land, and many nonforesters would agree.

Governments that lean heavily on taxes for public revenue do so because their culture teaches dependence on private enterprise for creating resources, and so the governments have relatively few funds of "their own." The economies they oversee feature private capitalism regulated by the market.

In the U.S., taxes levied by federal, state, and local governments total nearly a third of the gross national product. The tax total comprised the following broad categories in the mid-1980s:

Level of government and kind of tax	Percent of all taxes	
Federal taxes:		
Individual income tax	26	
Corporate income tax	5	
Other federal taxes	26	
All federal taxes		57
State taxes		17
Local taxes		26
All taxes		100

HOW INCOME TAX WORKS

The principle of income taxation is simple enough: A levy is made on the taxpayer's earnings during a period such as a year. The earnings figure is first reduced to allow for the cost of producing the earnings. Exemptions and deductions are also subtracted. Then the remainder **(taxable income)** is multiplied by a tax rate in order to arrive at the amount of the tax.

In practice, the simple principle is festooned with elaborations, exceptions, and special cases that fill long library shelves, so that if one wants to be a "tax expert," one must devote essentially one's whole working life to learning the procedure, practicing it, and keeping abreast of the continual changes.

The federal tax on corporate income is administered approximately as follows: The corporation maintains an office where records are kept for tax purposes. The services of an expert or team familiar with the Internal Revenue Code and its interpretations are brought to bear on the calculating as it progresses. During the corporation's fiscal year, which may be any 12-month period, the firm predicts its tax liability and submits "estimated tax payments" to the Internal Revenue Service (IRS), an arm of the U.S. Department of the Treasury. When the firm's tax office has closed its records for the year, it works up a tax report on forms supplied by IRS (Form 1120 is the central document), supplemented by detailed background evidence that may be voluminous to the point of constituting a set of books. The report is reviewed by IRS specialists in the industry (for example, in the Douglas-fir industry or in the southern timber industry), who commonly raise questions and meet with corporate representatives around the table. Points at issue may be resolved then and there or by seeking more formal IRS review or by going to court. Decisions reached find their way into the mounting lore and law that the experts feed on [1, 2, 5–8].

Although you and I may have no need or wish to be experts, we do have an obligation to understand a few basic points about the federal income tax, since it has a major bearing on forestry. Let us understand the meaning of two different kinds of income and two kinds of attendant cost. Tax rates and other specifics cited are those in effect for 1991 income [3].

1a Ordinary income. Payments received during the year for services, interest received, and the like, and notably income from the purchase (or manufacture) and sale of **stock in trade,** consisting of goods held for customers in the ordinary course of the taxpayer's trade or business. Corporations pay a 15-percent tax on the first $50,000 of taxable ordinary income, 25 percent on the next $25,000, and 34 percent on all beyond, except that an additional 5-percent "surtax" is paid on taxable income in the range $100,000–335,000.

The percentage tax rates just listed are the so-called **tax brackets.** They are the marginal tax rates, and the fact that they rise with income makes the federal corporate income tax a **progressive tax** up to $335,000. The average tax rate falls below the marginal in the second bracket, but in the course of the surtax bracket (to $335,000), the average rises to 34 percent, and thereafter both marginal and average rates are 34 percent. This is what is called **phasing out** the effect of initially low marginal rates for the high-income taxpayer. The general relation of average to marginal rates is explained on page 169.

1b Long-term capital gain. Income from the sale of capital assets held longer than 1 year (longer than 6 months in some cases). **Capital assets** are goods that are *not* stock in trade—for example, land sold by other than a developer or equipment sold by other than its manufacturer or dealer. Furthermore, Section 631 of the Internal Revenue Code allows long-term capital-gain treatment of income from *standing timber* under almost all circumstances where the time requirement is met: (1) Timber sold per unit of measure by its owner (Section 631b). (Lump-sum sales produce capital gains in any case, unless the seller is a dealer.) (2) Timber cut by its owner for sale as raw material (Section 631a). (3) Timber cut by its owner for use as raw material in the owner's manufacturing plant. (4) Timber cut for use as raw material by the holder of a contract cutting right—for instance, a buyer of public stumpage who has held the right for the requisite time period. In cases (3) and (4), which come under Section 631a, the gain is figured as though the timber-owning firm had sold the standing timber to itself; "fair market value" is estimated and taken to be the sales price.

2a Expensed cost. This is the cost of doing business that the taxpayer subtracts ("expenses") in full from current income. Outlays for nondurables are typically expensed; the cost, for instance, of wages, fees, travel, interest, insurance, raw materials, energy, some taxes, and forest protection and temporary roads.

2b Capitalized cost. This is the cost that the taxpayer enters in an "account" and then charges off against income over time. The charge-off can be made through depreciation—in the case, say, of durable equipment—"durable" meaning having a useful life of more than 1 year. Or the charge-off can be made through *depletion,* in the case of timber—or through sale, in the case of land and its durable improvements and nontemporary roads. In forest management, the taxpayer is required to capitalize the cost of durable property and of the improvement of this property, including site preparation and reforestation. Notice that *capitalize,* in the language of income taxation, has a meaning different from the one intended elsewhere in this book and first defined in Chap. 19.

> In general, if you had costs of forestry business chargeable against income either by expensing or by capitalizing, which method would you be likely to choose? Why?

INCOME FROM TIMBER LIQUIDATION

Whether timber income is reported to IRS as ordinary income or long-term capital gain can affect a forest owner's tax. If the owner is a corporation and has capital losses, it may charge these against capital gains, but not against ordinary income. Otherwise the capital-gain route may hold the disadvantage that the gain must (with certain fine-drawn exceptions, which we'll pass over here) be reported for the year when the timber is cut, even though money doesn't come in until later. Differences between ordinary income and long-term gain for the individual taxpayer are mentioned in Chap. 57.

In any case, the procedure for calculating income from the liquidation of a timber investment is the same regardless of how the income is reported. We want to sketch the procedure.

The basic idea of income from liquidating an investment is perfectly straightforward. For instance, you buy a home for $30,000. This figure is your **basis.** A few years later, you sell the place for $36,000, the **fair market value.** After paying the real estate agent $1,000 commission, your **cost of sale,** you have income of $5,000. This is the amount on which, after allowable subtractions, you pay tax at the required rate. Note that in tax calculations, a dollar is a dollar. Its purchasing power may have changed over time; this makes no difference.

But what about timber? When the trees are bought, the land comes with them. A sale or cutting may affect only part of the tract. Meanwhile the timber has grown, and the owner has incurred improvement costs. How is the income figured? Let's take the case of Dunstan Lumber Company's Bear Creek Block [4] in northwestern California. The company is working up its federal income tax return for 1991. In doing so, it is bringing forward its accounts from the close of 1990, when the company carried 832 million bd ft in its timber account—Table 56-1, line 3(i)—derived as explained there in parentheses. This timber represents an allowable investment claim (adjusted **basis for depletion**) of $7,046,100, line 3(h), so that the investment per M board feet **(unit depletion)** is $8.47 and is shown on the books at this figure 3(j).

The company makes a number of book changes in the Bear Creek account for 1991. First, it makes a cruise adjustment of nearly 10 percent, resulting in an increase in timber inventory at the beginning of the year, line 4(a). Second, there is timber planted in earlier years, for which a cost of $200,000, line 3(a), was capitalized; this timber, now of merchantable size, is moved into the timber account, line 4(c). The quantity of timber added is 167 million bd ft. Third, the company adds 4 million ft of estimated growth, lines 4(b) plus 4(d), since it made no such adjustment for 3 years. Of this growth, 3 million feet is in-growth of timber unmerchantable when the tract was bought and to which $10,000 of the purchase price, line 1(c), was allocated. Thus the company begins its 1991 tax estimates with 1,083 million bd ft of timber, line 4(e), in which the book investment is $7,256,100, line 4(f), or $6.70 per M board feet, line 4(g). Let's emphasize that this investment is in timber only: Land and its improvements such as roads are carried out in a land account, which is disregarded in Table 56-1.

During 1991, the company cuts 40 million bd ft of timber from the Bear Creek Block for lumber manufacture in its own mill, so that its basis for this transaction

TABLE 56-1
BEAR CREEK BLOCK ACCOUNT FOR FEDERAL INCOME TAXATION, 1991

Item	Timber quantity, M bd ft	Basis, depletion, etc., in dollars
1 1935 investment:		
(a) Land and timber		5,500,000.
(b) Allocated to land (30,000 acres)		90,000.
(c) Allocated to young growth		10,000.
(d) Basis for depletion (1a − 1b − 1c)		5,400,000.
2 1940 data:		
(a) Initial timber cruise	900,000	
(b) Unit depletion (1d ÷ 2a)		6.00
3 1940–1990, summary of highlights:		
(a) Investments in reforestation		200,000.
(b) Harvested before growth allowances	250,000	
(c) Depletion allowance (2b × 3b)		1,500,000.
(d) Revised basis for depletion (1d −3c)		3,900,000.
(e) Revised unit depletion (3d ÷ (2a − 3b))		6.00
(f) Adjustments for growth and harvesting, add	150,000	
(g) Acquired within the Block, 1979	32,000	3,146,100.[a]
(h) Ending basis (3d + 3g)		7,046,100.
(i) Ending timber inventory (2a − 3b + 3f + 3g)	832,000	
(j) Ending unit depletion (3h ÷ 3i)		8.47
4 1991 tax return, initial adjustments:		
(a) Timber cruise adjustment, add	80,000	
(b) From young-growth account (1c)	3,000	10,000.
(c) From reforestation account (3a)	167,000	200,000.
(d) Additional growth	1,000	
(e) Inventory entering 1991 (3i + 4a + 4b + 4c + 4d)	1,083,000	
(f) Revised basis for depletion (3h + 4b + 4c)		7,256,100.
(g) Revised unit depletion (4f ÷ 4e)		6.70
5 Timber cut for Dunstan mill, 1991:		
(a) Quantity; depletion allowance (Q × 4g)	40,000	268,000.
(b) Fair market value @ $110 per M bd ft		4,400,000.
6 Timber sold on scale basis, 1991:		
(a) Quantity; depletion allowance (Q × 4g)	10,000	67,000.
(b) Sales price minus cost of sale		960,000.
7 Timber lost by fire, 1991:		
(a) Quantity; depletion allowance (Q × 4g)	250	1,675.
8 Net income and tax, 1991:		
(a) Total depletion allowance (5a + 6a)		335,000.
(b) Total sales value (5b + 6b)		5,360,000.
(c) Net income (8b − 8a − 7a)		5,023,325.
(d) Tax (8c × .34)		1,707,930.[b]
9 At close of 1991 (entering 1992):		
(a) Timber inventory (4e − 5a − 6a − 7a)	1,032,750	
(b) Revised basis (4f − 5a − 6a − 7a)		6,919,425.
(c) Revised unit depletion (9b ÷ 9a)		6.70

[a]Excluding land value.
[b]Minus 34 percent of allowable expensed costs besides fire loss.

(depletion allowance), at $6.70 per M, is $268,000, line 5(a). Observe the reasoning behind this calculation: Since the company depleted 40/1083rds of its timber, it is entitled to a depletion allowance equal to 40/1083rds of its total investment of $7,256,100—that is, $268,000. The function of the unit depletion figure is to serve as a multiplier for easy arrival at such results.

The company calculates the "sales price" of the 40 million bd ft at a fair market value of $110 per M: $4,400,000, line 5(b). As you'll see, income is the excess of this figure over the $268,000 basis.

> The income amounts to $110.00 – $6.70, or $103.30 per M board feet. Why is it so huge? Why do you suppose the company does not report a lower fair market value and income?

Dunstan Lumber Company also disposes of 10 million ft of stumpage to other users, line 6(a). Here the company has an actual sales return, which it reduces, line 6(b), by selling cost in preparation for calculating income.

In the fall, a fire causes total loss of 250 M bd ft, representing an investment (at $6.70 per M) of $1,675, line 7(a). The timber is not insured. The company can expense this loss.

The total income from timber in the Bear Creek Block in 1991 is the sum of gains from the two harvesting operations minus the loss from fire. After this income has been reduced for expensed costs besides the fire loss, the tax is calculated as 34 percent of the remaining taxable income—section 8 in Table 56-1.

> Footnote b in Table 56-1 puts the tax at $1,707,930 minus 34 percent of expensed costs. What does this imply? Does it imply that the government pays a fraction of a taxpayer's costs: a fraction equal to the tax rate? Explain. Is this the average or the marginal rate if the two are different?

Section 9 lists the key data that the company carries from 1991 into 1992. Thus lines 9(a), 9(b), and 9(c) are 1-year updates of lines 4(e), 4(f), and 4(g) and 56-year updates of lines 2(a), 1(d), and 2(b). The tax record cycles along.

POSTSCRIPT

Before the 1986 law was passed, long-term capital gains had usually borne federal taxation at a lower rate than ordinary income, often at one-half this rate. The principle was that long-term holding of capital assets is in the social interest and should be encouraged and rewarded. In the early 1940s, Charles W. Briggs, general counsel to Weyerhaeuser Company, was instrumental in composing Section 117(k) of the Code. This section, which later became 631, provided that the bulk of income from coal and timber could be treated as a gain—even income that did not stem from the sale of a capital asset. When this provision became law, it was clear to all concerned that timber income was to be taxed at favorable rates.

But the dramatic further effect of Section 631 on forestry was not widely understood. Under 631a, a timber-owning and -processing firm could report selling raw material to itself, pay tax on the gain at a low rate, and expense the fair market value of the timber against current profits at the full ordinary-income rate. Thus the tax on ordinary income from wood manufacture was reduced or even extinguished. The excess gain-cost could be used to reduce taxable profit in any of the firm's enterprises: soft-drink bottling, motel operation, or what have you. Thus Section 631a stimulated forest-land acquisition, especially by firms that were or could become wood-using conglomerates. Some people say that the capital-gains feature of the federal income tax was the greatest single factor in the rise of private forest management in the U.S. after World War II.

With the 1986 law, a curtain fell on wood industry's four and a half notable decades. Section 631 and related sections still stand, but the distinction between capital gains and ordinary income in forestry is no longer the major consideration it used to be.

SUMMARY

The federal income tax on corporations is calculated under the Internal Revenue Code by taking stated percentages of (1) ordinary income less costs and deductions and (2) net long-term capital gains. These gains, primarily by virtue of Section 631 of the Code, include most net income related to timber or cutting rights held for more than a requisite period. Under the 1986 law, capital gains and ordinary income are taxed at the same rate, and the distinction between the two in forestry is no longer the major consideration it used to be.

REFERENCES

1 Commerce Clearing House: *Standard Federal Tax Reporter,* New York. Publication began in 1913 and is kept up to date with loose-leaf supplements.

2 Commerce Clearing House: *Tax Court Reporter,* New York. Publication began in 1965 and is kept up to date with loose-leaf supplements.

3 Commerce Clearing House: *Explanation of Tax Reform Act of 1986,* Chicago, 1986. Update, 1991.

4 The story of the Bear Creek Block and its tax data, all hypothetical, is a paraphrase of the exercise, "Federal Tax on Corporate Income," *In* Duerr, William A., and Neils B. Christiansen: *Exercises in the Managerial Economics of Forestry,* O.S.U. Book Stores, Corvallis, Oreg., 1973, pp. 245–247.
5 Prentice-Hall: *Federal Tax Handbooks.* Englewood Cliffs, N.J., annual.
6 Tax Research Institute of America: *Master Federal Tax Manual,* New York, annual.
7 U.S. Department of the Treasury, Internal Revenue Service. *Instructions for Form 1120, U.S. Corporation Income Tax Return,* Washington, annual.
8 U.S. Department of the Treasury, Internal Revenue Service: *Tax Guide for Small Business,* Washington, annual.

CHAPTER 57

FEDERAL TAX ON INDIVIDUAL INCOME

The question here is how an individual estimates the federal income tax for the past year in case there has been some revenue or cost connected with timber ownership or sale. The reply is that in general the procedure is just the same as for corporations (Chap. 56). Only some of the details are different, and the tax rates are different. Bearing the corporate case in mind, let's concentrate on the differences, talking about the rules in effect for 1991 income as a result of the Tax Reform Act of 1986 and further provisions of the law.

TAX FORMS AND RATES

Individuals report their income to IRS on Form 1040 and the various schedules and other forms that supplement 1040. For example, long-term capital gains from the sale or cutting of timber under Section 631 are reported first on Form 4797 and then, along with gains from other sources, on Schedule D of Form 1040.

For individuals, the 1986 Act provides just two tax brackets, 15 and 28 percent. However, as for corporations, there is a 5-percent surtax intended to phase out the effect of the lower bracket for higher-income taxpayers. And for such taxpayers the surtax holds through the successive increments of income until it has phased out, too, the effect of the **personal exemptions** to which they are entitled. Thus, like the corporate tax, the individual tax is progressive only through the top of the surtax bracket.

The 15-percent tax rate comes into play with the first dollar of taxable income (gross income minus all allowable subtractions). The breaking points between tax brackets depend on the taxpayer's status (married, unmarried, and so on). Since 1988, they depend also on the rate of inflation. They are indexed to inflation, and personal exemptions, too, are so indexed since 1989, which increases the tax concession to be phased out by the surtax. For example, on 1991 taxable income, the typical married couple with no dependents is allowed a total personal exemption of $3,900 and follows this schedule if they are filing jointly:

Taxable income, dollars	Marginal tax rate %
29,750 or less	15
19,751–71,900	28
71,901–171,090	33
171,091 or more	28

Anywhere within the uppermost bracket, the couple pays an average tax rate of 28 percent on the sum of its total taxable income plus $3,900.

> The income tax breaking points for married couples filing jointly, $29,750 and $71,900, are fixed for 1988 by law. The 5-percent level of surtax and the $3,900 personal exemption are also fixed. Given these numbers, calculate the top of a surtax bracket designed to phase out the 13-percent concession represented by the first bracket. Then calculate the top of the surtax bracket that will also phase out the $3,900 exemption.

The principle of a tax designed as the 1986 law provides is this: The fixed percentage (28 in our illustration) escapes the presumed penalty of a purely progressive tax, which is to constrain taxpayers' efforts to raise income and thus promote economic development. At the same time, the introduction of a low bracket for all incomes grasps the advantage of a progressive tax, in that it asks for more modest contributions from persons less well able to contribute. The indexing of the income breaking points to inflation escapes the presumed inequity of our traditionally fixed breaking points, that of letting inflation put taxpayers into higher tax brackets.

USING SECTION 631

Like the corporation, the individual is eligible for the use of Section 631a, recognizing fictitious sale of timber by the taxpaying firm to itself, for use as raw material in processing by the taxpayer or anyone else. Take the simple example of the individual forest owner who logs some timber and makes it into pulpwood for sale at the mill. If the time constraints are met, the owner can report a long-term gain under Section 631a as the difference between the (adjusted) basis for depletion and the fair market value of the *stumpage*. This fair market value then becomes one of the costs to be charged against receipts from the pulpwood sale.

Under Section 631b, a timber owner can claim a capital gain on stumpage sold per unit of mcasurc. And also likc thc corporation, thc individual can report capital gains from lump-sum sales where the stumpage qualifies as capital assets and is not stock in trade, as these terms are defined in Chap. 56. Chapter 49 points out the benefits of lump-sum stumpage selling that come from close timber utilization by the buyer and freedom from excessive supervision by the seller. Here one must recognize that if the timber owner does lump-sum selling at all regularly or repeatedly, IRS may look upon this owner as a dealer and the timber as stock in trade. IRS would then require that the net sales revenue be classed as ordinary income.

However, the usual pattern of timber sales, whether lump-sum or not, from nonindustrial private holdings is that they are widely spaced in time. Quite possibly a timber sale is a once-in-a-lifetime event for the owner. In such a case, the owner's taxable income may be uncommonly high in the year of the sale and thus may push the tax bracket to the top. The Internal Revenue Code provides for relief to the taxpayer in cases such as this. The taxpayer is allowed to retreat into a lower bracket by making installment sales and receiving payment over a period of years.

BENEFITS FROM CAPITAL-GAIN TREATMENT

The individual taxpayer may benefit in at least two principal ways by treating timber revenue as capital gain [3], even though the tax rates are the same as for ordinary income.

First, if there are capital losses, the individual may charge these losses against income. Against ordinary income, a maximum charge of $3,000 is allowed in any year. But against capital gains, there is no limit to the charge.

Second, sole proprietors whose timber enterprise constitutes a business are subject to a self-employment tax on ordinary income from the business. But capital gains from the business are not subject to this tax.

Still other benefits to the individual derive from capital-gains treatment. These are comparatively technical and uncommon benefits and will be bypassed here.

FAVORABLE TREATMENT OF REFORESTATION

It is well to mention just one other federal income-tax provision that affects primarily the individual timber owner. This is a provision for favorable tax treatment of reforestation.

Until 1980, landowners were required to capitalize into their timber account all direct costs connected with reforestation, such as site preparation, seedlings or seed, and the costs of labor, tools, and equipment depreciation. We have an example of the procedure in Table 56-1. Subsequently, the law was amended so as to carry three provisions favorable to individuals:

First, the taxpayer may claim an "investment tax credit" of 10 percent of the first $10,000 of reforestation costs in any year. The credit is subtracted directly from the tax.

Second, the taxpayer may amortize, over an 8-year period, 95 percent of the first $10,000 of reforestation costs per year. The taxpayer subtracts one-fourteenth of eligible costs from taxable income in the first year, one-seventh in each of the next 6 years, and the remaining one-fourteenth in the eighth year.

Third, the taxpayer who is granted cost-sharing subsidy under a federal or state incentives program to encourage reforestation may not need to report all or even any of the subsidy as taxable income. It is to the taxpayer's advantage to omit the reporting of subsidy if the taxpayer has not paid the maximum self-employment tax.

However, restrictions are imposed on taxpayers who do not participate actively in the management of their property.

POSTSCRIPT ON STATE TAXES

Though the main topic of this chapter is the federal tax, it does behoove us to glance at state taxes on individual income, recognizing that in some cases they may contribute significantly to determining a taxpayer's forest-related activity. Bettinger, Haney, and Siegel [1] have estimated the total income-tax load and the share of this total levied by 14 southern states on medium-income taxpayers who had a forest-management investment and made a timber sale in 1988. The state share ranged from 7 percent in Louisiana to 17 percent in North Carolina. The same authors [2] made analogous estimates for 13 western states. They found that the state share of total tax liability ranged from 12 percent in Arizona and Colorado to 21 percent in Hawaii.

SUMMARY

The gist of the case is this: Federal income-tax provisions affecting forest ownership and management are, in general, the same for individuals as for corporations (Chap. 56). Certainly the procedures for figuring forest income are the same. However, tax rates are different, and the individual is affected by certain special provisions, such as the treatment of the self-employment tax and reforestation costs.

States in the two principal timber regions of the U.S., west and south, are seen to cut a substantial figure in the individual taxpayer's liability and management program.

REFERENCES

1 Bettinger, Pete, Harry L. Haney, Jr., and William C. Siegel: " The Impact of Federal and State Income Taxes on Timber Income in the South Following the 1986 Tax Reform Act," *Southern Journal of Applied Forestry,* November 1989, pp. 196–203.

2 ———: "The Impact of Federal and State Income Taxes on Timber Income in the West Following the 1986 Tax Reform Act," *Western Journal of Applied Forestry,* January 1991, pp. 15–20.

3 Forest Industries Committee on Timber Valuation and Taxation: *Timber and the Federal Income Tax, Changes Made by the 1986 Tax Reform Act,* Washington, D.C., 1987.

CHAPTER **58**

TAXES ON FOREST PROPERTY

THE AD VALOREM PROPERTY TAX

The question now concerns the principal tax levied by local governments in the U.S.: the property tax. How does this tax bear upon timber ownership and management?

Let's start with the **ad valorem property tax,** the tax *on the value* of real property, and then go on to some modifications and substitutes designed for forest land and timber.

The ad valorem tax is an annual levy most appropriate, perhaps, to the farming community. The tax assessor comes around each year or so to review the value of land and improvements on each **tax parcel:** each holding or land subdivision within a holding. Commonly, the assessor estimates the "fair market value" of the property in its "highest and best use," or some specified fraction of this value. Page 386 offers a glimpse of assessors' use of transactions evidence for this purpose. After **assessment** comes **equalization,** the process, carried out occasionally for a state or major subdivision, of reviewing and, if necessary, adjusting individual assessors' estimates to a common standard. Meanwhile, the local government prepares a budget of proposed expenditures out of property-tax receipts in the coming year, and when this budget has been approved, its total is divided by the total assessed and equalized value of real property within the local government jurisdiction. The result is a **tax rate** which, if used to calculate each property owner's tax bill, will raise just enough revenue to fund the budget if everybody pays up. For example, in a farm county:

a Budget for the year	\$3.2 million
b Assessed and equalized property value	\$290 million
c Tax rate (1,000a ÷ b)	10.7 mills

This 10.7 **millage rate** is $10.70 per thousand dollars, or 1.07 percent. Each taxpayer is billed $10.70 for each $1,000 of assessed and equalized property value. The tax is paid out of receipts from the sale of last year's crops.

The ad valorem property tax is well suited for use by local government, primarily because it is a straightforward procedure, understandable, and easy to administer; and receipts from the tax are predictable (in contrast, for instance, to an income or sales tax).

> Why should receipts from a property tax be more predictable than those from an income or sales tax?

CONS AND PROS AS A FOREST TAX

In forestry circles, the ad valorem property tax has long had a bad name. People point out that an annual property tax may be fine as applied to cornfields, which supply a yearly income to meet the tax; but for woodland, it's quite a different matter. Years go by and the woods may yield no cash return, and meantime there is a tax to pay. Where will the money be found for paying the tax? The only thing the taxpayer can do, so the argument runs, is get out there and liquidate some timber to pay the tax: The tax *forces* the owner to liquidate, and thus the tax is at least partly responsible for the raggedy, depleted condition of small private forest holdings.

What's wrong with this argument? One thing that may strike you is that it disregards the unity of the firm. If a forest holding isn't large or efficient enough to provide the owner with an acceptable income, then the owner must have other income sources—a cornfield or a sawmill or a drugstore or unemployment relief—and it is the firm's function to judge all its enterprises together as a unit. The firm decides what enterprises to have and what size they should be, and presumably it does this in some sort of weighing of benefits and penalties. Taxes are one of the penalties. If, in the firm's view, they are worth bearing, then they will be paid—out of the firm's revenue, from whatever sources.

That there are some years when the woods contribute nothing to cash flow is irrelevant. What is relevant is whether total cash receipts will meet total cash costs, including taxes. If they won't, the firm may go to the woods for cash, but only if it regards this as the best alternative. Some of the relevant financial considerations are discussed in Part Three. If going to the woods entails a loss, it is unlikely to be the chosen alternative. If it entails a gain, then it's a desirable move, tax or no tax.

One should add that the tax bill on a partly wooded property may not separate the woods from the open land. To this extent, the tax procedure is consistent with the principle of the firm's unity.

VARIABLE-COST EFFECT

The unity principle notwithstanding, surely the ad valorem property tax may have some bearing on the behavior of a woods-owning taxpayer, and it behooves us to understand the bearing.

As you've often seen, the rational firm moves so as to capture rewards and escape penalties. The firm can move in this fashion if, by so doing, it can *change* net rewards—if net rewards are *variable* with the firm's alternatives. Suppose that nothing the firm does can change its tax bill; in this case, the tax has no bearing on the firm's behavior. On the other hand, suppose that the firm *can* change its tax bill so as to change its total net reward; then the tax *does* affect the firm's behavior: It offers the firm alternatives and causes the firm to choose its preferred alternative.

CASE (A):

Refer to the 50-year stand in Fig. 22-1. Recall that the problem is to estimate the best per-acre timber stocking of this Douglas-fir stand. Data are given on the total current annual growth of the stand as a function of stocking, and the firm's guiding rate of return is given as 4 percent. Both stock and growth are assumed to have the same value per M board feet. The procedure is to find the marginal (value) growth percent (marginal revenue): the extra growth obtainable from each addition to stock in the range studied. Where this percentage equals the firm's guiding rate (marginal cost), stocking is optimal. In the example, the optimum is found at 36 M bd ft per acre. Nothing is said about a property tax. The procedure is retraced in the first two columns and section (a) of Table 58-1, under the assumption that stock and growth value is $100 per M board feet and the property tax is zero. Total revenue (current annual growth) is recorded in the middle (a) column at each level of stock and then the marginal revenue is computed in the last (a) column. For example, in the first $400 stock interval, revenue shows a difference of $165 minus $142, or $23, which is 5.75 percent of the $400. Marginal value growth percent equals the 4-percent guiding rate at about 36 M bd ft of stock.

Now consider two contrasting cases.

CASE (B):

Suppose that a tax is imposed, amounting to $10 per acre. How will best stocking be affected? The tax will cause a $10 reduction in each of the previous total figures. The revised total revenue is given in the middle column of section (b) in Table 58-1. Because the successive differences in the column are unaffected by the constant $10 subtraction, the marginal value growth percentages are all the same as in the no-tax case, and best stocking is the same, about 36 M bd ft. The tax does not affect the taxpayer's behavior, for it is a fixed cost. There is no way to escape it and still hold the land in timber.

TABLE 58-1

BEST PER-ACRE STOCKING OF A 50-YEAR DOUGLAS-FIR STAND UNDER THREE ASSUMPTIONS ABOUT THE PROPERTY TAX: THAT IS (A) NIL, (B) $10 PER ACRE PER YEAR, AND (C) 10 MILLS PER YEAR ON THE VALUE OF THE STOCK

Total stock		(a) No tax			(b) $10 tax			(c) 10-mill tax		
			Annual net revenue			Annual net revenue			Annual net revenue	
Bd ft	Dollars	Tax, $	Total, $	Marginal, %	Tax, $	Total, $	Marginal, %	Tax, $	Total, $	Marginal, %
24,000	2,400	0	142		10	132		24	118	
				5.75			5.75			4.75
28,000	2,800	0	165		10	155		28	137	
				5.25			5.25			4.25
32,000	3,200	0	186		10	176		32	154	
				4.50			4.50			3.50
36,000	3,600	0	204		10	194		36	168	
				3.50			3.50			2.50
40,000	4,000	0	218		10	208		40	178	
				2.00			2.00			1.00
44,000	4,400	0	226		10	216		44	182	

Source: Fig. 22-1.

> The thought here is that marginal cost is unaffected by fixed cost. Where have we met with this thought before?

CASE (C):

But suppose there is a different sort of tax, literally an ad valorem tax, 1 percent of total stock value (10 mills). New changes now are introduced into Table 58-1. They are shown in the last three-column section, section (c). The tax is no longer constant: It is higher if stocking is higher. Consequently, the successive differences in total dollar revenue ($137 − $118 = $19; and so on) are less than before, and the marginal value growth percentages are all less than before. Best stocking lies somewhere below 32 M bd ft. This tax *does* influence the taxpayer's behavior, for it is a variable cost, variable with stocking. The form of the tax makes it possible for the firm to escape part of the tax without being hit by offsetting losses. The firm does this by carrying a lower timber stocking: following a less intensive forestry regime and producing a lower output.

The essential difference between cases (b) and (c) is that in (b), the assessor puts the same value on an acre of timber regardless of its stocking. The tax is fixed because the *assessment* is fixed. (Note that taxes will vary from year to year even with a constant assessment, but they will not vary in proportion to anything the taxpayer can do as a forest manager; they will be neutral toward forest management.) But in case (c), the assessment is variable, and thus the assessor penalizes higher stocking and output and rewards lower stocking and output, a reward that the taxpayer can accept by liquidating some stock. Let's emphasize that liquidation, in this case, is not undertaken to raise funds to pay the tax; it is undertaken to earn the reward offered by the tax system.

The tendency of the variable-cost property-tax assessment to exert a downward influence on timber stocking (and rotation, tree size, and so on) is here termed **variable-cost effect.** In recent years, the tendency has been called *time bias,* referring to the fact that commonly its effect arises when the assessor recognizes changes in a timber property *over time:* growth in timber quantity and value—or reduction in value through liquidation. However, the time-bias designation only encouraged a misunderstanding: that the issue at hand was the distinction between actual current income and "deferred income" [8], an issue logically extracted from the Forest Taxation Inquiry of 1935 [1], the classical technical work, addressed largely to questions beyond our concern here.

The simple remedy for the variable-cost effect is an assessment practice that once was denounced as slovenly and irresponsible: Don't change the assessments; let the assessor stay in his office and copy the same figures in the books from year to year. So far as variable-cost effect is concerned, the lazy assessor is the neutral assessor.

A very few states offer a constructive alternative to the lazy assessor. These states provide that a forest property owner may enter into a tax-stabilization contract with the local taxing authority. Vermont has such a provision [6].

CURRENT-USE ASSESSMENT

However, there are other and more widely endorsed ways of escaping the variable-cost effect. An approach introduced from Europe just a few decades ago and now widely adopted by the states is the "productivity assessment." The best way for us to view the procedure is as a subset of **current-use assessment.** So we'll have a look at current-use assessment and then return to productivity assessment.

A bias that may be introduced into the ad valorem tax on farm and forest property arises when an assessor anticipates land development and a consequent increase in values and raises assessments accordingly. The bias first came to public attention in farm areas on the urban fringe where residential, commercial, and industrial uses were pushing outward in a widening circle. Raising assessments (and thus taxes) in such fringe areas has the effect of hastening farm abandonment and land development, changes that the community may not desire, at least not at the pace generated by the heightened taxes. In any case, it puts the assessor in the position of swaying land use.

Since the 1950s, a revolution has been waged and widely won by advocates of current-use assessment for forest property. Back then, the current-use principle was being applied to some extent on farmland. Foresters saw a chance to protect forest land use from the ad valorem property tax. Many tax people viewed foresters' arguments as a bid for special subsidy.

But the current-use side won the support of legislators. Commencing with Maryland in 1957, some 35 states by 1991 had voted for current-use assessment of forest land [3]. These states instruct assessors to base their values on what they see today, not what they foresee for tomorrow. Here and there you are likely to find states reviewing their current-use legislation—some with a view to improving it, some, to repealing it.

The current-use policy can have an influence toward preservation of timberlands against encroachment, not only by residences, stores, or industry, but also by forest uses such as intensive recreation. On the other hand, current-use assessment is regarded by some as ineffective, as giving an unwarranted tax break to speculators, or as providing insufficient tax revenues.

Apropos of current-use assessment, you will find it helpful to review the paragraphs on long-run trends in forest use, pages 221 and 222; forecasting versus determining the future, page 244; and the case study of development versus current use, pages 419 to 421.

PRODUCTIVITY ASSESSMENT

Now the question arises, from where do the current-use states derive their forest property assessments. They could derive assessments from transactions evidence, which is advocated in Chap. 49 as the sensible way to make an appraisal. But if you turn back to Table 58-1, you will conclude that the second column lists values such as may come from transactions evidence, and a millage tax applied to these values exerts a variable-cost effect, as in section (c) of the table. How is a value set that doesn't climb as timber grows? Here is where **productivity assessment** enters [7].

Under the productivity approach, timberland is assessed at its forest expectation value: the capitalized value (net of all costs, including tax) of a property of equal site quality bearing or promising a satisfactory timber stock. This means that the assessor looks at the potential forest, not necessarily at the trees currently occupying the site.

By 1987, after just three decades of experience with productivity assessment in the U.S., the procedure had been adopted in pure or modified form by most current-use states. States commonly moderate their assessments of forest expectation value by using amply high interest rates and by making conservative assumptions about timber yields and unit values.

Why use "amply high interest rates"? What is the effect of doing this?

Productivity assessment has two outstanding advantages besides the defeat of the variable-cost effect. First is the ease of administration. The state can develop simple, reliable rules for identifying the major site-quality classes and can specify a per-acre value for each class. The assessor then classifies the site by the rules and is relieved of technical responsibility for estimating the quantity of timber. The result is uniform, equitable assessments that, by comparison with market-value assessments, are easy to equalize—or are already equalized. The second outstanding advantage of productivity assessment is that it encourages taxpayers to use their land efficiently. They gain by so doing, and the tax imposes no offsetting penalty. However, productivity assessment is regarded by some authorities as working to the disadvantage of the forestry sector because it burdens wooded properties that are distant in time from final yields.

YIELD AND SEVERANCE TAXES

The first popular approach to removing the variable-cost effect was to substitute an excise tax for the ad valorem tax on timber. This excise tax is called a **yield tax.** Sixteen states adopted the procedure during the years 1910–1930. A few thereafter dropped the yield tax, but then others took it up [2]. Today the count of yield-tax states is the same as at the 1930 high.

The most common yield-tax system is to place a flat assessment on the land, tax the land each year, and tax the timber only when there is a harvest, levying a flat percentage of the harvested stumpage value. Difficulties may show up. Tax receipts are relatively unpredictable. It is hard to arrive at a defensible yield-tax *rate,* one that produces a revenue comparable to the property-tax revenue. In any event, receipts fluctuate with harvests, and the state may need to set up a fund to tide local governments over the lean years. In some states, the yield tax is optional with the taxpayer and effective if the local authority approves [5]. You can picture some of the problems with an optional yield tax.

What do you suppose are the problems of having a yield tax that is optional? What advantages can you think of?

In any case, the yield tax is said to put the taxpayer at arm's length from the taxing authority, making the taxpayer less sensitive and responsive to the size and the contents of the public budget. Nevertheless, the yield tax is understandable and administratively straightforward. Some states have adopted it recently.

There are 12 states where *both* a property tax and an excise tax are levied on timber. In such cases, it is customary to refer to the excise tax as a **severance tax** [4]. Generally speaking, it is a mandatory **privilege tax,** originally justified as a charge for the privilege of extracting coal, oil, timber, or other free gifts of nature. The free-gift idea being no longer tenable, the timber severance tax today is commonly used to help defray public costs such as forest protection (as in Alabama), forest-product research (as in Oregon), and financial-assistance programs for nonindustrial private forest owners (as in Mississippi). Typically, the taxpayer is the logger or products manufacturer.

INCIDENCE OF A TAX

It behooves us now to consider the **incidence of a tax.** The question of incidence is this: On whom does the burden of a tax fall? Surely it is the taxpayer who forks over the levy to the government: The taxpayer pays the tax. But then is the taxpayer able to shift all or part of the burden to others? Is the incidence of the tax at least in some measure on people other than the immediate taxpayer? Under what circumstances? And why? Here is an aspect of the *equity* question (pages 162 and 163).

Recall the "principle of incidence" emphasized at the close of Chap. 44. It is, in effect, that burdens can be shifted from elastic firms to inelastic (and that benefits may shift from inelastic firms to elastic). How price-elastic is my demand for a service on which the seller must pay a tax? If it is inelastic ($0 > \epsilon > -1$), so that I am prepared to make a larger total outlay for the service at a high price than at a low price (page 92), then the seller may be able to pass the entire tax along to me in the form of a higher price.

So now, who bears the burden of a forest property tax? Is the taxpayer a nonindustrial private owner? To whom can the tax be passed? The timber buyer? Not likely. To be sure, if the price-demand curve for a stock of stumpage is vertical, the seller sets the price. But if, as *is* the case (page 217), it's supply that's price-inelastic, the incidence principle tips the advantage to the buyer.

Is the property taxpayer a wood-products manufacturer? This taxpayer will certainly want to pass the tax along to wood-products buyers. Whether the want is fulfilled depends on the market. If the market is competitive, as with lumber, the tax is apt to stop with the producer. If it is oligopolistic, as with paper, the tax may pass to the buyer and from there to the ultimate consumer.

> Why can the taxpayer pass the tax forward more readily into an oligopolistic market than into a competitive market?

What about a severance tax, paid by the severer or manufacturer? Here the same principles hold as for the property tax. Typically, a severance tax is part of conversion cost (page 387), deducted from stumpage price and thus borne by the stumpage owner. But if the severer is the timber owner and processor and has enough monopoly power in the market in which the wood product is sold, the incidence of the tax is on buyers and from there on ultimate consumers.

SUMMARY

The ad valorem property tax, then, in some respects the ideal tax for use by local government, has disadvantages when it comes to forest property. Most assessors have difficulty placing a value on timber. And if they change their assessments frequently to allow for timber growth and removal, they introduce a variable-cost bias into the tax. Two systems for escaping these disadvantages are the productivity assessment and the yield tax. Where land use is changing, present-use assessment can avoid undesirable abandonment of farms or timberland.

REFERENCES

1 Fairchild, Fred R., et al.: *Forest Taxation in the United States,* U.S. Department of Agriculture Misc. Publ. 218, 1935, pp. 525–527, elaborated on pp. 39–76.

2 Hickman, Clifford A.: *Emerging Patterns of Forest and Yield Taxes,* School of Forestry and Wildlife Resources, Virginia Polytechnic Institute and State University, Blacksburg, Publ. FWS-4-82, 1982, p. 55.

3 Hickman, Clifford A.: Information privately conveyed, 1991.

4 Hickman, Clifford A.: "Timber Severance Taxes, Current Status and Changing Role," *Forest Products Journal,* October 1989, pp. 31–34.

5 Kronrad, Gary D., Clifford A. Hickman, and William C. Siegel: *Optional Forest Yield Taxes in the United States,* U.S. Department of Agriculture, Forest Service Res. Pap. SO-227, 1985.

6 Sendak, Paul E., and Donald F. Dennis: *Vermont's Use Value Appraisal Property Tax Program, A Forest Inventory and Analysis,* U.S. Department of Agriculture, Forest Service Res. Pap. NE-627, 1989.

7 *Timber Tax Journal: Survey of State Forest Tax Laws,* annually, 1972 through 1984.

8 Trestrail, Richard W.: "Forest and the Property Tax, Unsound Accepted Theory," *National Tax Journal,* March 1969, pp. 347–356.

CHAPTER 59

SOCIAL PLANNING FOR THE TIMBER ECONOMY

INTRODUCTION TO PLANNING

At the outset of this book, the process of problem solving is outlined in four steps:

1 Identify the problem
2 Specify a model
3 Use model and data to solve the problem
4 Follow up

Then the first two steps are reversed, to display the decision process:

1 Specify decision topic and criteria
2 Identify alternative decisions
3 Use criteria and data to choose decision
4 Follow the decision chosen

And, having come this far, one realizes that deciding may be viewed as planning. True, planning underscores the future; but deciding is for the future, too. Here is **planning:**

1 Specify goal and criteria for planning
2 Identify alternative plans
3 Use criteria and data to choose a plan
4 Follow the plan, replanning at intervals

It is appropriate, then, that this chapter and the next, the last substantive chapters in the book, concern planning. The chapters ask how social planning for forestry is carried out and what some of the principal social plans for forestry in the U.S. have been. Why *social?* Because most of Parts Two through Five study questions raised by the individual firm: private planning. Part Six houses questions for society, including the planning question.

> The foregoing lists imply that "decision topic and criteria" (step 1 in decision making) and "goal and criteria for planning" (step 1 in planning) are both analogous to "a model" (step 2 in problem solving). Are they, in fact, models? Explain.

What does social planning betoken—socialism? At least a wisp of it. But not necessarily "central planning," to use FAO's phrase from Part One. All societies plan, and any plan may fail to bear fruit, even in a centrally planned society, let alone a "market" society. If a people lives, it plans, even without means for peaceful social accomplishment of the plan.

In forestry, the U.S. has produced several tons of social plans and a ton or two of positive results. Some of the most effective results, such as our systems of public forests, parks, and refuges, have flowed in part from formal plans. This chapter asks what some of the formal federal plans have been, with emphasis on the timber economy.

STATUS AND GOALS IN PLANNING

It will be helpful to think, all the while, about the four planning steps, so that the forestry plans described here can be evaluated in terms of their procedures and thus be better understood.

The first step is to specify the goal of planning—that is, the criteria for choosing one plan over another. At the start of Chap. 43 is a list of national economic goals. This list constitutes a general charter within which forestry may be expected to perform:

1 Full employment of resources
2 Price stability
3 Equity in income distribution
4 Progress in economic affairs
5 Economic efficiency

These goals were explained in ordinary language a good many years ago by President Franklin Roosevelt [2]:

> . . . to increase the security and the happiness of a large number of people in all occupations of life and in all parts of the country; to give them more of the good things of life; to give them a greater distribution, not only of wealth in the narrow terms but of wealth in the wider terms; to give them places to go in the summer time—recreation; to give them assurance that they are not going to starve in their old age; to give honest business a chance to go ahead and make a reasonable profit; and to give everyone a chance to make a living.

The second step in planning, to identify alternative plans, implies that the planner has already met an intermediate obligation, to specify how matters stand currently. If the community's goal is to have *n* acres of forest park and if it has none now, it must undertake the considerable task of fund raising to acquire *n* acres. This calls for alternative plans. But if it already has all but 5 percent of its goal, the remaining gap between ideal and real may take only a little enlarging of some other capital budget—or the project may be declared completed as is.

When it comes to planning for a wide forest region, describing the status quo may be a huge undertaking. And yet it may well be essential if the gap between ideal and real is to be made evident to all concerned. In fact, a good description of the current forest situation may have to forerun goal setting. Otherwise, people can't be expected to judge whether the proposed goals are realistic. Perhaps they reach so far beyond the existing situation that their accomplishment, though feasible in engineering and economic terms, is out of the question politically.

Thus one finds that many social plans begin, not by specifying goals, but by describing how things stand. The planner takes shelter in an easy beginning, and the plan's audience starts off on a familiar topic. All's well provided they don't let the familiar dominate the whole project.

HOUGH'S OPEN-ENDED GOALS, 1878–1882

One of the early federal plans for U.S. forestry is the work of Franklin Hough [1], who was appointed in the Department of Agriculture in 1876 to make a study, specially authorized by Congress, of

> . . . the annual amount of consumption, importation, and exportation of timber and other forest products; the probable supply for future wants; the means best adapted to the preservation and renewal of forests; the influence of forests upon climate; and the measures that have been successfully applied in various countries for the preservation and restoration or planting of forests.

Hough, working against his day's crippling deficiencies of data, produced a comprehensive, detailed, loosely assembled report addressed to the forestry problems of the time; the westward receding of the softwood timber industry from the depleted forests of the northeast; local timber shortages here and in the midwest; the supposed disturbances of climate attending forest removal; the presumed need for artificial forest regeneration for protective and timber purposes; the heavy drain on timber stocks for export; the deficiencies of national forestry policy in such matters as reserving public forest lands; the loss of timber to fire, insects, and disease; the lack of public education in forestry and of a forestry profession in the U.S.

And what goal did Hough hold in view in making his forestry recommendations? The goal that he implied was simple and open-ended: Grow more timber than we are growing.

What sort of goal is this, to produce more? *How much* is not specified, just *more.* It can be argued that this is no acceptable sort of goal at all [12]. It is not a target. Its advice is loose, unhelpful. It is misleading, too. It advocates a kind of allopathic formula, that if some is good, more is better. Yet anyone of right mind knows that this cannot be: There must be an optimum, and any quantity either smaller or larger is inferior. The rule is enforced by resource limitations. It is the fundamental economic rule.

But on second thought, is an open-ended forestry goal quite so childish and primitive as might be judged by a quantitative, computer-happy culture? Think of the production goals of private firms. If they are set as specific targets, they are set high, perhaps "too" high. And if they are in fact exceeded, few but rivals complain. If firms in the potato industry are efficient and aggressive, the public may find itself eating more potatoes than the "optimum" (however estimated) and proportionately less rice. And in the longer term, talents and tastes shift, consumers discipline producers, and pendulums swing.

In the longer term with which forestry is, by and large, concerned, which of the following two social policies is sounder? (1) For specified future dates, set goals in terms of acres, cubic feet, user-days, and so on and revise them continually as unforeseen wants and scarcities emerge on the horizon. (2) Set goals in terms of *more* or *less,* and let the economic and political market watch the horizon for unforeseen wants and scarcities and continually revise the rate at which changes in output are encouraged or allowed to occur.

An open-ended goal has these advantages: (1) It is realistic in being appropriate to a market economy rather than one centrally planned. (2) It escapes an energy-consuming, perhaps self-deluding, and possibly futile numbers exercise. (3) It leans on a well-tried economic principle, that supply gives rise to demand: that invention is the mother of necessity.

A specific goal has these advantages: (1) It satisfies our culture's want for quantitative statements. (2) It helps to make a long-term policy commitment, especially as between rival resources such as timber and wilderness acreages. (3) If set high enough and stated convincingly enough, it tends to recruit public support for constructive resource projects.

> Examine your own goals in life. Are they open-ended or targeted? Can you explain why? Are you satisfied with your goals? Do you find that your goal for a set of goals is more open-ended than the goals in the set? What is the significance of these questions for our social forestry goals?

BUREAU OF CORPORATIONS REPORT, 1911–1914

Be that as it may, the trend right from the outset of timber planning has been from open-ended goals toward specific, target-type goals. The 1897 Act establishing the national forests "to furnish a continuous supply of timber for the use and necessities of the citizens of the United States" presented the challenge of determining these necessities. In the early decades of the present century, as rough forest statistics came to be available, the practice grew of comparing timber growth with depletion, using the difference or ratio between them as a measure of the forest problem. Thus the next notable national planning output after Hough's, the Bureau of Corporations report of 1911–1914 [11], states that the rate of growth in timber forests of the U.S. was only about a third of the timber cut, implying that the two should be made equal (a comment on growth-depletion as a goal is that the two are equal on the Sahara Desert).

If you will look again at Fig. 45-1, you will see what an amazing price era the Bureau of Corporations was reflecting. The American industrial revolution, which had just been getting well under way in Hough's time, was gaining momentum and carrying the forestry economy with it into new problem spheres. The great national wave of softwood timbering had passed across the Lake States, had entered the southern pine region, and was heading toward the Far West. Everyone could see that the fading of the old timber surpluses was no longer merely a local thing, but was taking on nationwide significance.

The trend of lumber and timber prices, continuing to rise persistently, forced itself on the attention of the country. The spread of industrial combines had penetrated the area of wood manufacture and forest ownership: The widely accepted prospect of continued timber depletion and inflation had encouraged—and federal land-disposal policies had made possible—the concentration of much timberland in the hands of big operators and speculators. Congress asked the Bureau to find out why lumber prices were so high, and especially whether they were the result of any agreements among firms to restrain trade.

The Bureau produced one of the earliest nationwide inventories of the timber resource. It also carried out one of the first comprehensive studies of lumber production and distribution in this country. It found that although lumber manufacturing admitted to no such scale economies as were enjoyed in steelmaking and oil refining (exemplifying the great American trusts of the era), the industry had an avenue toward monopoly through the control of raw-material supplies. The development of such concentration was far advanced, three firms owning more than a tenth of all the privately held standing timber in the United States and 195 firms, nearly four-tenths. The Bureau described the vast speculative purchase and holding of timberland by the big owners, far ahead of use. It pointed to the enormous increases in the value of standing timber and in lumber prices and attributed them partly to the industry's concentration and its conspiracies to curtail output and push prices up. The theme of the plan was trust-busting, the theme of the day in all walks of industrial regulation.

FOUR FOREST SERVICE REPORTS, 1920–1958

While the Bureau of Corporations was reporting its anxiety over the industrial combination and concentration aspects of the timber problem, the Department of Agriculture's Forest Service was viewing with equal alarm the aspect of timber depletion. The Service believed that this aspect should not be permitted to suffer deemphasis, but should be given at least equal attention by the American public. There followed a series of nationwide planning analyses by the Forest Service, in which the emphasis was on timberland resources. The first two releases in the publications series were named for the U.S. senators whom the Service asked to call for reports. Later reports were produced under the Service's authority, from the McSweeney-McNary Forest Research Act of 1927, to assess the U.S. forest situation (notably the *timber* situation) on a continuing basis.

In recent years, too, multiresource reports have been coming out under the terms of the Forest and Rangeland Renewable Resources Planning Act of 1974 as amended by the National Forest Management Act of 1976, and under the terms, also, of the Forest and Rangeland Renewable Resources Research Act of 1978. Let's have a look at the timber series in this chapter and the multiresource group in Chap. 60, primarily in order to see how the issues and the approach to goals have changed over the years. One should emphasize that these are all exhaustive, painstaking analyses. The longest of the publications, Hough's and the Copeland report, run to about 1,600 pages each. but there isn't a lightweight in the lot. The typical reader, and certainly the nontechnical reader, may well feel overwhelmed.

The Capper Report of 1920 [3], representing principally the work of Earle Clapp of the Forest Service, contains one of the first plain statements of a goal for the U.S. timber economy. Under the heading "Future Trends in Requirements," and confining himself to lumber as the principal timber product, Clapp made a forecast based primarily on past trends in per-capita consumption in the U.S. and Europe.

> Even with large allowances for the substitution of other materials for timber, it seems hardly possible that our annual demand for lumber for years to come will fall below 35 billion [board] feet. . . . Any future lumber production falling below approximately 35 billion feet, unless we can make up the difference by imports, will result in hardship to many classes of consumers and to many industries. . . .

(In the last sentence quoted, you will recognize an explicit interpretation of the public interest.) The report held that forest depletion was indeed serious and that it was the fundamental problem underlying current scarcities and high prices.

The Copeland Report of 1933 [4], like the Capper report, stated a national goal in terms of requirements, but it broadened the goal to include all timber consumption and it offered a definition: **Requirements** are not irreducible needs, but a measure of use by consumers afforded a reasonable latitude of choice of materials. What's *reasonable* is, of course, the question, though the reply may well be that it doesn't really matter, so long as the figure chosen is high enough to be disturbing but not incredible. The Forest Service expressed itself as disturbed—over the nation's failure to realize forest potentialities. This was one of our major failures, the Service spokesman said, because forest land is more than a third of all land, because timber and the other services of the forest are part of the foundations of our national life, and because solving forestry ills gives employment, creates new income, and generally strengthens the American rural economy and therefore the economy at large. These were times of deep depression and mistrust of private enterprise.

Nearly all American forestry ills, the Copeland Report ventured, have grown out of private ownership. Nearly nine-tenths of all lightly stocked or denuded timberland is privately owned. More than 99 percent of the improvident cutting and 98 percent of the burning is on private forests. At the same time, four-fifths of the timber-growing land, with nine-tenths of the growth capacity, is privately owned. The ills of taxation, demoralized forest industry, impoverished forest communities, scanty timber growth in relation to requirements, and depleted watersheds and other nontimber resources have all, in the main, arisen from the mishandling of private lands. The proposed plan is: public forest acquisition; public controls over private practice; research and extension to further the public interest.

> It is generally conceded that the potentialities of the site are, as a whole in the U.S., significantly more fully attained in crop and pasture farming than in timber forestry. How do you explain the difference?

There followed two major Forest Service reviews of optimistic tone, contrasting with the dour work of depression days: the Reappraisal ("Forests and National Prosperity") of 1948 [5] and the Timber Resource Review, or TRR ("Timber Resources for America's future"), of 1958 [6]. In these releases, the Service put increasing emphasis on its planning goal, termed "potential requirements" in the Reappraisal and "projected timber demand" in the TRR. Targets were worked out painstakingly for each timber product, for dates extending as far out as the year 2000. The Service explained that

> Such estimates are not to be regarded as forecasts of actual future consumption of timber products. They are somewhat analogous to the potential demand estimates frequently made by large manufacturing concerns to serve as guidelines for planning their marketing and production facilities.

Nothing was said about the price at which the estimated consumption could take place, nor was the cost of meeting the goal analyzed.

The Reappraisal and TRR contrast with the Copeland Report in the forestry ills that are singled out for billing. What role are forests to perform in the abundant postwar world? The need is not to make jobs, as it was in the 1930s. Rather, it is to give forth a broad stream of services and to do so as efficiently as possible. For this purpose, there is plenty of forest land, but a scarcity of wood supplies, particularly in the preferred species, larger sizes, and high grades. Our forests are being operated in the red. But no longer are the big landowning corporations the bad guys. The major fault lies with the nonindustrial private owners, especially in the South. Nationwide, a fourth of the commercial forest land is less than 40 percent stocked with trees. A fourth of the timber cut is unused. Some essential measures: plant trees; build up stocks in the east; encourage the little fellow to intensify management; practice closer utilization in both woods and mill; make use of hardwood instead of softwood where possible; fight fire, insects, and disease.

FOUR MORE FOREST SERVICE REPORTS, 1965–1990

The last four members of the Forest Service timber series may be thought of as a unit. The reports are the Timber Trends Report of 1965 [7], the Outlook Report of 1973 [8], the Situation 1952–2030 Report of 1982 [9], and the Situation 1989–2040 Report of 1990 [10]. The last of these was released as a member of the Resources Planning Act series to be discussed in Chap. 60. The reports have many characteristics and lines of reasoning in common:

1 They are all written in a new era of U.S. forestry, the environmental era, which dates approximately from 1960. They show a significant and growing sensitivity to environmental-quality issues and a mounting awareness of a somewhat related trend in American culture, toward the preservation of wilderness and other nontimber single uses. A notable addition, in the last member of the series, is a pointed reference to global warming. The reference is termed "global forestry." What effect, the report asks, can warming be expected to have on timber management and output in the U.S.?

2 They represent a further and continuing elaboration of timber goals, with emphasis on calculating the goals and giving them the top spot in planning. Goals are termed *future,* or *projected, demands,* or sometimes just *consumption.* They are said to be, not forecasts, but statements of what will happen under assumed trends in population, economic activity, and materials availability.

3 Having brought "demands" into the story, the reports bow to the inevitable and introduce "supplies," keyed to the same target dates as "demands." These "supplies" are defined as timber growth on eastern timberlands plus allowable cut from western public forests plus actual cut from western private forests. Growth is not taken as a measure of output in the West because of the region's still substantial share of old-growth standing timber.

4 Given timber "demands" and "supplies," the prime criterion of success in U.S. timber management is made to order. The two should be equal. Since "demands" are consumption—that is, in essence, depletion—and "supplies" are growth or its proxy, one is tempted to inquire whether the supply-demand criterion is not simply the much-berated growth-depletion criterion in modern dress.

5 One finds a new brightness of outlook in the reports, a new note of encouragement to Americans on the status and immediate prospects for their forest resources. Although timber management is still not intensive enough, it has greatly improved. Cultural measures have been widely introduced. Fire has been checked. Timber growth has come to exceed depletion in some regions, and although the new additions to growing stock are in the traditionally less wanted categories of wood, timber utilization in woods and mills has improved and promises to continue to improve. Thus "supply-demand" prospects for the next few decades are good.

6 For the longer term, heavy consumption is in prospect:

Product	Measure	Consumption 1986	2040
Lumber	Billion bd ft	56.8	69.4
Panels	Billion sq ft	24.9	39.8
Paper	Million tons	81.7	173.0
Firewood	Billion cu ft	3.1	5.4
All roundwood	Billion cu ft	20.5	28.6

U.S. timber owners are probably not capable of supplying the 2040 consumption. Prices, says the Forest Service, are bound to rise. In terms of the analysis of the Z-formula and the price trends in Chap. 45, what does this mean? If we try to get very much *more* timber and wood products from U.S. forests (note the open-endedness of the goal!), we'll find ourselves running into higher costs and thus higher prices. If one can assume that advances in manufacturing technology will help to hold costs in check, then one must interpret *prices* to mean, in the first instance, stumpage prices.

7 However, says the Forest Service, measures can be taken in the near decades to soften later supply pinches. We can import more timber, softwood from Canada, hardwood from Asia. There are opportunities to intensify timber management, especially on nonindustrial private holdings in the South, that promise to return (at the anticipated high real prices) 4 percent or more on the investment. Timber utilization in woods and mill can still be tightened. Timber *consumption* can be made more efficient (the first mention of this!) so as to save materials.

> How do you explain the fact that the foresters have always stressed the desirability of more efficient production of forest outputs, but have had very little to say for more efficient consumption? In what ways could timber be conserved through more efficient consumption? How about wildlife? aesthetic values? developed recreation? wilderness?

SUMMARY

To be concise about it, this chapter has reviewed federal plans for the timber economy extending over a 100-year period. The plans represent a developmental sequence in such respects as these: More and more detailed and reliable resource statistics on which to base plans; goals for planning more elaborately devised and stated; an increasingly large, complex, and, all in all, promising timber economy for which to plan; increasingly detailed, specific, and constructive suggestions for implementation.

REFERENCES

1 Hough, Franklin B.: *Report upon Forestry,* U.S. Government Printing Office, Washington; vol. I, 1878; vol. II, 1880; vol. III, 1882.

2 *The New York Times,* June 8, 1935.

3 U.S. Department of Agriculture, Forest Service: *Timber Depletion, Lumber Prices, Lumber Exports, and Concentration of Timber Ownership,* Washington, 1920.

4 U.S. Department of Agriculture, Forest Service: *A National Plan for American Forestry,* 2 vols., Washington, 1933.

5 U.S. Department of Agriculture, Forest Service: *Forests and National Prosperity,* U.S. Department of Agriculture Misc. Publ. 668, 1948.

6 U.S. Department of Agriculture, Forest Service: *Timber Resources for America's Future,* Forest Resource Rep. 14, 1958.

7 U.S. Department of Agriculture, Forest Service: *Timber Trends in the United States,* Forest Resource Rep. 17, 1965.

8 U.S. Department of Agriculture, Forest Service: *The Outlook for Timber in the United States,* Forest Resource Rep. 20, 1973.

9 U.S. Department of Agriculture, Forest Service: *An Analysis of the Timber Situation in the United States 1952–2030,* Forest Resource Rep. 23, 1982.

10 U.S. Department of Agriculture, Forest Service: *An Analysis of the Timber Situation in the United States, 1989–2040,* General Tech. Rep. RM-199, 1990.

11 U.S. Department of Commerce and Labor, Bureau of Corporations: *Summary of Report of the Commissioner of Corporations on the Lumber Industry,* U.S. Government Printing Office, Washington, 1911. Part I, *Standing Timber,* 1913; Part II, *Concentration of Timber Ownership in Important Selected Regions,* 1914; Part III, *Land Holdings of Large Timber Owners,* 1914; Part IV, *Conditions in Production and Wholesale Distribution Including Wholesale Prices,* 1914.

12 Vaux, Henry J., and John A. Zivnuska: "Forest Production Goals, A Critical Analysis," *Land Economics,* November 1952, pp. 318–327.

CHAPTER **60**

SOCIAL PLANNING FOR FORESTRY

TRANSITION TOWARD MULTIRESOURCE PLANNING

Consider for another moment or two the remarkable series of documents sketched in Chap. 59. Two of the first reports in the series were the work of sole authors: Franklin Hough and Earle Clapp. Then came the Copeland Report, with Clapp as the director of a tiny group of Forest Service stalwarts from the Washington office and regional experiment stations. All these forester-analysts accepted the responsibility of identifying the nation's problems and recommending remedies. And on through the rest of the timber-planning series, the scholarly thinkers and writers were Service foresters recounting the tenets of their subculture to potential believers. These writers were totally informed, self-reliant, enthusiastic, unbending, and loyal. One may ask whether, as timber foresters, they were not at a disadvantage in seeing the nontimber aspects of their field, let alone the nonforestry considerations vital to their work. No, they were not, for they had a tenet of faith to sustain them: What's good for timber is good for all forestry—indeed, all society.

The undermining of the old order began imperceptibly in the mid-1940s. During the following decade or so, the radar screen was coming alive with blips of warning. The consumerists-environmentalists of our Chap. 10 were the chief question raisers. The conviction was spreading that the forestry subculture is approachable not only legislatively and administratively, but also judicially. The Forest Service began to find itself taken to court over its plans and practices, notably those flowing from old beliefs in the primacy of timber and the soundness of traditional silvicultural doctrines. The questioners were laypersons, members of a class long viewed skeptically in forestry circles. They were not the "general public." They belonged to interest groups. They were the public that was claiming seats around the planning table, the decision-making table.

In 1964, Ernest Gould, of the Harvard Forest, was drawing an analogy: Just as war has been said to be too serious a business to leave to generals, so it is being said that the future of forests is much too important to leave to foresters [6]. In the profession, this was a shocking, unpalatable thought. But Forest Service planners were now being confronted by the truth of it.

Between July 1, 1964, and June 30, 1974, 74 lawsuits were filed against the Forest Service [7]. Most of these suits charged the agency with environmental blunders or objected to the proposed use of Forest Service lands. During the 1970s, the suits characteristically alleged Forest Service violations of the National Environmental Policy Act. The most telling of the suits was brought in 1973 by the West Virginia Chapter of the Izaak Walton League, which claimed that clear-cuts on the Monongahela National Forest had violated provisions of the Organic Administration Act of 1897 establishing the national forests and describing their purpose and management. The Forest Service lost both the district court case and an appeal. Among very recent attacks upon the Forest Service for its handling of the timber resource, we can turn to a challenge leveled by the Foundation for Research on Economics and the Environment. "The Forest Service was intended to be a model of good government. . . . The experiment hasn't been . . . a success"[3].

Meantime, Senator Hubert Humphrey of Minnesota was coming to the rescue, to "get the Forest Service out of the courts and back into the woods." His vision was of new fundamental legislation that would attribute to the Forest Service a concern for all U.S. forest and rangeland resources: trees, grass, soil, water, and wildlife; wilderness and developed recreation areas; rocks and minerals; environmental values.

> The history sketched here of the agencies and persons active in the forestry policy developments of the times is vastly simplified [8]. What other agencies and individuals do you imagine were instrumental in shaping social planning policy in forestry?

The Service would take a broad social view of its field and would consider international influences on the field. The legislation would direct the agency, through the executive branch, to submit to the Congress a series of reports, drawn up in consultation with other agencies and the public, that would underpin a massive effort at stock taking, planning, and budgeting. Here is the gist of it:

1 A periodic *assessment* of the current and prospective forest and rangeland resource situation, which would give public and private landowners necessary perspective.

2 A periodic *recommended program* for the Forest Service, useful to the administration, the Congress, and interest groups in their planning and decision making.

3 A *statement of policy* to be submitted by the president to Congress every five years, at about the time when the recommended program was released by the secretary of agriculture. The statement is intended for use by Congress when it formulates the Forest Service budget.

4 An *annual report* from the Forest Service, showing progress toward program goals and a budget supportive of the agency's plans for land resource administration, for research, and for state and private forestry programs.

RPA, 1974; NFMA, 1976; RESEARCH ACT, 1978

Senator Humphrey's vision of assessment and planning was embodied in the Forest and Rangeland Renewable Resources Planning Act (RPA) of 1974, as amended by the National Forest Management Act (NFMA) of 1976 [1]. The latter provides for a periodic *national forest plan* for each of the forests or groups of forests. Also relevant, because it authorizes resource inventories: the Forest and Rangeland Renewable Resources Research Act of 1978 [2]. These acts are a mandate for the secretary of agriculture to prepare a Renewable Resources Assessment by December 31, 1975, with an update in 1979 and each tenth year thereafter. The assessment is to include "an analysis of present and anticipated uses, demand for, and supply of the renewable resources of forest, range, and other associated lands with consideration of the international resource situation, and an emphasis on pertinent supply, demand, and price relationship trends" [1, Section 3. (a)].

The acts, in effect, instruct the Forest Service to produce a torrent of documents and to accomplish this under the most trying conditions—namely, in consultation with the "public," that Hydra of many heads and many viewpoints [5].

Let us refer to the whole project as RPA and look in briefly on some of its results.

SOME RESULTS OF RPA: ASSESSMENTS

The first RPA *assessment,* a comparatively modest book of xvi plus 243 printed pages, was dated 1975 and published in 1977 [12]. It was followed by a 1979 assessment [13] and a 1984 supplement [14]. The next in the series, comprising 13 documents, appeared in 1988, 1989, and 1990 [18–30].

The assessments are organized into chapters or, in the 1988–1990 series, into separate reports on water, wildlife and fish, minerals, range forage, outdoor recreation and wilderness, and timber. The reports are introduced by a statement of assumptions about future U.S. population, income, other economic indicators, and an overview of forest, grassland, cropland, and related resources. They touch on the issues of our changing global climate. They contain a discussion of what the findings mean for Forest Service and other land managers, public and private.

Each resource is treated under somewhat the same outline as came to be adopted in the century-long series of timber assessments (Chap. 59). The first topic is "demand"—that is, consumption, use, depletion, as the case may be. Then, inevitably, comes "supply," couched in terms analogous to "demand"—that is, production, inventory, flow, growth. Both these estimates stretch ahead for some 50 years. "Demand" is foreseen to increase, "supply" to be pinched. If this were to happen, the result would be scarcity and deprivation. But countermeasures can be taken. These are discussed.

RECOMMENDED PROGRAMS

By 1992, four *recommended programs* had been produced under RPA: in 1975 [15], 1980, 1985 [17], and 1990. The 1980 update [16] is the weightiest of the documents, with some 680 pages. What, briefly, are its contents? How are they organized?

In its preface, the 1980 update is said to "respond to the projected renewable resource situation documented in its companion Assessment" (the "1979 assessment"). There follows an introduction in which 11 goals of the Forest Service are set out in some detail. These are wide-ranging, broad goals—for example, "Improve the welfare of underprivileged members of society. Encourage the development of forestry throughout the world."

Chapter 1 is the recommended program. Since most forest and rangeland is owned privately and by state and local governments, the federal government can only serve as a catalyst in land-resource improvement, the report says. Programs are recommended for each forest and rangeland resource and for each function (protection, etc.) respecting resources. Four features of the recommendations deserve our special thought:

1 The recommended plans cover the period 1981–2030: 50 years. The Forest Service has other planning efforts that are of shorter scope and more specific. For example, the forest land management plans look ahead 10 to 15 years.

2 Recommendations are stated as a range between a "high bound" and a "low bound." For example, national forest wilderness in 1985 is projected as 33 million acres at the low bound, 41 million at the high. For 2030, the respective acreages are 34 and 42 million. The rationale is to provide a range accommodated to changing resource emphasis and budget allocation.

3 A clear distinction appears not to be drawn between program inputs and program outputs. The wilderness recommendations just quoted are inputs. Outputs, which would be more appropriate to a consumer-oriented program, would be expressed in visitor-days or some such measure.

4 RPA gives the Forest Service broad responsibilities for nationwide "renewable-resource" planning. At the same time, the major share of the agency's year-to-year obligation is to the national-forest system: administering it, planning for the system as a whole, and planning for individual forests. How can the Service's duty to its "own" resources and its duty to nationwide resources be both wedded and held separate? The 1990 program does the best job so far of addressing both duties, for it puts a lot of emphasis on state and private forestry and research responsibilities.

Chapters 2 and following in the 1980 program concern, in essence, the methods that the Forest Service followed to produce the recommended program of Chap. 1. The methods obeyed rules proposed and widely approved in advance and then painstakingly respected. We can't go into the whole story. But we can touch upon just one fascinating part of it: how the Service went about *involving the public* in its planning.

As required by RPA, the public was involved in the 1980 planning effort (both public and Forest Service employees in the 1990 effort) and provided comments on the *draft* assessment and program that were used to prepare the *final* assessment and program. All interested publics had opportunities to learn and comment on the contents and selection of the assessment and program through numerous public briefings held throughout the country in the spring of 1979. In addition to these scheduled sessions, local Forest Service officials met and answered questions from interested groups, individuals, agencies, and officials.

Oral and written comments made during the meetings and briefings, and written comments received later, were reviewed and summarized. The Forest Service reviewed over 1,600 written comments during the public comment period, then systematically identified and summarized them. The recommendations voiced were used to improve the final documents in areas such as clarifying goals and targets, correcting data errors and omissions, making editorial improvements, and improving methodology [9]. An appendix of the 1980 report contains a detailed analysis of respondents by category and of responses by subject and viewpoint.

NATIONAL FOREST PLANS

NFMA provided that the *national forest plans* were to be finished by 1985. The assignment proved to be too gigantic for such a schedule. At the close of 1991, of the 123 plans under way for the units of the national forest system, some 116 were "final"—that is, drafted, reviewed, redrafted, and so on, and finally printed and released.

The national forest plans go through a process of public involvement analogous to that followed for the nationwide recommended programs. Then, after a plan is released, individuals and groups who disapprove of any of its provisions may make an appeal through the higher administrative levels of the Forest Service or Agriculture Department. Beyond that are the courts. Every one of the final forest plans has been appealed. The appellants were mostly environmental and wood-industry groups. At the beginning of 1992, all appeals were resolved for 72 of the forest plans. The Washington office of the Forest Service then was working on 325 appeals.

For an illustration of national forest planning, consider the Apache-Sitgreaves Forest, which occupied a bit more than 2 million acres in eastern Arizona. The forest extends along the New Mexico line 70–80 miles northward to the vicinity of Springerville, the supervisor's headquarters. The forest then continues westward for a similar distance.

Elevations—and, along with them, temperatures and precipitation—are widely variable. The lowlands are warm and commonly droughty. The high elevations enjoy a cooler and rainy or snowy climate. Vegetation ranges from ponderosa pine timber forest to semidesert grassland:

Vegetative group	Percent of area
Ponderosa pine	41
Pinyon-juniper	38
Mixed conifer	8
Mountain grassland	7
Semidesert grassland	3
Spruce fir	2
Riparian and aspen	1
All land	100

To characterize the Apache-Sitgreaves National Forest in briefest terms, it is a recreation-wildlife forest. Out of this and associated characteristics grow issues, concerns, and opportunities that the Forest Service and its publics stress—and around which the Service builds its planning.

Recreation Roads are essential to developed recreation—motoring and the use of campgrounds, summer homes, organization camps, downhill ski facilities, and the like—as well as to logging. The extent and location of roads are an issue. The use of off-road vehicles is common in the forest; it comes in conflict with other forms of recreation such as trail hiking and horseback riding. Construction and maintenance of trails and related facilities generate issues for the managers. More than 50,000 acres in four areas are wilderness designated by Congress. Another tract, of nearly 175,000 acres, is a primitive area. The forest has two candidates for admission to the Wild and Scenic Rivers System, but none had been formally established by 1992.

People Recreation management, like all management, involves people, and this means issues, concerns, and opportunities. Developed facilities tend to be overused; then they are closed at some times and rationed at other times by limiting users' length of stay. Drivers of off-road vehicles obtrude themselves on their environment, including gentler recreationists. Both the spirit and the letter of wilderness can be violated. The integrity of Native American culture (the Apache reservation adjoins the forest) can be encroached upon. Education and occasional policing are called for. This takes money: an appropriation that conforms to a budget that conforms to a plan. So far, as we'll see, such money is not to be had.

Wildlife The forest is potentially an ideal site for species of big game and for many kinds of fish. The bald eagle and peregrine falcon are found here, and the Apache trout—all threatened or endangered species. Forest managers and their publics are concerned that grazing, timber cutting, and associated activities will harm fish and wildlife habitat or adversely affect animal diversity and populations. Many species depend on riparian areas. There, conflicts are developing among range, recreation, road building, and water demands.

Timber What timber harvest level should be maintained? Reforestation involves problems in some areas. In others, slash is not being treated. The level of timber harvest affects the opportunities for management and use of other resources and the economic stability of nearby communities. There are views that the current timber harvest is too high, too low, and just right. Firewood is the big thing in the timber economy of this forest. "Demand" for many species of firewood may soon exceed "supply" in some areas. Commercial and personal-use cutters compete for the available firewood supply. Logging residue is available for firewood where timber has been cut; requirements should be placed on loggers to yard unmerchantable material for other uses. The public is clamoring for information about firewood.

HISTORY OF APACHE-SITGREAVES LAND-MANAGEMENT PROGRAM

Planning under RPA began for the national forest in Arizona in the spring of 1980. An interdisciplinary planning team was appointed. The team identified about 280 issues calling for planning effort and ultimately boiled these down to 18. The 18 were used as a tickler list in a "response booklet" distributed to about 22,000 people by mail and at workshops in behalf of the Arizona national forest. About 32,000 "interest response cards" were mailed to persons who had applied for Christmas tree permits the year before. Employees of the forest were briefed on the process. Meetings were held with state, county, and local officials and planners. Some 2,000 people made 20,000 comments, of which 7,000 were directed to the Apache-Sitgreaves planners. Meantime, the planners generated a "notice of intent" to draft an environmental impact statement (DEIS), the first official forest-planning statement under NFMA. The notice was entered in the Federal Register, and notices were placed in news media and posted in stores, post offices, and other gathering places.

Early in 1981, the planning team analyzed the responses. By the middle of the year, it had put together a statement of "final issues, concerns, and opportunities," which was sent to the regional forester for approval and to the public for comment and information. (A brief review of critiques of the entire RPA planning process was published by the Forest Service in 1989 [31].

The planners subdivided the Apache-Sitgreaves Forest into several hundred "management areas," which were traced on maps published as part of the project output. These areas fall into 13 categories, roughly the same as those listed on page 465, plus categories for developed recreation, natural areas, primitive areas, and wilderness. The areas served as the ultimate geographic subdivisions in the plans. A unique management prescription was written for each area.

In January 1986, the Forest Service issued a *Draft environmental impact statement, proposed Apache-Sitgreaves National Forest plan* and a *Summary of the draft environmental impact statement.* The draft statement describes six alternative plans for managing the forest resources and identifies one of them as chosen by the Forest Service. The planning period is said to be 200 years. Resource outputs for the first five decades are spelled out. Environmental consequences are described. Thirty-four persons are credited as joint authors; the academic background, work experience, and current position of each are given.

Also in January 1986, the Forest Service published a *Proposed Apache-Sitgreaves National Forest Plan.* The plan takes up the alternative identified in the impact statement as preferred by the Service. It enlarges on this alternative, describing the plans for individual management areas or groups of areas.

By August 1987, after further rounds of scrutiny by the publics and rewriting by the coauthors, the Forest Service was ready to release its Apache-Sitgreaves documents in "final" form: first, an *environmental impact statement for the Apache-Sitgreaves National Forest plan;* second, a *summary of the environmental impact statement;* third, a *records of decision,* announcing and explaining the Service's choice of one alternative as "the plan"; fourth, an *Apache-Sitgreaves National Forest plan;* fifth, *public comments and Forest Service response to the DEIS, proposed Apache-Sitgreaves National Forest plan.* The eight titles italicized in the foregoing run to more than 2,000 pages. Quite a heroic output for a unit of national forest, even one of 2 million acres!

What, indeed, is the optimal effort in planning? And the optimal result? Costs are hard to measure; benefits, impossible, for these benefits lie partly in a perpetually receding future. And they lie partly in the past, where they are an enigma, since they are the difference between what actually happened and an unknown, what would have happened if. What objects of expenditure alternative to forestry planning does society have for its resources? What alternative style and intensity of forestry planning could it light on? What success will social planners have in forecasting the trend of things in a changing world? What success will they have in making plans that can be carried out? The Forest Service has sponsored a thoughtful analysis of planning questions such as the foregoing [11].

BUDGET OUTCOME OF PLANNING

Carrying out social plans in the federal sphere is firstly a matter of federal budgeting. And this brings us to the last episode envisioned by Senator Humphrey, the translation of RPA plans into the *Forest Service budget.* The translation is anything but literal. Some critics (and we know by now what a wealth of them RPA possesses!) have stated that the elements of the act are intended to flow, each into the next, and that in fact they have turned out to be all too independent of one another. At any rate, the independence of the budget is undoubted. Here is the *Journal of Forestry* comment of a few years ago [10]:

> The president's 1984 federal budget proposal would reduce the USDA Forest Service agency budget by 7.9 percent to $1.8 billion. Compared with funding under the Resources Planning Act, the reduction is even more dramatic. The agency budget has been reduced from 75 percent of the recommended RPA level in fiscal year 1982 to 57 percent in 1984.
>
> Within the agency, forest research would receive 52 percent of the funding recommended in the RPA program, resulting in the closure of up to 19 research units and 4 research locations. State and Private Forestry would receive only 20 percent of the program needs suggested in RPA.
>
> The national forest system would have to operate at 58 percent of RPA's recommended

levels. Funding for reforestation and timber stand improvement would be 5 percent below fiscal year 1983. This would allow for the reforestation of only 134,000 acres, less than 13 percent of the more than 1 million national forest acres in need of trees.

Recreation, wildlife, range, soil, and water would be funded at slightly more than 50 percent of the RPA recommendations; fire protection and road maintenance would get 52 percent.

But we are obliged to stand back and view thoughtfully the character of the planning-budgeting connection. RPA seeks to tie national budget decisions to local production opportunities. Yet the budget is an act of national politics taken in a nationwide technical forestry context constrained by a century's tradition, whereas the forest's production opportunities are a technical conclusion reached in a largely local political context [4]. On-the-ground year-to-year forestry potentials are about as unchanging as can be. The manager's effort to fulfill them would thrive on budget stability. But will any federal administration surrender the prerogative of enjoying year-to-year budget flexibility? When 1988 opened, the Forest Service had for the first time submitted to the Department of Agriculture a budget request fully consistent with forest plans. But the planning-budgeting linkage remains a major debate item. Until there is evidence to the contrary, one is tempted to lament that in a market economy, to make a social plan is one thing; to see it carried out is another.

SUMMARY

This chapter looks briefly at the Forest and Rangeland Renewable Resources Planning Act (RPA) of 1974, as amended by the National Forest Management Act of 1976. The chapter sketches the background of the act, describes the act's principal provisions and output, takes up a case example of national forest planning, and closes with comments on the degree of our success in social planning for forestry.

REFERENCES

1 Act of August 17, 1974 (88 Stat. 476, as amended; 16 U.S.C. 1600-1614).

2 Act of June 30, 1978 (92 Stat. 353, as amended; 16 U.S.C. 1600, 1641-1647.

3 Baden, John: "Spare That Tree!" *Forbes,* December 9, 1991, pp. 229 ff.

4 Cortner, Hanna J., and Dennis L. Schweitzer: "Limits to Hierarchical Planning and Budgeting Systems, The Case of Public Forestry," *Journal of Environmental Management,* September 1983, pp. 191–205.

5 Gericke, Kevin L., Jay Sullivan, and J. Douglas Wellman: "Public Participation in National Forest Planning," *Journal of Forestry,* February 1992, pp. 35–38.

6 Gould, Ernest M.: "The Future of Forests in Society," *Forestry Chronicle,* December 1964, pp. 431–444.

7 Hunt, Frances A.: "We Asked for It, or Did We?" *American Forests,* October 1986, p. 20.

8 LeMaster, Dennis C.: "Decade of Change, The Remaking of Forest Service Statutory Authority During the 1970s," *Forest History Society,* 1984.

9 The paragraph to this point is a paraphrase of the introduction to the public-involvement topic in the 1980 program update.

10 Petersen, Philip V.: "National Highlights, Forest Service Budget Out of Line with RPA," *Journal of Forestry,* July 1983, p. 429.

11 Teeguarden, Dennis E.: *National Forest Planning Under RPA/NFMA: What Needs Fixing?* USDA Forest Service, June 1990, iii + 66 pages.
12 U.S. Department of Agriculture, Forest Service: *The Nation's Renewable Resources, An Assessment, 1975,* Forest Resource Rep. 21, 1977.
13 U.S. Department of Agriculture, Forest Service: *An Assessment of the Forest and Rangeland Situation in the United States,* Forest Resource Rep. 22, 1981.
14 U.S. Department of Agriculture, Forest Service: *America's Renewable Resources, A Supplement to the 1979 Assessment of the Forest and Rangeland Situation in the United States,* FS-386, 1984.
15 U.S. Department of Agriculture, Forest Service: *Final Environmental Statement and Renewable Resources Program, 1977 to 2020,* Washington, 1976.
16 U.S. Department of Agriculture, Forest Service: *A Recommended Renewable Resources Program, 1980 Update,* FS-346, 1980.
17 U.S. Department of Agriculture, Forest Service: *A Recommended Renewable Resources Program, 1985–2030, 1985 update,* FS-400, 1986.
18 U.S. Department of Agriculture, Forest Service: *Forest and Rangeland Resource Interactions,* Gen. Tech. Rep. RM-156, 31 pages, 1988.
19 U.S. Department of Agriculture, Forest Service: *RPA Assessment of the Forest and Rangeland Situation in the United States, 1989,* Forest Resource Rep. 26, 72 pages, 1989.
20 U.S. Department of Agriculture, Forest Service: *Basic Assumptions,* Gen. Tech. Rep. RM-174, 11 pages, 1989.
21 U.S. Department of Agriculture, Forest Service: *The Evolving Use and Management of the Nation's Forests, Grasslands, Croplands, and Related Resources,* Gen. Tech. Rep. RM-175, 66 pages, 1989.
22 U.S. Department of Agriculture, Forest Service: *A Description of Forest Service Programs and Responsibilities,* Gen. Tech. Rep. RM-176, 45 pages, 1989.
23 U.S. Department of Agriculture, Forest Service: *An Analysis of the Water Situation in the United States, 1989–2040,* Gen. Tech. Rep. RM-177, 178 pages, 1989.
24 U.S. Department of Agriculture, Forest Service: *An Analysis of the Wildlife and Fish Situation in the United States, 1989–2040,* Gen. Tech. Rep. RM-178, 146 pages, 1989.
25 U.S. Department of Agriculture, Forest Service: *An Analysis of the Minerals Situation in the United States, 1989–2040,* Gen. Tech. Rep. RM-179, 38 pages, 1989.
26 U.S. Department of Agriculture, Forest Service: *An Analysis of the Range Forage Situation in the United States, 1989–2040,* Gen. Tech. Rep. RM-180, 136 pages, 1989.
27 U.S. Department of Agriculture, Forest Service: *An Analysis of the Land Base Situation in the United States, 1989–2040,* Gen. Tech. Rep. RM-181, 76 pages, 1989.
28 U.S. Department of Agriculture, Forest Service: *Climate Change and America's Forests,* Gen. Tech. Rep. RM-187, 12 pages, 1990.
29 U.S. Department of Agriculture, Forest Service: *An Analysis of the Outdoor Recreation and Wilderness Situation in the United States, 1989–2040,* Gen. Tech. Rep. RM-189, 112 pages, 1990.
30 U.S. Department of Agriculture, Forest Service: *An Analysis of the Timber Situation in the United States, 1989–2040,* Gen. Tech. Rep. RM-199, 268 pages, 1990.
31 U.S. Department of Agriculture, Forest Service: *Review of Critiques of the USDA Forest Service Land Management Planning Process,* Gen. Tech. Rep. RM-170, 15 pages, 1989.

CHAPTER 61

HIGH POINTS IN PART SIX

1 Social institutions are laws, customs, and articles of faith that vary among cultures. The forestry examples studied in Part Six are the status of labor, regulation of land use, and federal financial policies (one chapter each); taxation (three chapters); and social planning (two chapters).

2 The 2½ percent of all U.S. workers who are employed in the forestry economy are, generally speaking, disadvantaged by comparison with nonforestry labor with respect to such matters as working conditions, safety, turnover, education, pay rates, employment of women, and unionization. But the status of workers in pulp and paper manufacture and land management is superior, and poorer conditions are found only as one moves down through the pulverization hierarchy: furniture and fixtures, panels, lumber, and logging. Furthermore, there is an upward trend in status in all forestry sectors.

3 Land resources comprise a bundle of useful and scarce attributes. Using land is exercising rights to these attributes. Some of the rights to a land parcel may be held by a private landowner; some are held by the public. Society has two principal means for transferring rights from private to public hands and thus regulating land use: the police power (for example, zoning) and acquisition (of easements or of all rights).

4 Federal fiscal and monetary policies affect forestry through their bearing on interest and monetary-inflation rates, economic trends and cycles, public expenditures on specific projects, the value of the dollar in international trade, and income taxation.

5 Income taxes represent well over half of all taxes levied by the federal government in the U.S. Today's federal income tax derives from the Tax Reform Act of 1986. The Act provides a progressive tax for low-income taxpayers, but a fixed-rate tax for those of high income. Ordinary income and long-term capital gains are taxed at the same rates, a reform that largely removes the favorable treatment formerly accorded to timber revenue.

6 A property tax administered as a cost directly variable with changes in timber value has a bias that works against intensive timber management. The bias is defeated by holding assessments constant or putting them out of the taxpayer's reach, as in a yield or a productivity tax.

7 Social planning as applied to U.S. timber resources has a 100-year history. Since 1910, a plan has appeared during every decade, and the plans have been continually more elaborate, based on steadily improving data, and offering increasingly detailed, specific, and constructive programs. The necessity for social planning in forestry in a multiresource sense became apparent during the 1950s and 1960s and gave birth to the Forest and Rangeland Renewable Resources Planning Act of 1974, as amended by the National Forest Management Act of 1976. The weight the act places on the Forest Service for communicating with the public has produced heroic responses from the Service and unbelievable interchange with unbelievers.

A parting shot: If it is true, as heralded from the very beginning of this book, that everything depends on everything else, what justification have we for recognizing forestry economics—or even forestry—as a separate field? On what grounds is a boundary line drawn around the subject of this book?

INDEX

[1]Numbers set in bold face designate pages on which terms are defined or their meaning is implied. This index thus doubles as a glossary.